S0-BLI-582

Herbert Baur

Thermophysics of Polymers I

Springer

Berlin
Heidelberg
New York
Barcelona
Hong Kong
London
Milan
Paris
Singapore
Tokyo

Herbert Baur

Thermophysics of Polymers I

Theory

With 79 Figures and 2 Tables

Springer

PD Dr. rer. nat. Herbert Baur
Sonnenwendstraße 41
D-67098 Bad Dürkheim

ISBN 3-540-65046-6 Springer-Verlag Berlin Heidelberg New York

Library of Congress Cataloging-in-Publication Data
Baur, Herbert.
Thermophysics of polymers I : theory / Herbert Baur.
p. cm.
Includes bibliographical references and index.
ISBN 3-540-65046-6 (alk. paper)
1. Polymers--Thermal properties. I. Title.
QC173.4.P65B38 1999
620.1'92--dc21 98-53819

Coverdesign: Design & Production GmbH, Heidelberg
Typesetting: Fotosatz-Service Köhler GmbH, Würzburg
SPIN: 10114932 2/3020 – 5 4 3 2 1 0 – Printed on acid-free paper

Preface

"Physics is physics" a physicist may think and be surprised that there is supposed to be a "polymer physics". However, physics lives to a large extent from abstractions. The Newtonian laws of mechanics could only be formulated and start their triumphant march after one had learned to disregard friction (the dissipative effects which occur during motion). The physicist likes to regard particles, *e.g.* molecules, as mass points or rigid spheres. Many "laws" are based on this abstraction. However, if one builds a polyethylene molecule using the Stuart-Brigleb models, one obtains a flexible chain with a diameter of about 5 cm (2 inches), which can attain a length of 2 km (1.24 miles). Even on the true microscopic scale, such a molecule is certainly not a mass point, nor is it a structureless rigid sphere. The flexibility and the enormous length of the threadlike molecules induce properties in polymers which do not usually occur in physics, or which are considered to be negligible. In this respect, it is, therefore, certainly possible to speak of a polymer-specific physics, a "polymer physics". The physical peculiarities, which are caused by the very long flexible chain molecules, are mainly of an entropic nature. One can even attribute an entropy to an individual polymer molecule. These molecules usually respond dissipatively to an external perturbation, *i.e.* with a production of entropy. For this reason, thermodynamics is the basis of the polymer-specific physics. "Polymer physics" is at its roots a thermophysics of polymers, as indicated by the title of this book.

Experimental aspects will be treated by B. Wunderlich in Vol. II of this book. The objective of the present volume is to provide a simple description of the theoretical framework which constitutes the basis of polymer physics. According to the statements above, it is clear that this framework must be established with the help of thermodynamics. Unfortunately, a complete theory does not exist to date.

The problems of polymer physics can be formulated with an almost arbitrary complexity. But this is not our objective. In order to illustrate the theoretical skeleton, we have only treated simple, easily comprehensible problems of polymer physics, but, in detail and with all the necessary information on the assumptions used. The description of dissipative processes within the framework of thermodynamics shows clearly that these are generally non-linear phenomena. Nevertheless, we will also confine ourselves here mainly to the simpler linearized cases. The so-called glass transition, for example, is found to be a typical dissipative process which cannot be linearized.

At this point I would like to thank once again: Mrs. C. Paciello (Bad Dürkheim) who very carefully translated the originally German text into English, Prof. Dr. B. Wunderlich (Knoxville, TN) who critically perused the entire text, Prof. Dr. G. Kanig (Ludwigshafen/Rhein) who kindly gave me the original of the Fig. 8.3, and Mr. F. Maron (BASF AG, Ludwigshafen/Rhein) who prepared most of the figures. I also wish to thank Dr. M. Hertel (Springer-Verlag, Heidelberg), whose unremitting patience I greatly appreciate.

Bad Dürkheim, May 1998 H. Baur

Contents

CHAPTER 1

Theoretical Aspects of Polymer Thermodynamics

Theoretical macroscopic-phenomenological physics is usually divided into three areas: continuum mechanics (hydrodynamics, theory of elasticity), electrodynamics and thermodynamics. Thermodynamics deals with thermal properties of materials, the most prominent representative of which is the heat capacity.

A strict tripartition, however, can only be made with certain idealizations. There are classes of matter in which the other areas are decisively affected by thermodynamics. A pure mechanics of gases, for example, cannot be formulated. The mechanical properties of a gas are essentially dependent upon its entropy. Their explanation is, therefore, mainly a thermodynamic problem.

Entropy can also have a major influence on the mechanical properties of polymers. The best-known example is the rubber elasticity found in cross-linked or long-chain, entangled polymers. In the case of polymers, however, yet another reason blurs the division into three classical areas. Polymers usually behave dissipatively when responding to an external perturbation, *i.e.*, the external perturbation is linked to a non-negligible entropy production. For example, when one measures Young's modulus $E'(\omega)$ or the dielectric constant $\varepsilon'(\omega)$ of a polymer as a function of frequency ω of an external sinusoidal disturbance, these mechanical or electrical quantities are accompanied by equally measurable, so-called loss quantities $E''(\omega)$ and $\varepsilon''(\omega)$ which are of thermodynamic origin and character. The loss quantities are a measure of the entropy produced per half period of the perturbation. According to the Kramers-Kronig relations, however, the reactive quantities $E'(\omega)$ and $\varepsilon'(\omega)$ cannot be separated from the dissipative quantities $E''(\omega)$ and $\varepsilon''(\omega)$ (see Sect. 7.5). Rather, the functions $E'(\omega)$ and $\varepsilon'(\omega)$ can be determined unequivocally from the functions $E''(\omega)$ and $\varepsilon''(\omega)$ (and *vice versa*). In this case also, the supposedly mechanical or electrical problem proves to be a thermodynamic problem.

Thus, the thermodynamics of polymers has to do more than simply explain the purely thermal behaviour of this class of substances. It assumes the role of a comprehensive thermophysics of polymers. In the following, however – as is common in classical equilibrium thermodynamics – we will confine ourselves to homogeneous or heterogeneous systems at rest that are not exposed to a locally variable external force field. Transport phenomena, such as heat conduction, viscosity, diffusion, and electrical conductivity, will be ignored or only mentioned in passing when it seems appropriate. We will consider the thermo-mechanical phenomena of polymers in detail, but will forgo consideration of the theory of thermoelectric and thermomagnetic behaviour, which can be developed in analogy to thermomechanics.

Equilibrium thermodynamics is certainly the basis and at the same time an important component of polymer thermophysics. Its basic principles can be most simply described by means of a homogeneous, fluid, one-component system (Chap. 2). In practice, however, equilibrium thermodynamics can only be applied in this simple form to non-cross-linked polymers with a low molecular mass and very narrow molecular mass distribution. As the polymerization process leads to different lengths in the chain molecules, a hundred percent fractionation is impossible with longer chain molecules, and there are often low-molecular-mass impurities or deliberate admixtures present, polymers almost always constitute homogeneous mixtures (Chap. 3). It is a typical feature of polymers that they are incapable of forming perfect mixtures. The perfect mixture, used as a reference system for low-molecular-mass substances, is replaced by an ideal-athermal mixture in the case of polymers (Sect. 3.4). Polymers in a segregated or partially crystalline state, as well as filled or reinforced polymers, constitute heterogeneous systems (Chap. 4). Furthermore – in contrast to fluid media – polymers in a partially crystalline or crystalline state, in a glass- or rubber-like state, as well as melts of high-molecular-mass polymers in certain non-equilibrium states, resist not only a change in volume, but also a change in shape. Polymers are, thus, often elastic systems (Chap. 5). In this connection, it is important to differentiate between energetic and entropic elasticity, which, among other things, explains the polymer-specificity of rubber elasticity (Sect. 5.3).

The specific properties of polymers are mainly due to the fact that the long chain-like molecules of which they are composed are capable of producing a multitude of conformational isomers. The thermal excitation of these isomers results in coiled polymer molecules which constitute an equilibrium state with maximal entropy (Sect. 6.2). The simplest model for a polymeric liquid is the two-level system (Sect. 6.3). Despite the absence of the characteristics due to thermally excitated conformational isomerism in an ideal polymer crystal consisting of extended chain molecules, there are still polymer-specific effects. These are caused by the pronounced hybrid interaction between the lattice units. Strong covalent bonds, that cannot be treated as central forces, act in the chain direction. In contrast to this, predominantly weak van-der-Waals- or dipol-forces act perpendicular to the chain direction (Sect. 6.4 and 6.5).

Equilibrium thermodynamics requires that all changes of state occur so slowly that the microscopic internal degrees of freedom of the system under consideration are capable of adjusting their equilibrium in response. Due to their conformational degrees of freedom, the polymers have a class of internal degrees of freedom which attain equilibrium relatively slowly. This severely restricts the applicability of equilibrium thermodynamics to the problems of polymer physics. If the rate of external perturbation of a thermodynamic system is of the same order of magnitude as the rate with which an internal degree of freedom of the system tries to adjust to equilibrium, equilibrium thermodynamics is no longer valid. The perturbation process becomes an irreversible, dissipative sequence of non-equilibrium states, whose character is determined by the macroscopic relevance of the internal degree of freedom. In the following, we proceed from the thermodynamics of irreversible processes to describe the resulting phenomena (Chap. 7). The relevant internal degrees of freedom are directly considered in the thermodynamics of irreversible processes by assignment of macroscopic internal variables. This makes the thermodynamics of irreversible processes appear much less formal and much more suitable for a direct understanding of the phenomena than other theories [for example, rational thermodynamics; *cf.* Truesdell (1969)], which describe the effect of internal degrees of freedom using so-called memory functions or memory functionals but conceal the internal degrees of freedom.

Relaxation processes are the simplest dissipative processes which take place in polymers. In general, these are non-linear phenomena with feedback. To a good approximation, however, relaxation processes can often be described by simple linear differential equations (Sect. 7.4). At the same level of approximation, the question about the reactions of a polymer to periodic mechanical perturbations leads to the linear theory of response, to the concept of memory functions, as well as to the phenomena of anelasticity or viscoelasticity (Sects. 7.5 and 7.6).

Relaxation times prove to be a collective property of the internal degrees of freedom in systems with several intercoupled internal degrees of freedom. The relaxation times can only be assigned individually to the so-called normal modes, whose composition of internal variables changes with temperature and mechanical boundary conditions. Decomposition of a macroscopic process into many elementary normal modes is, however, somewhat problematic (Sect. 7.7).

Under normal circumstances, polymers comprised of long chain molecules crystallize only partially – if at all. Therefore, they usually contain a glassy solidified portion at lower temperatures. Polymers consisting of irregularly assembled flexible chain molecules do not crystallize. They form a glass at lower temperatures. The transition from the liquid state into the glassy state is – at least in all practical cases – also a dissipative process. Without a change in the molecular degree of order, exactly those internal degrees of freedom of the liquid freeze which substantially determine the temperature of the liquid. Therefore, one needs to differentiate between a dynamic and a static temperature during the glass transition (Sect. 7.8).

Finally, the phase transitions (Chap. 8) have a major influence on the thermal and mechanical properties of polymers – especially crystallization, which occurs only partially under normal circumstances, as well as melting, which may occur over a broad temperature range. A special feature of long flexible chain molecules is that they necessarily develop amorphous non-autonomous boundary phases during crystallization (Sect. 8.2). A special phase that may develop from a polymer crystal on the basis of conformational isomerism is the so-called condis crystal (conformationally disordered crystal) (Sect. 8.3).

CHAPTER 2

Equilibrium Thermodynamics of Simple Fluids

2.1
Fundamental Equations and Equations of State

The basis for a thermodynamic description of a physical system is Gibbs' fundamental equation. It connects the so-called Gibbs function (or the Gibbs potential) with the mutually independent variables of the system. In thermodynamic equilibrium, the Gibbs function allows a complete and unequivocal description of the system.

The number of mutually independent variables corresponds to the number of macroscopic degrees of freedom of a system. In thermodynamics, the variables of every macroscopic field theory, *e.g.*, the variables of the theory of elasticity, of hydromechanics, or electrodynamics, can appear as variables. In addition, the temperature and the entropy always appear as variables in thermodynamics.

As physical systems rarely manifest themselves as such, but rather through their changes in response to external actions, the differentials and differential coefficients of the Gibbs fundamental equation are particularly important. A precondition for all changes in the thermodynamic equilibrium is that the value which the Gibbs function assumes for a given state is independent of the path by which the state was achieved. Mathematically, this means that the differential of the Gibbs function is a total differential that can always be integrated. It also means that in the second derivatives of the Gibbs function, the order of the differentiations can be changed with respect to the variables.

In order to introduce the so-called Gibbs formalism, we will confine ourselves to simple fluid media for the present. These are homogeneous, isotropic gases or liquids composed of chemically pure and inert substances which are electrically and magnetically neutral and react to mechanical stress only by elastic changes in volume. It is possible – at least in principle – to understand polymers in a liquid state of aggregation as simple fluid media, if they are not cross-linked and have a uniform molecular mass (or at least a very narrow molecular mass distribution). It is, however, always uncertain whether longer-chain polymers achieve internal equilibrium during a real external perturbation, which necessarily occurs with a finite velocity (*cf.* Sect. 2.4 and Chap. 7). In practice, the applicability of the equilibrium thermodynamics of simple fluid media with respect to polymers is in most cases limited to non-cross-linked substances with low and uniform molecular masses.

In a state of equilibrium, simple fluid media possess three macroscopic degrees of freedom, a thermal, a mechanical, and a material degree of freedom. The temperature T, the volume V, and the mass M or the mole number N can be assigned as variables to these degrees of freedom (for a definition of the mole number see Sect. 3.1). Gibbs fundamental equation is then

$$F = F(T, V, N)\,, \tag{2.1}$$

in which the free energy F (or Helmholtz free energy) takes over the role of the Gibbs function. The differential form of the fundamental equation (2.1) is

$$\mathrm{d}F = \left(\frac{\partial F}{\partial T}\right)_{V,N} \mathrm{d}T + \left(\frac{\partial F}{\partial V}\right)_{T,N} \mathrm{d}V + \left(\frac{\partial F}{\partial N}\right)_{T,V} \mathrm{d}N\,. \tag{2.2}$$

The coefficients of this equation, which are a measure of the change in the Gibbs function if one of the variables changes, are labelled by individual symbols and names:

$$-\left(\frac{\partial F}{\partial T}\right)_{V,N} \equiv S = S(T, V, N) \tag{2.3}$$

is referred to as the entropy and

$$\left(\frac{\partial F}{\partial N}\right)_{T,V} \equiv \mu = \mu(T, V, N) \tag{2.4}$$

as the chemical potential.

$$-\left(\frac{\partial F}{\partial V}\right)_{T,N} \equiv p = p(T, V, N) \tag{2.5}$$

is identical to the hydrostatic pressure known from hydromechanics. The equations (2.3–2.5) are referred to as the "equations of state" of the system. They describe the equilibrium states of the system as completely and unequivocally as the fundamental equation (2.1). Using (2.3–2.5) one can replace (2.2) with

$$\mathrm{d}F = -S\mathrm{d}T - p\mathrm{d}V + \mu\mathrm{d}N\,. \tag{2.6}$$

Among the variables, one has to distinguish between extensive and intensive variables (see Tab. 2.1). Extensive variables are assigned to quantitative properties. On multiplication of a system, the values of the extensive variables, *e.g.*, the mass, are multiplied by the same factor. Extensive variables can be balanced (Sect. 2.2) and expressed in the form of specific quantities and densities. Intensive variables, on the other hand, are assigned to properties which are not affected by multiplication of a system.

F, V, N are extensive variables, whereas T is an intensive variable. Thus, with an arbitrary factor λ

$$\lambda F = F(T, \lambda V, \lambda N) \tag{2.7}$$

Table 2.1. Selected extensive and intensive properties

Extensive properties	intensive properties
Energy, enthalpy	
Entropy	Temperature
Mass, mole number	Chemical potential
Electric charge	Electrical potential
Momentum	Velocity
Volume	Pressure
Mechanical strain	Mechanical stress
Dielectric displacement	electric field strength

must apply. As a result, the fundamental equation (2.1) must also be representable in the form

$$F = -pV + \mu N \tag{2.8}$$

[Euler's theorem on homogeneous functions of the first degree; as proof, one differentiates (2.7) with respect to λ and subsequently sets $\lambda = 1$]. From (2.8) follows

$$\mathrm{d}F = -p\mathrm{d}V - V\mathrm{d}p + \mu\mathrm{d}N + N\mathrm{d}\mu \,.$$

The comparison with (2.6) shows that, in addition, the Gibbs–Duhem relation

$$S\mathrm{d}T - V\mathrm{d}p + N\mathrm{d}\mu = 0 \tag{2.9}$$

must always be fulfilled. Relation (2.9) expresses clearly that only two of the three intensive variables T, p, μ of a simple fluid medium are independent of each other.

In the case of heterogeneous systems which are composed of several homogeneous domains (phases), the value of an extensive quantity is given by the sum of the values of the individual phases. The values of the intensive quantities, on the other hand, cannot be added together. In equilibrium, heterogeneous systems are homogeneous with respect to the values of their intensive variables as long as the individual phases are not separated from each other by insulating walls (*cf.* Chap. 4).

The equilibrium properties of thermodynamic systems can be unequivocally and completely described by a Gibbs fundamental equation. The reverse, however, is not true: a thermodynamic system is by no means described by only a single Gibbs fundamental equation. Other representations of Gibbs fundamental equation can be obtained from the F-representation (2.1) of Gibbs fundamental equation using special, purely mathematical transformations, the so-called Legendre transformations. Legendre transformations are contact transformations, which have to do with the fact that, for example, every smooth curve $y(x)$ can also be described by the family of its enveloping tangent lines

$$y = ax + b\,, \quad \text{with} \quad a \equiv \mathrm{d}y/\mathrm{d}x \quad \text{and} \quad b \equiv y - (\mathrm{d}y/\mathrm{d}x)x\,.$$

If the variable pairs y, x and b, a are coupled with each other by the given relations, the same information can be found in the functions $y(x)$ and $b(a)$.

T, p, N are in most cases given experimentally as the independent variables of a simple fluid medium. The Legendre transformation

$$V \to -\left(\frac{\partial F}{\partial V}\right)_{T,N} ; \quad F \to G = F - \left(\frac{\partial F}{\partial V}\right)_{T,N} V ,$$

i.e., according to (2.5)

$$V \to p ; \quad F \to G = F + pV ,$$

then transforms the F-representation (2.1) into the G-representation

$$G = G(T, p, N) \tag{2.10}$$

of Gibbs fundamental equation, in which the free enthalpy (or Gibbs free energy) G takes over the role of the Gibbs function. Equations (2.1) and (2.10) contain the same information about the medium as long as F and G are coupled *via*

$$G = F + pV \tag{2.11}$$

Comparison of (2.11) with (2.8) shows that the Gibbs fundamental equation must have the form

$$G = \mu N \tag{2.12}$$

in the G-representation. Equation (2.12) expresses the fact that G and N are extensive, whereas T and p are intensive variables, so that with an arbitrary factor λ

$$\lambda G = G(T, p, \lambda N)$$

must also be valid in place of (2.10).

During the transformation from the F- to the G-representation, (2.11) is expressed at full length by

$$G(T, p, N) = F(T, V(T, p, N), N) + pV(T, p, N) .$$

By differentiation of this relation with respect to the new independent variables T, p, N, one obtains *via* (2.3–2.5)

$$-\left(\frac{\partial G}{\partial T}\right)_{p,N} \equiv S = S(T, p, N) , \tag{2.13}$$

$$\left(\frac{\partial G}{\partial p}\right)_{T,N} \equiv V = V(T, p, N) , \tag{2.14}$$

$$\left(\frac{\partial G}{\partial N}\right)_{T,p} \equiv \mu = \mu(T, p, N) , \tag{2.15}$$

as equations of state in the G-representation. Thus, the differential form of Gibbs fundamental equation (2.10) is

$$\mathrm{d}G = -S\mathrm{d}T + V\mathrm{d}p + \mu\mathrm{d}N\,. \tag{2.16}$$

(2.16), however, is only in agreement with (2.12) if the Gibbs–Duhem relation (2.9) is once again fulfilled.

It should be noted that the differentiation between extensive and intensive variables also induces particular forms in the intensive functions. For example, as the chemical potential μ is an intensive variable,

$$\mu = \mu(T, p, \lambda N)$$

must also apply with every arbitrary factor λ in place of (2.15). This means, however, that μ cannot be dependent on N in the G-representation (Euler's theorem on homogeneous functions of the zeroth degree).

$$\mu = \mu(T, p)$$

holds, corresponding to the Gibbs–Duhem relation (2.9), according to which only two of the intensive variables T, p, μ are independent of each other.

The Legendre transformation

$$T \to S\,; \quad F \to U = F + TS\,,$$

transforms (2.1) into the U-representation

$$U = U(S, V, N) \tag{2.17}$$

of the Gibbs fundamental equation. U is the internal energy of the system, which corresponds to the total energy of the system if the system is at rest and not exposed to external force fields (*cf.* Sect. 2.2). U and F are coupled *via* the relation

$$U = F + TS\,. \tag{2.18}$$

Hence, according to (2.8), we have

$$U = TS - pV + \mu N\,. \tag{2.19}$$

In the U-representation, the equations of state are given by

$$\left(\frac{\partial U}{\partial S}\right)_{V,N} \equiv T = T(S, V, N)\,, \tag{2.20}$$

$$-\left(\frac{\partial U}{\partial V}\right)_{S,N} \equiv p = p(S, V, N)\,, \tag{2.21}$$

$$\left(\frac{\partial U}{\partial N}\right)_{S,V} \equiv \mu = \mu(S, V, N)\,, \tag{2.22}$$

the differential form of the Gibbs fundamental equation by

$$dU = T dS - p dV + \mu dN . \tag{2.23}$$

In the U-representation, all the mutually independent variables are extensive. The equations of state, on the other hand, only apply to intensive variables. The Gibbs–Duhem relation (2.9) is also in this case responsible for consistency.

Using Legendre transformations, one can also replace several variables simultaneously in (2.1). The double transformation,

$$T \to -\left(\frac{\partial F}{\partial T}\right)_{V,N} ; \quad V \to -\left(\frac{\partial F}{\partial V}\right)_{T,N} ,$$

$$F \to H = F - \left(\frac{\partial F}{\partial T}\right)_{V,N} T - \left(\frac{\partial F}{\partial V}\right)_{T,N} V ,$$

i.e., the transformation,

$$T \to S ; \quad V \to p ; \quad F \to H = F + TS + pV$$

for example, leads to the H-representation of the Gibbs fundamental equation

$$H = H(S, p, N) \tag{2.24}$$

(H: enthalpy or heat function of the system) with the equations of state

$$\left(\frac{\partial H}{\partial S}\right)_{p,N} \equiv T = T(S, p, N) , \tag{2.25}$$

$$\left(\frac{\partial H}{\partial p}\right)_{S,N} \equiv V = V(S, p, N) , \tag{2.26}$$

$$\left(\frac{\partial H}{\partial N}\right)_{S,p} \equiv \mu = \mu(S, p, N) , \tag{2.27}$$

the differential fundamental equation

$$dH = T dS + V dp + \mu dN \tag{2.28}$$

and the relation

$$H = F + TS + pV \tag{2.29}$$

or according to (2.8)

$$H = TS + \mu N . \tag{2.30}$$

The different representations of Gibbs fundamental equation obtained from (2.1) can obviously also be interconnected by Legendre transformations. Furthermore, the Gibbs function in a specific representation can, of course, also be expressed as a function of the independent variables of another representation. Moreover, by resolution of the equation (2.19) according to S,

$$S = \frac{1}{T} U + \frac{p}{T} V - \frac{\mu}{T} N, \tag{2.31}$$

we can obtain an S-representation

$$S = S(U, V, N) \tag{2.32}$$

of the Gibbs fundamental equation, with the equations of state

$$\left(\frac{\partial S}{\partial U}\right)_{V,N} \equiv \frac{1}{T} = \frac{1}{T}(U, V, N), \tag{2.33}$$

$$\left(\frac{\partial S}{\partial V}\right)_{U,N} \equiv \frac{p}{T} = \frac{p}{T}(U, V, N), \tag{2.34}$$

$$-\left(\frac{\partial S}{\partial N}\right)_{U,V} \equiv \frac{\mu}{T} = \frac{\mu}{T}(U, V, N) \tag{2.35}$$

and the differential form of the fundamental equation

$$\mathrm{d}S = \frac{1}{T}\,\mathrm{d}U + \frac{p}{T}\,\mathrm{d}V - \frac{\mu}{T}\,\mathrm{d}N\,. \tag{2.36}$$

2.2 Balance Equations

The explicit forms of the fundamental equations and the equations of state are characteristic for a specific material. The formats of the equations are dependent on the atomic and molecular composition of the substances and on the material structure, *e.g.*, on the state of aggregation. Fundamental equations and equations of state in a closed mathematical form over a broader range of values of the independent variables, however, can only be given for a few simple materials. The Gibbs formalism alone – to the extent that it was described in the previous section – is, therefore, often of little use. These approaches become more powerful when connected to further physical observations. Primarily these include balances of the extensive properties, which are independent of the type of material. For example, balances are extracted from the equations of motion in hydromechanics or Maxwell's equations in electrodynamics. In thermodynamics, the balance of mass (principle of conservation of mass), the balance of internal energy (first law), and the balance of entropy (second law) are particularly important.

The change $\mathrm{d}E$ of every arbitrary extensive variable E (with the exception of volume) can be split into two additive terms [1]:

$$\mathrm{d}E = \mathrm{d}_a E + \mathrm{d}_i E . \tag{2.37}$$

The first term $\mathrm{d}_a E$ describes the change in E caused by the exchange of a quantity of E between the observed system and its surroundings. The second term $\mathrm{d}_i E$ describes the change in E resulting from creation or annihilation of a quantity of E in the interior of the system. The volume has to be excluded since the surroundings and interior of the system are defined by means of the volume itself. Hence, the following balance equations are valid:

$$\mathrm{d}N = \mathrm{d}_a N + \mathrm{d}_i N , \tag{2.38}$$

$$\mathrm{d}U = \mathrm{d}_a U + \mathrm{d}_i U , \tag{2.39}$$

$$\mathrm{d}S = \mathrm{d}_a S + \mathrm{d}_i S . \tag{2.40}$$

It should be noted that, during changes of an equilibrium, the value which an extensive variable E, like the value of a Gibbs function, reaches for a specific new state of equilibrium is independent of the path by which this state is achieved. This means that $\mathrm{d}E$ represents a total differential. In contrast, $\mathrm{d}_a E$ and $\mathrm{d}_i E$ generally do not represent total differentials. Therefore, expressions in which $\mathrm{d}_a E$ or $\mathrm{d}_i E$ appear individually, in particular integrals over these differentials, are only well-defined if the path over which the changes occur is also given. If $\mathrm{d}E$ is a total differential, $\mathrm{d}_a E$ is, of course, also a total differential if $\mathrm{d}_i E = 0$.

The principle of conservation of mass states that in the interior of a system mass can neither be created nor annihilated (as long as processes induced by subatomic particles and effects of the theory of relativity are excluded). The mass of a system can therefore only be changed by adding mass from or removing it to the exterior. Hence, it is always valid that:

$$\mathrm{d}N = \mathrm{d}_a N ; \quad \mathrm{d}_i N = 0 . \tag{2.41}$$

In equilibrium thermodynamics, the first law also represents a conservation law and at the same time gives a detailed balance of the internal energy. For simple fluid media, the first law is

$$\mathrm{d}_a U = \mathrm{d}Q - p\,\mathrm{d}V + h\,\mathrm{d}_a N , \tag{2.42a}$$

$$\mathrm{d}_i U = 0 . \tag{2.42b}$$

1 In the English literature $\mathrm{d}_a E$ is usally designated as $\mathrm{d}_e E$. In the following, however, we wish to reserve the index "e" for the internal equilibrium.

Equally, the internal energy can neither be produced nor consumed in the interior of the system during changes in equilibrium. The internal energy, however, can be changed by three different exchanges with the surroundings: 1. by heat exchange ($dQ > 0$ describes the heat the system receives from the exterior), 2. by the performance of work (under the condition $d_a N = 0, p\,dV > 0$ describes the work performed by the system), and 3. by mass exchange ($h = u + pv$ is the enthalpy for one mole[2]), so that $h\,d_a N$ describes the heat which is transferred by an exchange of mass).

If the system is at rest and not exposed to external force fields (which, as already mentioned, is generally assumed), (2.42b) also holds for changes during non-equilibrium processes. The system then possesses neither kinetic nor potential energy, so that the internal energy U is equal to the total energy of the system. Under these circumstances, (2.42) expresses nothing more than the universally valid law of conservation of energy.

The quantity dQ contains the fraction of the change dU of the internal energy which is based purely on thermal changes. This fraction is due to excitations of the atomic and molecular internal degrees of freedom of the system concealed in the macroscopic equilibrium thermodynamics. These internal degrees of freedom include for example, the degrees of freedom of random (diffusive) translational motion of the atoms or molecules in fluids, the degrees of freedom of the collective lattice vibrations (phonons) in solids, the rotational and vibrational degrees of freedom of the individual molecules, the degrees of freedom of mutual arrangement and orientation of the molecules, and especially – with respect to polymers – the conformational degrees of freedom.

The first part of the second law states that for all changes in the thermodynamic equilibrium, we have:

$$d_a S = \frac{1}{T}\,dQ + s\,d_a N\,, \tag{2.43a}$$

$$d_i S = 0\,. \tag{2.43b}$$

The entropy is also conserved during changes in equilibrium. In the case of closed systems (these are systems which do not exchange mass with their surroundings), (2.43) directly links the exchange of entropy with the exchange of heat. According to (2.40) and (2.43), when $d_a N = 0$, $dS = dQ/T$.

The second part of the second law refers to non-equilibrium processes, *i. e.*, to processes which cannot be described explicitly by equilibrium thermodynamics.

Equilibrium thermodynamics is a static theory, as it only applies to states of equilibrium. It only allows a description of processes that can be mapped onto sequences of points in the space of equilibrium states spanned by the Gibbs function. Processes which fully correspond to such simply geometrically order-

2 The specific extensive quantities relative to a mass unit or one mole are marked by the corresponding small letters.

ed sequences of points are designated as quasi-static processes. These processes can only be approximatively realized, as they require infinitesimal driving forces and infinitely long process durations. Real processes, self-driven equilibration processes, or processes caused by finitely large external stresses, on the other hand, represent a time-ordered succession of non-equilibrium states which are traversed with finite velocity.

Among the possible processes, one distinguishes between reversible and irreversible ones. All processes which are invariant on time reflection ($t \to -t$) are reversible. These include all quasi-static processes. Purely mechanical processes in ideal Hookean solids or in ideal, frictionless fluids are also reversible. Since reversible processes always require thermal equilibrium (homogeneity with respect to the temperature; *cf.* Chap. 4), and since there are always some ideal conditions attached that only can be approximated, they are generally not classed with real processes in thermodynamics. The class of real processes is, in turn, identified with the class of irreversible processes. Typical irreversible processes are transport processes – which will not be treated in detail here – such as heat conduction (transport of energy), diffusion (transport of mass), viscous flow (dissipative transport of momentum), or electrical conduction (charge transport). Moreover, all processes are also irreversible in which the statistical equilibrium of the microscopic internal degrees of freedom, concealed in equilibrium thermodynamics, is disturbed (Chap. 7).

In its second part, the second law states that (2.43 a, b) hold for all reversible processes. For a state of equilibrium A, however, which is transformed into a state of equilibrium B by an irreversible process, the following is valid:

$$S(\mathrm{B}) - S(\mathrm{A}) > \int_{\mathrm{A}}^{\mathrm{B}} \mathrm{d}_a S \tag{2.43 c}$$

Here, $\mathrm{d}_a S$ describes the entropy taken up or released by the system during the process. Thus, $-\mathrm{d}_a S$ is the entropy released or taken up by the surroundings. (2.43 c) may be interpreted fully in terms of equilibrium thermodynamics when it is assumed that the surroundings are in equilibrium at the beginning of the process and large enough so that this equilibrium is not disturbed during the course of the process (infinitely large contact reservoir). If we have irreversible processes in which an entropic value can be unequivocally attributed to each of the non-equilibrium states traversed, then (2.43 c) is replaced by the inequality

$$\mathrm{d}S > \mathrm{d}_a S \,,$$

which is valid for each infinitesimal step of the process. Then, according to (2.40), (2.43 a) can be supplemented by the simple statement that

$$\mathrm{d}_i S > 0 \tag{2.43 d}$$

always holds during the course of an irreversible process. Irreversible processes always involve an entropy production in the interior of the system. There are no entropy sinks. Nevertheless (2.43 c, d) do not imply that every irreversible process leads to an increase in the entropy of the system. If $\mathrm{d}_a S < 0$, *i.e.*, if the system releases entropy to its surroundings during the process, the process can most certainly also involve a decrease in entropy. Irreversible processes only necessarily lead to an increase in entropy if they occur in adiabatically closed systems. For such adiabatically closed systems it holds that $dQ \equiv 0$ and $d_a N \equiv 0$, *i. e.*, according to (2.43 a), $d_a S \equiv 0$ and thus, according to (2.43 c) or (2.43 d) and (2.40),

$$S(\mathrm{B}) - S(\mathrm{A}) > 0 \quad \text{or} \quad dS = d_i S > 0\,, \tag{2.43 e}$$

respectively.

2.3 Response Functions

The variables defined by the equations of state are more accessible to experiments than the Gibbs function. Since these variables are given by the first derivatives of the Gibbs function with respect to the independent variables, the stimulation of a macroscopic degree of freedom in thermodynamic equilibrium manifests itself particularly in the second derivatives of the Gibbs function. The second derivatives of the Gibbs functions with respect to the independent variables provide the so-called response functions of the systems. After appropriate normalization, they are also given individual symbols and names.

The mechanical, thermo-mechanical and thermal response functions which can be derived from the F-representation (2.1 – 2.5) are: the isothermal bulk modulus

$$k_T \equiv V \frac{\partial^2 F}{\partial V^2} = -V \left(\frac{\partial p}{\partial V}\right)_{T,N}\,, \tag{2.44}$$

the coefficient of thermal pressure

$$\beta \equiv -\frac{1}{p}\frac{\partial^2 F}{\partial V\,\partial T} = \frac{1}{p}\left(\frac{\partial S}{\partial V}\right)_{T,N} = \frac{1}{p}\left(\frac{\partial p}{\partial T}\right)_{V,N} = \left(\frac{\partial \ln p}{\partial T}\right)_{V,N} \tag{2.45}$$

and the heat capacity at constant volume

$$C_V \equiv -T\frac{\partial^2 F}{\partial T^2} = T\left(\frac{\partial S}{\partial T}\right)_{V,N}\,. \tag{2.46}$$

The G-representation (2.10, 2.13 – 2.16) results in the response functions: isothermal compressibility

$$\kappa_T \equiv -\frac{1}{V}\frac{\partial^2 G}{\partial p^2} = -\frac{1}{V}\left(\frac{\partial V}{\partial p}\right)_{T,N} = -\left(\frac{\partial \ln V}{\partial p}\right)_{T,N}\,, \tag{2.47}$$

coefficient of thermal expansion (thermal expansivity)

$$\alpha \equiv \frac{1}{V}\frac{\partial^2 G}{\partial T\,\partial p} = -\frac{1}{V}\left(\frac{\partial S}{\partial p}\right)_{T,N} = \frac{1}{V}\left(\frac{\partial V}{\partial T}\right)_{p,N} = \left(\frac{\partial \ln V}{\partial T}\right)_{p,N}\,, \tag{2.48}$$

and heat capacity at constant pressure

$$C_p \equiv - T \frac{\partial^2 G}{\partial T^2} = T \left(\frac{\partial S}{\partial T}\right)_{p,\,N} . \tag{2.49}$$

Based on the condition of integrability for the differentials dF and dG, the order of the differentiations can be exchanged in the second derivatives of the Gibbs function as used in (2.45) and (2.48).

Of particular interest in the following are (as a result of the U- and H-representations of the Gibbs fundamental equation): the adiabatic (isentropic) bulk modulus

$$k_S \equiv V \frac{\partial^2 U}{\partial V^2} = - V \left(\frac{\partial p}{\partial V}\right)_{S,\,N} , \tag{2.50}$$

and the adiabatic (isentropic) compressibility

$$\kappa_S \equiv - \frac{1}{V} \frac{\partial^2 H}{\partial p^2} = - \frac{1}{V} \left(\frac{\partial V}{\partial p}\right)_{S,\,N} = - \left(\frac{\partial \ln V}{\partial p}\right)_{S,\,N} . \tag{2.51}$$

The physical meaning of the response functions (2.44, 2.45) and (2.47, 2.48) is obvious: the isothermal bulk modulus is a relative measure for the change in pressure with the volume, the coefficient of thermal pressure is a relative measure for the change in pressure with temperature *etc.* The significance of the heat capacity becomes evident upon transformation of the defining equations (2.46) and (2.49). Using the balance equations (2.38 – 2.43) under the condition that $V, N =$ const. and the assumption that the changes of state are reversible, (2.46) results in

$$C_V \equiv \left(\frac{\partial U}{\partial T}\right)_{V,\,N} = \left(\frac{\mathrm{d}Q}{\mathrm{d}T}\right)_{V,\,N} . \tag{2.52}$$

Correspondingly, for reversible changes of state, (2.49) with (2.38 – 2.43), (2.28) and $p, N =$ const. leads to

$$C_p \equiv \left(\frac{\partial H}{\partial T}\right)_{p,\,N} = \left(\frac{\mathrm{d}Q}{\mathrm{d}T}\right)_{p,\,N} . \tag{2.53}$$

Thus, in the case of reversible changes of state, the heat capacity is a measure of the change in the heat content of a system with temperature (in the case of irreversible changes of state, see Sect. 7.3). With regard to measurement techniques, the adiabatic response functions (2.50) and (2.51) refer to idealized limiting cases which can be approximatively realized with rapidly proceeding changes of state, *e.g.*, by periodic stress with a small amplitude and a high frequency.

There are several purely mathematical relations between the response functions (2.44 – 2.51). To start with, the relations:

$$k_T = 1/\kappa_T , \tag{2.54}$$

$$k_S = 1/\kappa_S \tag{2.55}$$

become directly apparent from the defining equations (2.44, 2.47) and (2.50, 2.51). Furthermore, the equation of state (2.5) results in

$$\mathrm{d}p \equiv \left(\frac{\partial p}{\partial T}\right)_{V,N} \mathrm{d}T + \left(\frac{\partial p}{\partial V}\right)_{T,N} \mathrm{d}V + \left(\frac{\partial p}{\partial N}\right)_{T,V} \mathrm{d}N \tag{2.56}$$

and from this, when one maintains p and N constant

$$\left(\frac{\partial p}{\partial T}\right)_{V,N} \mathrm{d}T + \left(\frac{\partial p}{\partial V}\right)_{T,N} \mathrm{d}V = 0 \,.$$

However, when $p, N = \text{const.}$

$$\left(\frac{\mathrm{d}V}{\mathrm{d}T}\right) \equiv \left(\frac{\partial V}{\partial T}\right)_{p,N} ,$$

i.e.,

$$\left(\frac{\partial p}{\partial T}\right)_{V,N} + \left(\frac{\partial p}{\partial V}\right)_{T,N} \left(\frac{\partial V}{\partial T}\right)_{p,N} = 0 \tag{2.57a}$$

or

$$\left(\frac{\partial p}{\partial T}\right)_{V,N} \left(\frac{\partial T}{\partial V}\right)_{p,N} \left(\frac{\partial V}{\partial p}\right)_{T,N} = -1 \tag{2.57b}$$

is valid. According to (2.45), (2.47), and (2.48), this corresponds to the relation

$$\alpha = p\,\kappa_T\,\beta \,. \tag{2.58}$$

Similarly, one obtains from the differential form of the equation of state (2.3)

$$\mathrm{d}S = \left(\frac{\partial S}{\partial T}\right)_{V,N} \mathrm{d}T + \left(\frac{\partial S}{\partial V}\right)_{T,N} \mathrm{d}V + \left(\frac{\partial S}{\partial N}\right)_{T,V} \mathrm{d}N$$

when $p, N = \text{const.}$

$$\left(\frac{\partial S}{\partial T}\right)_{p,N} = \left(\frac{\partial S}{\partial T}\right)_{V,N} + \left(\frac{\partial S}{\partial V}\right)_{T,N} \left(\frac{\partial V}{\partial T}\right)_{p,N} .$$

Substitution with the expressions given by (2.45, 2.46) and (2.48, 2.49) leads to the relation

$$C_p = C_V + T p V \alpha \beta \tag{2.59}$$

or with (2.58) and (2.54) to

$$C_p = C_V + TVk_T\alpha^2 \,. \tag{2.60}$$

The relations (2.59, 2.60) allow an interconversion of C_p and C_V if the coefficient of thermal expansion and the coefficient of thermal pressure or the bulk modulus are known. The possibility of conversion is particularly important for materials in a condensed state, as C_p is experimentally and C_V theoretically more accessible in this state (with regard to solids, see Sect. 5.2).

From the equation of state (2.21) it follows that:

$$\mathrm{d}p = \left(\frac{\partial p}{\partial S}\right)_{V,N} \mathrm{d}S + \left(\frac{\partial p}{\partial V}\right)_{S,N} \mathrm{d}V + \left(\frac{\partial p}{\partial N}\right)_{S,V} \mathrm{d}N$$

and from this, provided that p, N = const.,

$$\left(\frac{\partial p}{\partial S}\right)_{V,N} \left(\frac{\partial S}{\partial V}\right)_{p,N} + \left(\frac{\partial p}{\partial V}\right)_{S,N} = 0 \tag{2.61 a}$$

or

$$\left(\frac{\partial p}{\partial S}\right)_{V,N} \left(\frac{\partial S}{\partial V}\right)_{p,N} \left(\frac{\partial V}{\partial p}\right)_{S,N} = -1 \,. \tag{2.61 b}$$

With the help of (2.57) and (2.61), we obtain, according to (2.51) and (2.47), the ratio

$$\begin{aligned}
\frac{\kappa_S}{\kappa_T} &= \left(\frac{\partial V}{\partial p}\right)_{S,N} \Big/ \left(\frac{\partial V}{\partial p}\right)_{T,N} \\
&= \left(\frac{\partial p}{\partial T}\right)_{V,N} \left(\frac{\partial T}{\partial V}\right)_{p,N} \Big/ \left(\frac{\partial p}{\partial S}\right)_{V,N} \left(\frac{\partial S}{\partial V}\right)_{p,N} \\
&= \left(\frac{\partial S}{\partial p}\right)_{V,N} \left(\frac{\partial p}{\partial T}\right)_{V,N} \Big/ \left(\frac{\partial S}{\partial V}\right)_{p,N} \left(\frac{\partial V}{\partial T}\right)_{p,N} \\
&= \left(\frac{\partial S}{\partial T}\right)_{V,N} \Big/ \left(\frac{\partial S}{\partial T}\right)_{p,N} \,.
\end{aligned}$$

Thus, according to (2.54, 2.55) and (2.46, 2.49), one gets the relation

$$\frac{\kappa_S}{\kappa_T} = \frac{k_T}{k_S} = \frac{C_V}{C_p} \,. \tag{2.62}$$

Additional relationships can, of course, be derived from the given interrelations between the response functions. For example, (2.60) and (2.62) result in

$$C_p = \frac{TV\alpha^2}{\kappa_T - \kappa_S} . \tag{2.63}$$

Corresponding to the relations between the response functions, there are also relations between the functional dependencies of the response functions with respect to the independent variables.

With T, p, N as independent variables (G-representation) we obtain, *e.g.*, by differentiation of (2.47) with respect to T and of (2.48) with respect to p, the relation

$$\left(\frac{\partial \kappa_T}{\partial T}\right)_{p,N} = -\left(\frac{\partial \alpha}{\partial p}\right)_{T,N} . \tag{2.64}$$

That is, if the isothermal compressibility increases with temperature at constant pressure, the coefficient of expansion decreases with increasing pressure at constant temperature. By differentiation of (2.49) with respect to p and of (2.48) with respect to T, we obtain with (2.14)

$$\left(\frac{\partial C_p}{\partial p}\right)_{T,N} = -TV\left[\alpha^2 + \left(\frac{\partial \alpha}{\partial T}\right)_{p,N}\right] . \tag{2.65a}$$

For solids with a very small coefficient of thermal expansion, one can write to a good approximation

$$\left(\frac{\partial C_p}{\partial p}\right)_{T,N} \approx -TV\left(\frac{\partial \alpha}{\partial T}\right)_{p,N} . \tag{2.65b}$$

If the expansivity increases with temperature at constant pressure, the heat capacity at constant pressure decreases with increasing pressure at constant temperature. Correspondingly, the F-representation (with T, V, N as independent variables) yields, for example, with respect to (2.44–2.46)

$$\left(\frac{\partial k_T}{\partial T}\right)_{V,N} = k_T\beta - pV\left(\frac{\partial \beta}{\partial V}\right)_{T,N} . \tag{2.66}$$

and

$$\left(\frac{\partial C_V}{\partial V}\right)_{T,N} = -Tp\left[\beta^2 + \left(\frac{\partial \beta}{\partial T}\right)_{V,N}\right] . \tag{2.67}$$

On the other hand, statements regarding the temperature dependence of the heat capacity, the pressure dependence of the compressibility, and the volume dependence of the bulk modulus cannot be obtained from the purely mathematical framework. Quantum-mechanical treatments reveal (*e.g.*, see Guggenheim 1959) that α and β vanish when $T \to 0$ in the case of condensed media in an undisturbed stable or metastable state of internal equilibrium (so-called third law of thermodynamics). Equally, C_p and C_V vanish with temperature (Sect. 6.4.1). The compressibilities κ_T and κ_S, or rather the moduli k_T and k_S, on the other hand, become equal as $T \to 0$. They retain, however, a finite value. Apart from these considerations, the pressure dependence of the compressibility matters only for very high pressures. The response functions can, of course, in the vicinity of every arbitrary temperature outside the range of thermal transformations, be developed into suitable power series as function of temperature. Only a few terms are often sufficient at higher temperatures in order to achieve good approximation of the experimental data over a sometimes quite large interval (see also Sect. 6.5.4).

2.4 Equilibrium and Stability Conditions

In Chap. 2 we confine ourselves to the consideration of homogeneous simple fluids. The homogeneity assumed with respect to temperature, pressure, and chemical potential means that the medium is, by definition, in a stable thermal, mechanical, and material equilibrium (see also below, as well as Chap. 4). In such media, non-equilibrium states and instabilities can only be induced by the macroscopic relevance of atomic or molecular internal degrees of freedom, which are generally concealed in equilibrium thermodynamics and only become globally apparent in dQ. With regard to polymers, particularly the degrees of freedom of conformation, of mutual arrangement, and mutual orientation of the chain molecules should be mentioned. In the macroscopic-phenomenological theory, these internal degrees of freedom can, of course, only be described if they can be associated with macroscopic variables, the so-called internal variables, such as the concentration of conformational isomers, the degree of crystallization, or the mean degree of orientation with respect to a particular direction. In a state of non-equilibrium (particularly true for polymers), more than three macroscopic degrees of freedom are, therefore, attributed to a homogeneous fluid thermodynamic system. Only the restrictive equilibrium conditions cause the direct macroscopic effect of these additional degrees of freedom to disappear (*cf.* further below).

In the following, we assume that the Gibbs formalism, as far as it has been described up to now, also remains valid in non-equilibrium (for further comments on this assumption see Sect. 7.1). In particular, we assume that an entropy can also uniquely be attributed to states of non-equilibrium. Furthermore, we assume (as before) that U describes the total energy of the systems, so that according to the principle of conservation of energy, $\mathrm{d}_i U = 0$ also holds for non-equilibrium states.

The change in the free energy of a homogeneous fluid system during an arbitrary process is then, according to (2.18), given by

$$\mathrm{d}F = \mathrm{d}U - T\mathrm{d}S - S\mathrm{d}T\,.$$

With (2.39), (2.40), (2.42) and (2.43 a), we obtain

$$dF = -S\,dT - p\,dV + (h - Ts)\,d_aN - T\,d_iS$$

and with (2.30) and (2.41)

$$dF = -S\,dT - p\,dV + \mu\,dN - T\,d_iS\,. \tag{2.68}$$

This equation describes the change in free energy during an irreversible process. With Eq. (2.43 b), Eq. (2.68) transforms, of course, again into Eq. (2.6), which is valid for changes in equilibrium or for reversible processes.

If one holds temperature, volume, and mass constant and leaves the system to itself, (2.68) reduces to

$$dF = -T\,d_iS\,. \tag{2.69}$$

Under these circumstances, the second law (2.43) says that during all irreversible changes in non-equilibrium:

$$dF < 0 \tag{2.70a}$$

and during all reversible changes in equilibrium it is valid that:

$$dF = 0\,. \tag{2.70b}$$

The inequality (2.70 a) means that the free energy of a system left to itself in a state of non-equilibrium always decreases, provided that T, V, N = const. Consequently, if the process should ever come to a standstill, the system strives to achieve a state of equilibrium "e", which possesses a minimal free energy F_e.

Thus, $F > F_e$ must hold for every non-equilibrium state neighbouring the attained state of equilibrium "e". In the Taylor-expansion of the free energy F about the equilibrium value F_e:

$$F = F_e + (dF)_e + \frac{1}{2}\,(d^2F)_e + \ldots \tag{2.71}$$

the linear and quadratic terms must necessarily fulfil the respective conditions (minimum conditions):

$$(dF)_e = 0 \quad \text{and} \quad (d^2F)_e > 0\,. \tag{2.72}$$

With a greater distance from the equilibrium, the higher terms of expansion which were not given above are, of course, also of importance. If the inequality $F > F_e$ is maintained on consideration of the higher terms, the state of equilibrium is designated as (absolutely) stable (Fig. 2.1 A). If this is not the case, the state of equilibrium is designated as metastable (Fig. 2.1 B). On the contrary, if $(d^2F)_e < 0$ holds in addition to the equilibrium condition $(dF)_e = 0$, we obtain $dF < 0$, *i. e.*, $F < F_e$, after an infinitesimal perturbation. According to (2.70), a real process is initiated. The state of equilibrium is unstable (Fig. 2.1 C). If $(\mathrm{d}F)_e = 0$ as well as $(\mathrm{d}^2F)_e = 0$ because $F = F_e = \text{const.}$, we have a neutral state of equilibrium (Fig. 2.1 D).

The equilibrium and stability conditions (2.72) require that the function F is differentiable at the position "e". If this is not the case, *e.g.*, because F follows two different analytic functions left and right of the position "e" or because no real physical states correspond to the values of the variables on one side of "e" (*e.g.*, compare Fig. 8.4), then (2.72) must be replaced by the condition

$$(\mathrm{d}F)_{e+0} > 0 \quad \text{or} \quad (\mathrm{d}F)_{e-0} > 0\,, \tag{2.73}$$

whereby $(\mathrm{d}F)_{e\pm 0}$ defines the differential on the right or left side of F at the position "e", respectively. (2.73), like (2.72), expresses that for a state of equilibrium "e", an infinitesimal perturbation cannot induce a real process for which $\mathrm{d}F < 0$ would have to apply.

Furthermore, if we assume a closed system which is not subject to any mass exchange with its surroundings, then $\mathrm{d}N \equiv 0$ or $\mathrm{d}M \equiv 0$, and the mass variable vanishes. The extensive variables can now be referred to unit of mass as specific quantities and are written by the appropriate small letters: $f \equiv F/M$ = specific free energy; $v \equiv V/M$ = specific volume, etc. If the non-equilibrium states are due to the macroscopic relevance of a single molecular internal degree of freedom, and if the macroscopic internal variable ζ is attributed to this degree of freedom, the complete Gibbs fundamental equation of the system in the F-representation is

$$f = f(T, v, \zeta)\,. \tag{2.74}$$

The equilibrium and stability conditions (2.72) then become explicitly

$$(\mathrm{d}f)_e = \left(\frac{\partial f}{\partial \zeta}\right)_e \mathrm{d}\zeta = 0 \tag{2.75}$$

and

$$(\mathrm{d}^2 f)_e = \left(\frac{\partial^2 f}{\partial \zeta^2}\right)_e (\mathrm{d}\zeta)^2 > 0\,. \tag{2.76}$$

Since the variation of ζ is completely arbitrary, it follows for the equilibrium states:

$$\left(\frac{\partial f}{\partial \zeta}\right)_e = 0\,. \tag{2.77}$$

In addition,

$$\left(\frac{\partial^2 f}{\partial \zeta^2}\right)_e > 0 \tag{2.78}$$

holds particularly for a stable or metastable equilibrium.

$$\left(\frac{\partial^2 f}{\partial \zeta^2}\right)_e < 0 \quad \text{or} \quad \left(\frac{\partial^2 f}{\partial \zeta^2}\right)_e = 0\,, \tag{2.79}$$

respectively, is valid for an unstable and a neutral equilibrium. The equilibrium characterized by (2.77) is designated as an internal equilibrium. Since $(\partial f/\partial \zeta)_e$ is a function of the variables T, v, ζ, the equilibrium condition (2.77) fixes a given variable as a function of the others, *e.g.*,

$$\zeta = \zeta_e(T, v)\,. \tag{2.80}$$

Although ζ is variable in internal equilibrium, it is not an independent variable. The internal variable can be eliminated by means of the equation (2.80). Then only T and v explicitly appear as variables in equilibrium thermodynamics, the internal variable remains hidden (see Sect. 6.3 for an example).

For the change in free energy, when $T, V, N = \text{const.}$,

$$\mathrm{d}f = \left(\frac{\partial f}{\partial \zeta}\right) \mathrm{d}\zeta\,,$$

is generally given in a homogeneous, closed system with a macroscopically relevant internal degree of freedom. Thus, the equilibrium condition (2.70b) is also fulfilled if

$$\mathrm{d}\zeta = 0\,, \quad \textit{i.e.,} \quad \zeta = \text{const.}\,. \tag{2.81}$$

The equilibrium described by (2.70b) and (2.81) is referred to as an arrested equilibrium (with respect to the internal variable ζ). Only T and v are variable in an arrested equilibrium, ζ can be ignored. A typical arrested state of equilibrium is the vitreous state, in which the internal degrees of freedom of the diffusive translational motions of the molecules are arrested (frozen) (Sect. 7.8).

The path of the free energy $F(T, v, \zeta)$ in the vicinity of the different states of equilibrium is represented in Fig. 2.1. On the right (E), a state of non-equilibrium with $(\partial f/\partial \zeta) \neq 0$ and $f \neq f_e$ was assumed to be an arrested state of equilibrium. It should be noted that stable, metastable, unstable, or neutral states of equilibrium can, of course, also appear as arrested states of equilibrium, as long as (2.81) is valid.

If several, mutually independent internal degrees of freedom are macroscopically relevant, and if the internal variables $\zeta_1 \ldots \zeta_n$ are attributed to these degrees of freedom,

$$(\mathrm{d}f)_e = \sum_i \left(\frac{\partial f}{\partial \zeta_i}\right)_e \mathrm{d}\zeta_i = 0 \tag{2.82a}$$

holds in place of the equilibrium condition (2.75). As the ζ_i are mutually independent and can be varied at will, the equilibrium condition for each individual

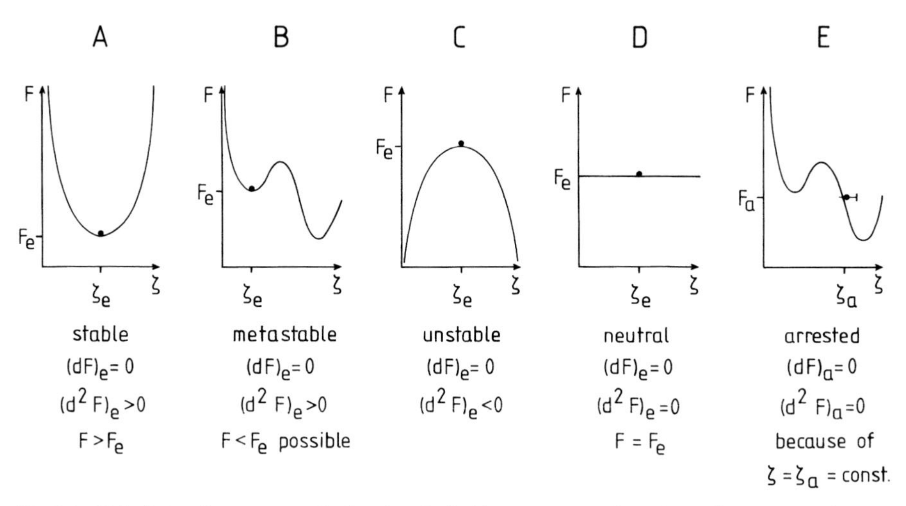

Fig. 2.1. Helmholtz free energy F of a simple fluid system as a function of the internal variable ζ in the neighbourhood of various internal equilibrium states "e". $T, V, N =$ const. was assumed. A non-equilibrium state "a" is represented in the right as an arrested equilibrium state

degree of freedom becomes

$$\left(\frac{\partial f}{\partial \zeta_i}\right)_e = 0\,, \qquad i = 1 \ldots n\,. \tag{2.82b}$$

The internal variables $\zeta_i(T, v)$ can, in turn, be eliminated from the equilibrium thermodynamics by means of these n equations. The stability condition is now

$$(\mathrm{d}^2 f)_e = \sum_i \sum_k \left(\frac{\partial^2 f}{\partial \zeta_i \partial \zeta_k}\right)_e \mathrm{d}\zeta_i \mathrm{d}\zeta_k > 0\,. \tag{2.83}$$

The partial derivatives are interchangeable in the coefficients of this inequality, as they refer to the equilibrium. Hence, the matrix of the coefficients is symmetrical. Thus, the inequality (2.83) expresses that the matrix of the coefficients $(\partial^2 f / \partial \zeta_i \partial \zeta_k)_e$ must be positive definite in stable and metastable equilibria.

A symmetric matrix is called positive definite if all of its diagonal elements are positive and if their determinant and all principal minors are positive or zero. The principal minors of a matrix are obtained by eliminating an arbitrary number of rows and columns with the same ordinal number in the matrix and by forming the determinant from the remaining elements of the matrix. A symmetric matrix is called negative definite if their principal minors of odd order are negative and their principal minors of even order are positive. Thus, the matrix

$$\begin{pmatrix} M_{11} & M_{12} \\ M_{21} & M_{22} \end{pmatrix} \qquad \text{with } M_{12} = M_{21}$$

is positive definite if

$$M_{11} > 0 \,; \quad M_{22} > 0 \,; \quad M_{11}M_{22} - M_{12}^2 \geq 0$$

and negative definite if

$$M_{11} < 0 \,; \quad M_{22} < 0 \,; \quad M_{11}M_{22} - M_{12}^2 > 0 \,.$$

For the change in free enthalpy G, internal energy U, and entropy S of a homogeneous fluid system in non-equilibrium, we obtain corresponding to equation (2.68)

$$\mathrm{d}G = -\,S\mathrm{d}T + V\mathrm{d}p + \mu\mathrm{d}N - T\,\mathrm{d}_i S \,, \tag{2.84}$$

$$\mathrm{d}U = \quad T\mathrm{d}S - p\mathrm{d}V + \mu\mathrm{d}N - T\,\mathrm{d}_i S \,, \tag{2.85}$$

$$\mathrm{d}S = \frac{1}{T}\,\mathrm{d}U + \frac{p}{T}\,\mathrm{d}V - \frac{\mu}{T}\,\mathrm{d}N + \mathrm{d}_i S \,. \tag{2.86}$$

Because of $\mathrm{d}_i U = 0$, (2.85) is, of course, merely a different form of the conservation law of energy (2.42) – as can easily be confirmed by substitution with (2.40) and (2.43 a) under consideration of (2.30). From (2.84) and (2.85) it follows that the second law (2.43), provided that $T, p, N =$ const., is reduced to a minimum principle for the free enthalpy

$$\mathrm{d}G = -\,T\,\mathrm{d}_i S \leq 0 \tag{2.87}$$

with the equilibrium and stability conditions

$$(\mathrm{d}G)_e = 0 \,; \quad (\mathrm{d}^2 G)_e > 0 \,, \tag{2.88}$$

and provided that $S, V, N =$ const., to a minimum principle for the internal energy

$$\mathrm{d}U = -\,T\,\mathrm{d}_i S \leq 0 \tag{2.89}$$

with the equilibrium and stability conditions

$$(\mathrm{d}U)_e = 0 \,; \quad (\mathrm{d}^2 U)_e > 0 \,. \tag{2.90}$$

For energetically closed systems with $U, V, N =$ const., (2.86) reduces to the maximum principle (2.43 e) with the equilibrium and stability conditions

$$(\mathrm{d}S)_e = 0 \,; \quad (\mathrm{d}^2 S)_e < 0 \,. \tag{2.91}$$

In addition, these minimum and maximum principles and the resulting equilibrium and stability conditions apply equally to mixtures (Chap. 3) and to heterogeneous systems (Chap. 4). In heterogeneous systems, the external intensive variables can also appear as non-equilibrium variables. Heterogeneity (or inhomogeneity) with regard to the temperature means thermal non-equilibrium, heterogeneity (or inhomogeneity) with regard to the pressure (or with regard to the components of the stress tensor; see Chap. 5) means mechanical non-equilibrium, and heterogeneity (or inhomogeneity) with regard to the chemical potential means material non-equilibrium. It should be noted that the conditions $T, V, N =$ const. posed for (2.70, 2.72) refer to the total system in the case of heterogeneous systems, and are in detail $T =$ const., $V = \sum V_k =$ const., $N = \sum N_k =$ const. (V_k, N_k: volumes and mole numbers of the homogeneous domains from which the heterogeneous system is composed; Chap. 4). Hence, the conditions certainly allow a variation of the extensive variables V_k, N_k in the domains of the system. The constancy of the intensive variable T, on the other hand, requires constancy (homogeneity) in the entire system. Equations (2.70, 2.72), therefore, only allow statements about mechanical, material, and internal equilibria, but no statements about thermal equilibrium. Correspondingly, only statements about material and internal equilibria can be made on the basis of (2.87, 2.88) due to the constancy of the intensive variables T and p. Equations (2.89–2.91), on the other hand, as the subsidiary conditions refer only to extensive variables, yield information about the thermal and mechanical as well as material and internal equilibria (*cf.* Chap. 4).

CHAPTER 3

Homogeneous Mixtures

The material properties of a thermodynamic system are essentially determined by the mass, shape, size, and interaction of the molecules present in the system if chemical reactions are excluded. The molecular mass acts as a unit of measurement for the amount of material present in the system. In the macroscopic theory, the amount of material acts directly as a variable, whereas the shape, size, and interaction of the molecules contribute to the chemical potential. The composition of a system of several, different substances leads to a considerable expansion in the variety of physical behaviour. At the same time, this makes the finding of solutions to theoretical problems considerably more difficult.

With regard to polymers, one observes that macromolecules with a uniform mass, shape, and size commonly originate only during biosyntheses. Most polymers represent mixtures of molecules of different masses, shapes, and sizes, as well as a complex diversity of possible interactions. An exact description of such mixtures is not possible. Relatively simple, physically transparent approximations can only be obtained in the case of homologous polymer mixtures, *i. e.*, mixtures composed of long linear molecules with the same chemical composition, but different degrees of polymerization. An unequivocal description of mixtures composed of more complicated polymer molecules, such as branched polymers or copolymers, is not only extremely difficult theoretically, but also experimentally. The behaviour of such complicated systems, however, can be quite similar to that of simple systems. In the case of cross-linked polymers, even the specification of a molecular mass no longer makes sense.

3.1 Mass Variables

We will examine a homogeneous mixture composed of different components. Each component is assumed to contain a certain (very large) number of molecules which are in every respect mutually identical, but different from the molecules of the other components. The individual components are numbered with $i = 1 \ldots r$.

For a characterization of the relative proportions of the masses, an introduction of the masses

$$M_i = Z_i m_i^+ \tag{3.1}$$

of the individual components is obvious, and certainly always possible (Z_i: number of molecules of the species i; $m_i^\dagger$: mass of the individual molecule of the species i). If the total mass

$$M = \sum_i M_i \tag{3.2}$$

of the system is constant, it is often advantageous to use the mass concentrations (the so-called mass or weight fractions) as mass variables in place of the masses M_i:

$$c_i \equiv M_i/M \quad \text{with} \quad \sum_i c_i = 1\,. \tag{3.3}$$

In thermodynamics, however, the mole numbers N_i are preferred to the masses M_i. These allow a substance-independent formulation for several fundamental correlations (*e.g.*, the equations of state of ideal gases and the entropic aspects of ideal mixtures). One mole of a substance contains as many grams of a substance as its molecular weight indicates. One mole always contains, independent of the substance,

$$N_A = 6.022 \cdot 10^{23}$$

molecules (Avogadro's or Loschmidt's number). From a logical point of view, N_A is a dimensionless pure number. In the following, however, we will follow the practice in physical chemistry of attributing the dimension mol^{-1} to this number. The mole number (with the dimension mol) is then defined by

$$N_i \equiv Z_i/N_A\,. \tag{3.4 a}$$

As a result of the dimensioning of Avogadro's number with mol^{-1}, the factor mol^{-1} must be added to the usual dimensions of the physical quantities if they refer to one mole.

If $m_i = N_A m_i^\dagger$ stands for the mass of a mole (in g/mol or kg/mol) of the component i, there exists for the mass M_i the relationship

$$M_i = m_i N_i\,. \tag{3.4 b}$$

The total number of moles in the system is

$$N = \sum_i N_i\,.$$

The mole fractions

$$x_i \equiv N_i/N \quad \text{with} \quad \sum_i x_i = 1 \tag{3.5}$$

can be introduced as mass variables corresponding to the weight fractions (3.3). According to (3.1), (3.3), and (3.4), the weight fractions and mole fractions

(3.5) are linked *via*

$$c_i = \frac{m_i x_i}{\sum\limits_i m_i x_i} = \frac{m_i^{\dagger} x_i}{\sum\limits_i m_i^{\dagger} x_i} \tag{3.6}$$

In addition to the weight and mole fractions, the so-called volume fractions are also used as mass variables. The volume fractions φ_i of the components are defined by

$$\varphi_i = \frac{\bar{v}_i N_i}{\sum\limits_i \bar{v}_i N_i} = \frac{\bar{v}_i x_i}{\sum\limits_i \bar{v}_i x_i} \quad \text{with} \quad \sum_i \varphi_i = 1\,, \tag{3.7}$$

whereby

$$\bar{v}_i = \left(\frac{\partial V}{\partial N_i}\right)_{T,\,p,\,N_{k \neq i}} \tag{3.7}$$

is the partial molar volume of the component i [*cf.* Eq. (3.17) in Sect. 3.2].

In the case of polymers whose molecules consist of long, non-branched chains of repeating molecular units, it has often proved useful not to proceed from the mole numbers N_i of the molecules, but from the mole numbers N_i^* of the mutually identical repeating units. This implies, however, that the chains are long enough so that their ends, which are always different from the internal links, can be neglected. If P_i is the degree of polymerization, *i.e.*, the number of repeating units in the chain molecules of the components i, the mole fractions

$$x_i^* = \frac{N_i^*}{\sum\limits_i N_i^*} = \frac{P_i N_i}{\sum\limits_i P_i N_i} = \frac{P_i x_i}{\sum\limits_i P_i x_i} \quad \text{with} \quad \sum_i x_i^* = 1 \tag{3.8}$$

can be introduced as mass variables. Disregarding extreme pressure conditions, the volume of a polymer chain is practically proportional to the degree of polymerization. The mole fractions (3.8) are, therefore, often identified with the volume fractions (3.7). This is an exact identification in the case of ideal-athermal polymer mixtures (Sect. 3.4) and generally then, when shape, size, and interaction of the molecules are not dependent on pressure [*cf.* Sects. 3.3. and 3.5].

According to (3.4a) and (3.5),

$$x_i = \frac{Z_i}{\sum\limits_i Z_i}\,, \tag{3.9}$$

is also valid for the mole fractions. If the molecules in the mixture are statistically distributed, x_i has the significance of being the probability of finding a molecule of the species i when one selects one molecule out of the entire population. By means of x_i, average values can then be determined. For example,

$$\langle P \rangle = \sum_i x_i P_i \tag{3.10}$$

is the mean degree of polymerization and

$$\langle m\rangle_n = \sum_i x_i m_i \tag{3.11}$$

the mean molar mass. Equations (3.10) and (3.11) represent average values referenced to the number of molecules. The weight averages, which can correspondingly be determined using the weight fractions (3.3), are often also of importance. For example, the weight average of the molar mass is:

$$\langle m\rangle_w = \sum_i c_i m_i\,. \tag{3.12}$$

Substitution of (3.6) into (3.12) results in

$$\langle m\rangle_w = \frac{\sum\limits_i x_i m_i^2}{\sum\limits_i x_i m_i}\,. \qquad (3.12)$$

Hence, the product

$$\langle m\rangle_n \langle m\rangle_w = \sum_i x_i m_i^2$$

is equal to the second moment of the number average. Higher moments are also of importance experimentally. The so-called centrifuge or z-average:

$$\langle m\rangle_z = \frac{\sum\limits_i x_i m_i^3}{\sum\limits_i x_i m_i^2} \tag{3.14}$$

is, for example, of importance when determining the mass distribution with an ultracentrifuge.

3.2 Fundamental Equations, Subsidiary and Stability Conditions

The thermodynamics of mixtures can, as long as the nature of the problem permits, best be described on the basis of the G-representation. The G-representation not only takes into account the usually experimentally given fact that T and p should be considered as independent variables. Even more so, it is particularly suitable for a description of the conditions caused by the presence of different substances, since the mole numbers are the only extensive variables in this representation.

In the G-representation for every arbitrary extensive quantity E (*e.g.*, for the energy, the entropy, the volume, or the heat capacity) the following is valid if r mutually independent components are found in the mixture:

$$E = E(T, p, N_1 \ldots N_r) \tag{3.15}$$

or, with an arbitrary number λ,

$$\lambda E = E(T, p, \lambda N_1 \ldots \lambda N_r) .$$

Thus, according to Euler's theorem on homogeneous functions of the first degree, every arbitrary, extensive quantity E referring to a mixture of r components must be representable in the form:

$$E = \sum_i \bar{e}_i N_i \tag{3.16}$$

The quantities

$$\bar{e}_i (T, P, N_1 \ldots N_r) \equiv \left(\frac{\partial E}{\partial N_i} \right)_{T, p, N_{k \neq i}} \tag{3.17}$$ [3]

are defined as partial molar values of the quantity E. They describe the components' partial contribution to the quantity E in the mixture (which is dependent on the mixing ratio). In general, they do not directly refer to the molar quantities of the pure substances. The $\bar{e}_i$, however, can always be normalized in such a way that at the limit

$$\lim_{N_i \to N} \bar{e}_i = e_i^o (T, p)$$

(where $N_i \to N$ means at the same time $N_{k \neq i} \to 0$), they transform into the molar values $e_i^o = E/N$ relative to one mole of the pure substances [*e.g.*, compare equation (3.60) with (3.49b)]. These limiting values can sometimes have a somewhat hypothetical character, for example, when the substance i under pressure p and temperature T occurs in the mixture in a fluid state, but as a pure substance in a solid state.

A comparison of the differentials of (3.15) and (3.16) shows that a generalized form of the Gibbs–Duhem relation must be valid:

$$\left(\frac{\partial E}{\partial T} \right)_{p, N_i} \mathrm{d}T + \left(\frac{\partial E}{\partial p} \right)_{T, N_i} \mathrm{d}p - \sum_i N_i \mathrm{d}\bar{e}_i = 0 . \tag{3.18}$$ [4]

Hence, of the $r + 2$ intensive variables $T, p, \bar{e}_i$, only $r + 1$ are independent of each other. If specifically $T, p =$ const. is valid, only $r - 1$ of the r partial molar quantities $\bar{e}_i$ are independent of each other. Differentiation of (3.17) with respect to N_k results in (if the mixture is in equilibrium)

$$\frac{\partial \bar{e}_i}{\partial N_k} = \frac{\partial^2 E}{\partial N_k \, \partial N_i} = \frac{\partial^2 E}{\partial N_i \, \partial N_k} = \frac{\partial \bar{e}_k}{\partial N_i} . \tag{3.19}$$

3 The subscript $N_{k \neq i}$ in (3.17) means that except for N_i all N_k must be held constant for differentation.

4 The subscript N_i in (3.18) means that all N_i must be held constant for determination of the partial derivatives.

The matrix of the quantities of state

$$e_{ik}(T, p, N_1 \dots N_r) \equiv \frac{\partial \bar{e}_i}{\partial N_k} \tag{3.20}$$

is symmetrical. Since the N_i are independent of each other, differentiation of (3.16) with respect to N_k results with (3.17) and (3.20) in:

$$\bar{e}_k = \sum_i e_{ik} N_i + \bar{e}_k \, .$$

This means, however, that the linear system of equations holds (Duhem-Margules relations):

$$\sum_i e_{ki} N_i = 0 \, , \quad k = 1 \dots r \tag{3.21}$$

This system of equations only has non-trivial solutions $N_i \neq 0$ if the determinant of the coefficients vanishes:

$$\begin{vmatrix} e_{11} \dots e_{1r} \\ \vdots \quad\quad \vdots \\ e_{r1} \dots e_{rr} \end{vmatrix} = 0 \, . \tag{3.22}$$

The matrix of the functions of state (3.20) possesses r^2 elements and r diagonal elements, *i.e.*, $r(r-1)$ non-diagonal elements. On the basis of the symmetry relations (3.19), half, *i.e.*, $r(r-1)/2$, of the non-diagonal elements are dependent on the other elements. In addition, the r equations (3.21) are valid between the e_{ik}. Therefore, only

$$r^2 - r(r-1)/2 - r = r(r-1)/2 \tag{3.23}$$

of the e_{ik} can be independent of each other.

Gibbs' fundamental equation in the G-representation for a mixture of r components is:

$$G = G(T, p, N_1 \dots N_r) \, . \tag{3.24}$$

The $r+2$ equations of state are given by

$$\left(\frac{\partial G}{\partial T}\right)_{p, N_i} = - S(T, p, N_1 \dots N_r) \, , \tag{3.25}$$

$$\left(\frac{\partial G}{\partial p}\right)_{T, N_i} = V(T, p, N_1 \dots N_r) \, , \tag{3.26}$$

$$\left(\frac{\partial G}{\partial N_i}\right)_{T, p, N_{k \neq i}} = \mu_i(T, p, N_1 \dots N_r) \, . \tag{3.27}$$

The chemical potentials μ_i of the components are identical with the partial molar free enthalpies $\bar{g}_i$ of the components [compare (3.27) with (3.17)]. Hence, the

μ_i must fulfil the subsidiary conditions (3.18), (3.19), (3.21), and (3.22). Corresponding to (3.16),

$$G = \sum_i \mu_i N_i \tag{3.28}$$

is valid. The equilibrium thermodynamics of a mixture is completely determined if its composition and the chemical potentials of the individual components in the mixture are known. Therefore, the main problem of the theory lies in the determination of suitable representations for the functions $\mu_i\,(T, p, N_1 \ldots N_r)$.

By differentiation of the equations of state (3.25) and (3.26) with respect to N_i, we obtain with (3.17)

$$\bar{s}_i = -\frac{\partial \mu_i}{\partial T}\,; \quad \bar{v}_i = \frac{\partial \mu_i}{\partial p}\,. \tag{3.29}$$

Furthermore, we obtain in the G-representation from the universally valid relations

$$U = G + TS - pV \tag{3.30}$$

and

$$H = G + TS \tag{3.31}$$

[see Eqs. (2.12), (2.19), and (2.30)] by differentiation with respect to N_i with (3.29), the relations

$$\bar{u}_i = \mu_i - T\frac{\partial \mu_i}{\partial T} - p\,\frac{\partial \mu_i}{\partial p} \tag{3.32}$$

and

$$\bar{h}_i = \mu_i - T\frac{\partial \mu_i}{\partial T}\,. \tag{3.33}$$

If one considers a mixture of r components under the boundary condition T, p, $N =$ const., only $r - 1$ of the mole numbers are still independent of each other. When we number the dependent variable with 1 and proceed to the mole fractions (3.5), it follows for the dependent variable:

$$x_1 = 1 - \sum_i{}' x_i\,; \quad \mathrm{d}x_1 = -\sum_i{}' \mathrm{d}x_i\,. \tag{3.34}$$

The prime at the symbol of summation denotes that summation only occurs from $i = 2 \ldots r$. According to (3.16),

$$g = \sum_i \mu_i x_i = \mu_1 + \sum_i{}' (\mu_i - \mu_1)\,x_i \tag{3.35}$$

holds for the molar Gibbs free enthalpy $g \equiv G/N$. The change in molar free enthalpy is given by

$$dg = \sum_i \mu_i \, dx_i + \sum_i x_i \, d\mu_i \, .$$

According to the Gibbs–Duhem relation (3.18), however, provided that $T, p =$ const., one finds that:

$$\sum_i x_i \, d\mu_i = 0 \, .$$

Thus, if $T, p, N =$ const., we have

$$dg = \sum_i \mu_i \, dx_i = \sum_i{}' (\mu_i - \mu_1) \, dx_i \, . \tag{3.36}$$

If we consider g as a function solely of the independent variables $x_2 \ldots x_r$, we obtain

$$\left(\frac{\partial g}{\partial x_i}\right)_{T,\, p,\, x_{k \neq i}} = \mu_i - \mu_1 \quad (i = 2 \ldots r) \, . \tag{3.37}$$

Questions about the stability of a homogeneous mixture are synonymous with questions about the separation of a mixture under the boundary condition $T, p, N =$ const. and if chemical reactions are excluded. Demixing is caused by diffusion processes and induces a decomposition of the homogeneous into a heterogeneous system (Chap. 4). In order to simplify matters, we assume in the following that a decomposition into a 2-phase system is potentially possible, that each component is present in each phase, and that the mole fractions are bilaterally variable (*cf.* Sect. 2.4). We mark the molar or rather partial molar quantities, which refer to the individual potential phases, with a prime and a double prime. Since the total mole numbers do not change during the demixing process, we obtain the boundary conditions for the molar or rather partial molar quantities:

$$x_i' + x_i'' = \text{const.} \, ; \qquad dx_i' = - \, dx_i'' \, . \tag{3.38}$$

The molar free enthalpy of the total system is given by

$$g = \sum_i \mu_i' x_i' + \sum_i \mu_i'' x_i'' \, . \tag{3.39}$$

As the Gibbs–Duhem relation (3.18) must hold for each phase individually (*cf.* Chap. 4), the following is valid:

$$\sum_i x_i' \, d\mu_i' = \sum_i x_i'' \, d\mu_i'' = 0 \, ,$$

and hence, with (3.38):

$$dg = \sum_i \mu_i' dx_i' + \sum_i \mu_i'' dx_i'' = \sum_i (\mu_i' - \mu_i'')\, dx_i' . \tag{3.40}$$

The equilibrium condition $dg = 0$ [*cf.* (2.88)] thus implies that:

$$\mu_i' = \mu_i'' . \tag{3.41}$$

With respect to the chemical potentials, a mixture is definitely homogeneous in equilibrium. With regard to questions about the stability of the equilibrium, we obtain with (3.20) and (3.38)

$$d^2g = \sum_{i,k} \mu_{ik}' dx_i' dx_k' + \sum_{i,k} \mu_i'' dx_i'' dx_k'' = \sum_{i,k} (\mu_{ik}' + \mu_{ik}'')\, dx_i' dx_k' .$$

If the mixture is homogeneous from the start, *i.e.*, also regarding the extensive quantities, and especially regarding the mole numbers N_i, we have $\mu_{ik}' = \mu_{ik}'' = \mu_{ik}$. Hence, according to (2.88), a homogeneous mixture is stable or metastable if it is valid that:

$$d^2g = 2 \sum_{i,k} \mu_{ik}\, dx_i\, dx_k > 0 . \tag{3.42}$$

A homogeneous mixture is stable when the matrix μ_{ik} is positive definite. Demixing sets in when the matrix μ_{ik} is negative definite.

According to (3.23), only a single μ_{ik} is independent of the others in the case of a binary mixture. As a stability condition, for example, it is then sufficient that:

$$\mu_{11} = \frac{\partial^2 g}{\partial x_1^2} > 0 . \tag{3.43}$$

3.3 Activities, Standard and Mixing Terms

The chemical potentials μ_i of the components of a mixture can always be described by

$$\mu_i = RT \ln \lambda_i \tag{3.44}$$

This is, a purely mathematical transformation which introduces the so-called absolute activities

$$\lambda_i = \lambda_i(T, p, x_1 \ldots x_r)$$

in place of the chemical potentials. Correspondingly,

$$\mu_i^o = RT \ln \lambda_i^o \quad \text{with} \quad \lambda_i^o = \lambda_i^o(T, p) \tag{3.45}$$

must hold for the chemical potentials $\mu_i^o(T, p)$ of the pure substances. From (3.44) and (3.45) follows that:

$$\mu_i = \mu_i^o + RT \ln a_i \tag{3.46}$$

whereby

$$a_i(T, p, x_1 \dots x_r) \equiv \lambda_i / \lambda_i^o \tag{3.47}$$

are referred to as the relative activities. The relative activity is a measure of the change of the chemical potential of a substance on mixing the substance with other substances. Such a mixing effect even occurs in the absence of any interactions during the statistical intermixing of particles of the same shapes and sizes. This simplest, purely entropic mixing effect is described by

$$a_i = x_i \tag{3.48}$$

(*cf.* Sects. 3.4 and 6.1). The starting equations (3.44–3.47) also acquire a physical background if one generally writes the activities as

$$a_i = x_i \gamma_i \, . \tag{3.49a}$$

The activity coefficients

$$\gamma_i = \gamma_i(T, p, x_1 \dots x_r)$$

contain those mixing effects that are due to the different shapes and sizes, and the interactions of the molecules. Because of (3.47), γ_i must satisfy the normalization

$$\lim_{x_i \to 1} \gamma_i = 1 \, . \tag{3.49b}$$

For the chemical potentials of the components of a mixture, one can generally set up the expression:

$$\mu_i = \mu_i^o(T, p) + RT \ln (x_i \gamma_i) \tag{3.50}$$

Via (3.28) and (3.5), it then follows as the Gibbs fundamental equation of the mixture in the G-representation that:

$$G = N \sum_i \mu_i^o x_i + RTN \sum_i x_i \ln (x_i \gamma_i) \, . \tag{3.51a}$$

Thereby

$$G_o \equiv N \sum_i \mu_i^o x_i \tag{3.51b}$$

is referred to as the standard free enthalpy of the mixture. According to (3.9), the molar standard free enthalpy is the mean value over the chemical potentials of the pure substances:

$$\langle \mu_i^o \rangle = g_o = G_o/N .$$

The deviations from this mean value are given by the free enthalpy of mixing

$$G_M \equiv RTN \sum_i x_i \ln(x_i \gamma_i) . \tag{3.51 c}$$

Likewise, the entropy (3.25) and the volume (3.26) can be split into a standard term and a mixing term. Differentiating (3.51 a) with respect to T, we obtain with (3.25) and (3.29)

$$S = S_o + S_M , \tag{3.52 a}$$

$$S_o \equiv - N \sum_i \frac{\partial \mu_i^o}{\partial T} x_i = - N \sum_i s_i^o x_i , \tag{3.52 b}$$

$$S_M \equiv - RN \left[\sum_i x_i \ln(x_i \gamma_i) + T \sum_i x_i \frac{\partial \ln \gamma_i}{\partial T} \right] . \tag{3.52 c}$$

Differentiating (3.51 a) with respect to p, one gets with (3.26) and (3.29)

$$V = V_o + V_M , \tag{3.53 a}$$

$$V_o \equiv N \sum_i \frac{\partial \mu_i^o}{\partial p} x_i = N \sum_i v_i^o x_i , \tag{3.53 b}$$

$$V_M \equiv RNT \sum_i x_i \frac{\partial \ln \gamma_i}{\partial p} , \tag{3.53 c}$$

where s_i^o and v_i^o, as well as u_i^o and h_i^o, in the following, denote the molar quantities of the pure substances. Furthermore, from (3.30–3.33) and (3.51–3.53) we obtain for the internal energy U and the enthalpy H

$$U = U_o + U_M , \tag{3.54 a}$$

$$U_o \equiv N \sum_i \left(\mu_i^o - p \frac{\partial \mu_i^o}{\partial p} - T \frac{\partial \mu_i^o}{\partial T} \right) x_i = N \sum_i u_i^o x_i , \tag{3.54 b}$$

$$U_M \equiv - RNT \left[p \sum_i x_i \frac{\partial \ln \gamma_i}{\partial p} + T \sum_i x_i \frac{\partial \ln \gamma_i}{\partial T} \right] , \tag{3.54 c}$$

and

$$H = H_o + H_M , \tag{3.55a}$$

$$H_o \equiv N \sum_i \left(\mu_i^o - T \frac{\partial \mu_i^o}{\partial T} \right) x_i = N \sum_i h_i^o x_i , \tag{3.55b}$$

$$H_M \equiv - RNT^2 \sum_i x_i \frac{\partial \ln \gamma_i}{\partial T} . \tag{3.55c}$$

If the activity coefficients γ_i are not dependent on temperature and pressure, the mixing term only appears in the entropy, and with the entropy in the free enthalpy $G = H - TS$ and the (not explicitly mentioned) free energy $F = U - TS$. In this case, all the other extensive quantities are free from mixing terms, *i.e.*, their molar quantities represent mean values of the quantities relative to the pure substances.

Differentiating (3.51a) twice with respect to T and p, one gets as response functions of the mixture *via* (2.47–2.49) the isothermal compressibility

$$\kappa_T = \kappa_T^o + \kappa_T^M , \tag{3.56a}$$

$$\kappa_T^o \equiv - \frac{1}{v} \sum_i \frac{\partial^2 \mu_i^o}{\partial p^2} x_i = \sum_i \kappa_{T,i}^o x_i , \tag{3.56b}$$

$$\kappa_T^M \equiv - \frac{RT}{v} \sum_i x_i \frac{\partial^2 \ln \gamma_i}{\partial p^2} , \tag{3.56c}$$

the thermal expansivity

$$\alpha = \alpha_o + \alpha_M , \tag{3.57a}$$

$$\alpha_o \equiv \frac{1}{v} \sum_i \frac{\partial^2 \mu_i^o}{\partial T \partial p} x_i = \sum_i \alpha_i^o x_i , \tag{3.57b}$$

$$\alpha_M \equiv \frac{R}{v} \left[\sum_i x_i \frac{\partial \ln \gamma_i}{\partial p} + T \sum_i x_i \frac{\partial^2 \ln \gamma_i}{\partial T \partial p} \right] , \tag{3.57c}$$

and the heat capacity

$$c_p = c_p^o + c_p^M , \tag{3.58a}$$

$$c_p^o \equiv - NT \sum_i \frac{\partial^2 \mu_i^o}{\partial T^2} x_i = N \sum_i c_{p,i}^o x_i , \tag{3.58b}$$

$$c_p^M \equiv - RNT \left[2 \sum_i x_i \frac{\partial \ln \gamma_i}{\partial T} + T \sum_i x_i \frac{\partial^2 \ln \gamma_i}{\partial T^2} \right] . \tag{3.58c}$$

In (3.56) and (3.57), $v = V/N$ denotes the molar volume of the mixture. The response functions are also free from mixing terms if the activity coefficients are neither dependent on temperature nor on pressure. A mixing term does not even appear in the compressibility if the activity coefficients are linearly dependent on pressure.

We should remember here that the heat capacity, which is an extensive quantity, can, in addition to (3.58), also be described with respect to (3.16), (3.17) by

$$C_p = N \sum_i \bar{c}_{p,i} x_i \,. \tag{3.59}$$

A comparison of the coefficients in (3.58) and (3.59) results in

$$\bar{c}_{p,i} = c^o_{p,i} - RT \left[2 \frac{\partial \ln \gamma_i}{\partial T} + T \frac{\partial^2 \ln \gamma_i}{\partial T^2} \right] . \tag{3.60}$$

Corresponding expressions can easily be derived for all other partial molar quantities (3.17). The partial molar free enthalpies $\bar{g}_i$ of the mixture are expressed by the equations (3.50), our starting point.

It should finally be noted that (3.50) and (3.20) result in

$$\mu_{ik} = RT \left(\frac{\partial \ln x_i}{\partial N_k} + \frac{\partial \ln \gamma_i}{\partial N_k} \right) = \frac{RT}{N} \left(\frac{\delta_{ik}}{x_i} - 1 \right) + RT \frac{\partial \ln \gamma_i}{\partial N_k} ,$$

where δ_{ik} is the Kronecker delta; *i.e.*, $\delta_{ik} = 1$ for $i = k$ and $\delta_{ik} = 0$ for $i \neq k$. The activity coefficients must, therefore, comply with the Duhem–Margules relations in the form

$$\sum_i N_i \frac{\partial \ln \gamma_i}{\partial N_k} = 0 \,, \quad k = 1 \ldots r \,, \tag{3.61}$$

3.4 Classes of Mixtures, Excess Quantities

A mixture in which Eq. (3.48) is fulfilled for all components is referred to as an ideal mixture. If Eq. (3.48) is fulfilled in the whole range of mole fractions x_i, the mixture is also referred to as a perfect mixture. Mixtures of ideal gases are always perfect. According to van't Hoff's law [see any physical chemistry textbook, *e.g.*, Moelwyn-Hughes (1961)], the substances k in a highly diluted solution ($x_i \approx 1, x_{k \neq i} \ll 1$) behave like an ideal gas. This means that sufficiently dilute solutions also exhibit ideal behaviour. Ideal or perfect behaviour can generally be expected for a mixture if all molecules have roughly the same shape and size; the molecules are statistically distributed throughout the mixture, and the molecular interaction is negligible. It is not the strengths of the binding energies, but rather the energy differences between the different types of binding that are decisive with regard to questions about the negligibility of interactions.

The molecular interactions in a mixture are negligible when the binding energy w_{ik} between two unequal molecules of the species i and the species k is equal to the arithmetic mean of the binding energies w_{ii} and w_{kk} between molecules of the same kind, *i.e.*, if

$$w_{ik} = \frac{1}{2}\,(w_{ii} + w_{kk})$$

or rather

$$w_{ik} - w_{ii} = w_{kk} - w_{ik}$$

is valid (*cf.* Sect. 3.5).

According to (3.51) with $\gamma_i = 1$, Gibbs fundamental equation for ideal mixtures is

$$G = N \sum_i \mu_i^o x_i + RTN \sum_i x_i \ln x_i\,. \tag{3.62}$$

The equations of state are given by

$$S = -\,N \sum_i s_i^o x_i - RN \sum_i x_i \ln x_i\,, \tag{3.63}$$

$$V = N \sum_i v_i^o x_i\,, \tag{3.64}$$

$$\mu_i = \mu_i^o + RT \ln x_i \tag{3.65}$$

[see Eqs. (6.37–6.43) in Sect. 6.1 for a statistical explanation of the equations for ideal mixtures]. Equation (3.64) expresses that the volume of an ideal mixture without a mixing effect is the sum of the partial volumes of the pure components. In ideal mixtures, this additivity applies to all extensive quantities except for the entropy and the free energy and free enthalpy which are connected with the entropy. In particular, there is no heat of mixing for an ideal mixture [compare (3.55c) with $\gamma_i = 1$].

According to (3.5), (3.20), and (3.65) (with $N = \sum_i N_i$), the matrix μ_{ik} characteristic for the stability of the mixtures is given by

$$\mu_{ik} = \frac{RT}{N}\left(\frac{\delta_{ik}}{x_i} - 1\right).$$

Corresponding to Eq. (3.22), the determinant of this matrix vanishes. Apart from that, the matrix is positive definite in the range $0 < x_i < 1$; *i.e.*, regardless of its composition, an ideal mixture is always stable. A separation of the mixture is not possible.

The ideal mixture is often used as a reference system for the characterization of more complicated types of mixtures. Deviations from the ideal mixture are referred to as excess quantities. By comparing Eqs. (3.50–3.58) with (3.62–3.65), we obtain for these excess quantities, which we indicate with "E":

$$G_E = RTN \sum_i x_i \ln \gamma_i \,, \tag{3.66}$$

$$S_E = -RN \left[\sum_i x_i \ln \gamma_i + T \sum_i x_i \frac{\partial \ln \gamma_i}{\partial T} \right], \tag{3.67}$$

$$\mu_i^E = RT \ln \gamma_i \tag{3.68}$$

$$H_E = H_M \,, \qquad U_E = U_M \,, \qquad V_E = V_M \,, \tag{3.69}$$

$$\kappa_T^E = \kappa_T^M \,, \quad \alpha_E = \alpha_M \,, \qquad C_p^E = C_p^M \,. \tag{3.70}$$

Corresponding to (3.31), the relation

$$G_E = H_E - TS_E \tag{3.71}$$

holds for the excess free enthalpy G_E. The mixture is called a regular mixture if the deviations from the ideal mixture are solely due to the heat of mixing, *i.e.*, if

$$G_E = H_E \,; \qquad S_E = 0 \tag{3.72}$$

is valid. From (3.67) and (3.72) it follows for the regular mixture that:

$$\sum_i x_i \left[\ln \gamma_i + T \frac{\partial \ln \gamma_i}{\partial T} \right] = 0 \,.$$

If this equation holds for every arbitrary mixing ratio, it further follows that

$$\frac{\partial}{\partial T} (T \ln \gamma_i) = 0 \,.$$

i.e.,

$$\mu_i^E = RT \ln \gamma_i = \text{const.} \tag{3.73a}$$

The excess chemical potential of a regular mixture is not dependent on the temperature:

$$\mu_i^E = \mu_i^E (p, x_1 \ldots x_r) \,. \tag{3.73b}$$

The logarithm of the activity coefficients is inversely proportional to the temperature:

$$\ln \gamma_i \sim 1/T \,. \tag{3.73c}$$

If the deviations from the ideal mixture are solely due to the entropy of mixing, *i.e.*, if

$$G_E = -TS_E; \qquad H_E = 0 \tag{3.74}$$

is valid, the mixture is called athermal. From (3.55c), (3.69) and (3.74),

$$\sum_i x_i \left(\frac{\partial \ln \gamma_i}{\partial T}\right) = 0$$

follows for the athermal mixture, and

$$\frac{\partial \ln \gamma_i}{\partial T} = 0\,,$$

if this equation remains valid for every arbitrary mixing ratio. The activity coefficients are independent of the temperature:

$$\gamma_i = \gamma_i\,(p, x_1 \ldots x_r)\,. \tag{3.75}$$

There is no heat of mixing, and the heat capacity (3.58) is given by the mean value of the heat capacities of the pure substances.

Because of the different sizes of the molecules, polymers with a more or less broad distribution of chain lengths, or polymers dissolved in a low-molecular-mass substance are fundamentally not capable of forming an ideal mixture. However, if the units or segments of the chain molecules are all of roughly the same shape and size, the Gibbs fundamental equation (3.62) of an ideal mixture can be replaced by the generalized Flory–Huggins equation

$$G = G_o + RTN \sum_i x_i \ln x_i^* \tag{3.76}$$

[*e.g.*, compare Kurata (1982)]. The mole fraction (3.5) must merely be replaced by the mole fraction (3.8) in the logarithmic term in (3.62). A mixture which fulfils the Gibbs fundamental equation (3.76) is referred to as an ideal-athermal polymer mixture. In the case of polymers, the ideal-athermal polymer mixture replaces the ideal mixture. However, it should be emphasized here that (3.76), as opposed to (3.62), does not exactly reflect the statistical distribution of the mutually independent molecules, but is rather subject to some approximations [*cf.* Flory (1953)]. Excess quantities which have been experimentally determined and defined with respect to the fundamental equation (3.76) can, therefore, also contain parts which merely correct these approximations.

From (3.17) with (3.76), (3.8), and (3.10), one obtains for the chemical potentials of the components of an ideal-athermal polymer mixture

$$\mu_i = \mu_i^o + RT\left(\ln \frac{P_i x_i}{\langle P\rangle} + 1 - \frac{P_i}{\langle P\rangle}\right). \tag{3.77}$$

A comparison with (3.50) shows that in this case the activity coefficients are given by

$$\gamma_i = \frac{P_i}{\langle P \rangle} \exp\left(1 - \frac{P_i}{\langle P \rangle}\right) \tag{3.78}$$

The activity coefficients of an ideal-athermal polymer mixture are independent of the pressure and temperature. They only take into account that the molecules of a polymer-homologous series are different in length. For a mixture composed of chain molecules of the same length ($P_i \equiv P$ for all i), Eqs. (3.76–3.78) transform into the corresponding equations for the ideal mixture. Due to the pressure and temperature independence of the activity coefficients, one observes, as in the ideal mixture, additivity of the extensive quantities with the exception of the entropy, free energy, and free enthalpy.

For a binary solution of a polymer (subscript "2" with $P_2 \equiv P$, an integral multiple of P_1) in a low-molecular-mass substance (subscript "1" with $P_1 \equiv 1$), we obtain from (3.76) and (3.77) the Flory–Huggins equations in their original forms [Flory (1941), Huggins (1941)]:

$$G = G_o + RT\left[N_1 \ln \frac{N_1}{N_1 + PN_2} + N_2 \ln \frac{PN_2}{N_1 + PN_2}\right], \tag{3.79a}$$

$$\mu_1 = \mu_1^o + RT\left[\ln \frac{N_1}{N_1 + PN_2} + \frac{(P-1)N_2}{N_1 + PN_2}\right], \tag{3.79b}$$

$$\mu_2 = \mu_2^o + RT\left[\ln \frac{PN_2}{N_1 + PN_2} - \frac{(P-1)N_1}{N_1 + PN_2}\right], \tag{3.79c}$$

or with the mole fractions (3.8) as variables

$$\mu_1 = \mu_1^o + RT\left[\ln x_1^* + \left(1 - \frac{1}{P}\right) x_2^*\right], \tag{3.80a}$$

$$\mu_2 = \mu_2^o + RT\left[\ln x_2^* - (P-1)\, x_1^*\right]. \tag{3.80b}$$

3.5 Simple Non-ideal Mixtures

The different shapes and sizes of the molecules as well as intermolecular interactions – as already mentioned – lead to sometimes considerable deviations from the ideal behaviour of mixtures.

To simplify matters we will examine a binary mixture of molecules of the species 1 and 2 with roughly the same shape and size. We assume the binding energy between two molecules of the same species to be w_{11} or w_{22} and the binding energy between two molecules of different species to be w_{12}. Mixing of the two substances 1 and 2 causes a part of the [11]- and [22]-bonds to break and leads to the formation of [12]-bonds. As a result, two [12]-pairs develop from one [11]-pair and one [22]-pair. The energy change connected with this process amounts to

$$2\Delta w_{12} = 2w_{12} - w_{11} - w_{22} \,.$$

If N_{12} [12]-pairs develop during mixing, the total energy change amounts to

$$\Delta G = N_{12}\Delta w_{12} \,. \tag{3.81 a}$$

An ensemble of N molecules is composed of $N/2$ molecule pairs if each molecule has one nearest neighbour and of $zN/2$ molecule pairs if each molecule has z nearest neighbours (z: coordination number). Hence, the probability x_{12} of finding a [12]-pair in the mixture is given by

$$x_{12} = N_{12}/(zN/2) \,.$$

Thus, the energy change (3.81 a) induced by the change in binding conditions upon mixing can also be described by

$$\Delta G = zN\Delta w_{12} x_{12}/2 \,. \tag{3.81 b}$$

With $\Delta w_{12} < 0$, the molecules tend to be surrounded by unequal partners (solvation); with $\Delta w_{12} > 0$, the molecules tend to be surrounded by equal partners (aggregation). Solvation or aggregation tendencies, which both result from attempts at energy minimization, can possibly lead to a considerable deviation from the statistical distribution of the molecules. If this is not the case, the deviation from the ideal mixture is already completely described by (3.81), as we assumed roughly the same shape and size for the molecules, and we can set $\Delta G = G_E$. With a statistical distribution of the molecules, the probability x_{12} of finding a [12]-pair in the mixture is equal to the probability of finding a molecule 1 and a molecule 2, or a molecule 2 and a molecule 1. According to (3.9), we obtain with a statistical distribution of the molecules

$$x_{12} = x_1x_2 + x_2x_1 = 2x_1x_2 \,.$$

The excess free enthalpy of the mixture is then:

$$G_E = zN\Delta w_{12} x_1 x_2 = zN\Delta w_{12} x_1(1 - x_1) \,. \tag{3.82}$$

Guggenheim (1959) designated mixtures with this excess free enthalpy as simple mixtures. Here, Δw_{12} must generally be regarded as a function of temperature and pressure. Of course, $\Delta w_{12} = 0$ leads back to the case of an ideal mixture.

From (3.82) one obtains *via*

$$\mu_i^E = \frac{\partial G_E}{\partial N_i} = RT \ln \gamma_i$$

for the activity coefficients of a simple mixture

$$\ln \gamma_1 = z\Delta w_{12} x_2^2 / RT , \tag{3.83a}$$

$$\ln \gamma_2 = z\Delta w_{12} x_1^2 / RT , \tag{3.83b}$$

and from this with (3.67), (3.69), (3.53), and (3.55)

$$S_E = - zNx_1x_2 \frac{\partial \Delta w_{12}}{\partial T} , \tag{3.84}$$

$$H_E = zNx_1x_2 \left(\Delta w_{12} - T \frac{\partial \Delta w_{12}}{\partial T} \right) , \tag{3.85}$$

$$V_E = zNx_1x_2 \frac{\partial \Delta w_{12}}{\partial p} . \tag{3.86}$$

If Δw_{12} is independent of temperature, a simple mixture behaves regularly ($S_E = 0$). If Δw_{12} is proportional to temperature, a simple mixture behaves athermally ($H_E = 0$) [*cf.* Eqs. (3.71–3.75)]. If Δw_{12} is a linear function of temperature, i.e., if Δw_{12} can be represented in the form

$$\Delta w_{12} = \Delta w_{12}^h - T\Delta w_{12}^s \quad \text{with} \quad \Delta w_{12}^h , \Delta w_{12}^s = \text{const.} , \tag{3.87}$$

it follows that:

$$S_E = zN\Delta w_{12}^s x_1 x_2 \; ; \quad H_E = zN\Delta w_{12}^h x_1 x_2 .$$

It should be mentioned here that (3.87) does not mean that binding energies can be split into an enthalpy term and an entropy term. Equation (3.87) merely indicates a special dependence of the binding energies on temperature. In general, binding energies as such are not capable of contributing to the entropy. Rather, they are only capable of contributing to the entropy of the system *via* their temperature dependence. According to (3.86), the volume remains additive if Δw_{12} does not depend on pressure. The excess volume in condensed systems is usually only significant under extreme pressure conditions.

For the thermo-mechanical response functions of a simple mixture we obtain from (3.82), (3.70), and (3.56–3.58)

$$\kappa_T^E = - \frac{z}{v} x_1x_2 \frac{\partial^2 \Delta w_{12}}{\partial p^2} , \tag{3.88}$$

$$\alpha_E = \frac{z}{v} x_1x_2 \frac{\partial^2 \Delta w_{12}}{\partial T \partial p} , \tag{3.89}$$

$$C_p^E = - zNx_1x_2 T \frac{\partial^2 \Delta w_{12}}{\partial T^2} . \tag{3.90}$$

Deviations from the ideal system are only observed in the compressibility and heat capacity when the binding energies depend non-linearly on pressure or temperature, respectively. The intermolecular interaction contributes only to the excess quantity of the coefficient of thermal expansion when the binding energies are dependent on both pressure as well as temperature.

A simple mixture, like the ideal mixture, requires that the molecules are of equal size and shape. Polymers are, therefore, not capable of forming a simple mixture. However, if the units or segments of the chain molecules are of roughly the same size and shape, one can also arrive – in analogy to the transition from an ideal mixture to an ideal-athermal polymer mixture – at a "simple polymer mixture". In the case of the binary solution of a polymer in a low-molecular-mass substance, examined at the end of Sect. 3.4, (3.82) can be approximatively replaced by

$$\Gamma = z(N_1 + PN_2)\,\Delta w_{12}x_1^* x_2^* = zN\,\langle P\rangle\,\Delta w_{12}x_1^* x_2^* \,.$$

Γ is the excess free enthalpy related to the fundamental equation (3.79 a) of the ideal-athermal polymer solution. Using the dimensionless Huggins' interaction parameter [Huggins (1943)]:

$$\chi = z\Delta w_{12}/RT\,. \tag{3.91}$$

one obtains

$$\Gamma = RTN\chi\,\langle P\rangle\,x_1^* x_2^* = RTN\chi\,\frac{P}{\langle P\rangle}\,x_1 x_2\,. \tag{3.92}$$

According to (3.79a) and (3.92), the Gibbs fundamental equation for one mole of the simple binary polymer solution is

$$g = \frac{G}{N} = g_o(T, p) + RT\,[x_1 \ln x_1^* + x_2 \ln x_2^* + \chi\,\langle P\rangle\,x_1^* x_2^*\,]\,. \tag{3.93}$$

The chemical potentials are given by

$$\mu_1 = \mu_1^o + RT\left[\ln x_1^* + \left(1 - \frac{1}{P}\right)x_2^* + \chi x_2^{*2}\right], \tag{3.94a}$$

$$\mu_2 = \mu_2^o + RT\,[\ln x_2^* + (P-1)\,x_1^* + P\chi x_1^{*2}]\,. \tag{3.94b}$$

For the entropy, enthalpy, and the volume of the simple polymer solution, we obtain

$$s = s_o - R\,(x_1 \ln x_1^* + x_2 \ln x_2^*) - R\left(\chi + T\frac{\partial\chi}{\partial T}\right)\langle P\rangle\,x_1^* x_2^*\,, \tag{3.95}$$

$$h = h_o - RT^2\,\frac{\partial\chi}{\partial T}\,\langle P\rangle\,x_1^* x_2^*\,, \tag{3.96}$$

$$v = v_o + RT \frac{\partial \chi}{\partial p} \langle P \rangle x_1^* x_2^* , \tag{3.97}$$

and for the thermo-mechanical response functions

$$\kappa_T = \kappa_T^o - \frac{RT}{v} \frac{\partial^2 \chi}{\partial p^2} \langle P \rangle x_1^* x_2^* , \tag{3.98}$$

$$\alpha = \alpha_o + \frac{R}{v} \left(\frac{\partial \chi}{\partial p} + T \frac{\partial^2 \chi}{\partial p \, \partial T} \right) \langle P \rangle x_1^* x_2^* , \tag{3.99}$$

$$c_p = c_p^o - RT \left(2 \frac{\partial \chi}{\partial T} + T \frac{\partial^2 \chi}{\partial T^2} \right) \langle P \rangle x_1^* x_2^* , \tag{3.100}$$

The somewhat changed structure of the excess quantities in the equations (3.95–3.100) as compared to the equations (3.84–3.86, 3.88–3.90), in which, however, the structure of the excess quantities (3.66–3.70) can once more be recognized, is due to the fact that the factor $1/RT$ is carried along in the Huggins interaction parameter (3.91). The observation χ = const. means $\Delta w_{12} \sim T$. Hence, the simple polymer solution with a constant interaction parameter χ behaves athermally, and the one with $\chi \sim 1/T$, regularly.

According to (3.43), the simple polymer solution (3.93) is stable, provided that T, p, N = const., and if

$$\frac{\partial^2 g}{\partial x_1^2} = RT \left[\frac{1}{x_1 x_2} + \frac{(1-P)^2}{\langle P \rangle^2} - 2\chi \frac{P^2}{\langle P \rangle^3} \right] > 0 , \tag{3.101}$$

Thus, the solution is definitely stable for all mixing ratios if $\chi < 0$, *i.e.*, $\Delta w_{12} < 0$, *i.e.*, if there is a tendency towards solvation. The (g, x_1)-plane is generally divided into two ranges in which the solution is either stable or unstable. The limiting curve which separates these two ranges is determined by the equation

$$g''(\chi, x_1) = RT \left[\frac{1}{x_1 x_2} + \frac{(1-P)^2}{\langle P \rangle^2} - 2\chi \frac{P^2}{\langle P \rangle^3} \right] = 0 . \tag{3.102}$$

There exists a critical value χ_c of the interaction parameter, with which for $\chi < \chi_c$ the curve $g(x_1)$ runs entirely in the stable range, whereas for $\chi > \chi_c$ the free enthalpy $g(x_1)$ also runs through the unstable range. There must be a critical point x_1^c for $\chi = \chi_c$ at which the curve $g(x_1)$ and the limiting curve given by (3.102) touch (see Fig. 3.2). The common tangent of both curves at the critical point yields the additional condition

$$\frac{\partial^3 g}{\partial x_1^3} = -RT \left[\frac{x_2 - x_1}{x_1 x_2} + \frac{2(1-P)^3}{\langle P \rangle^3} - 6\chi \frac{P^2 (1-P)}{\langle P \rangle^4} \right] = 0 . \tag{3.103}$$

After a longer but simple calculation, one obtains from (3.102) and (3.103) with $\langle P \rangle = x_1 + Px_2$ and $x_1 + x_2 = 1$ the critical values

$$x_2^c = 1 - x_1^c = \frac{1}{1 + P\sqrt{P}}, \tag{3.104}$$

$$\chi_c = \frac{1}{2}\left(1 + \frac{1}{\sqrt{P}}\right)^2. \tag{3.105}$$

Thus, with an increasing degree of polymerization P, the critical interaction parameter drops from $\chi_c = 2$ (for $P = 1$) to $\chi_c = 0.5$ (for $P \to \infty$). The critical mole fraction of the polymer then decreases from $x_2^c = 0.5$ (for $P = 1$) to $x_2^c = 0$ (for $P \to \infty$). For a finite degree of polymerization and $\chi > \chi_c$, there are two concentration ranges where miscibility is given: a larger area $0 < x_1 < x_1^c$ (solution of the low-molecular-mass substance in the polymer) and a very small area near $x_1^c < x_1 \approx 1$ (solution of the polymer in the low-molecular-mass substance). Figures 3.1 and 3.2 give an example with $P = 30$, *i.e.*, $\chi_c = 0.6992$ and $x_1^c = 0.9939$. The value of Huggins' interaction parameter for ordinary solvents is mostly between 0.2 and 0.8 at room temperature and under normal pressure [Brandrup, Immergut (1989)].

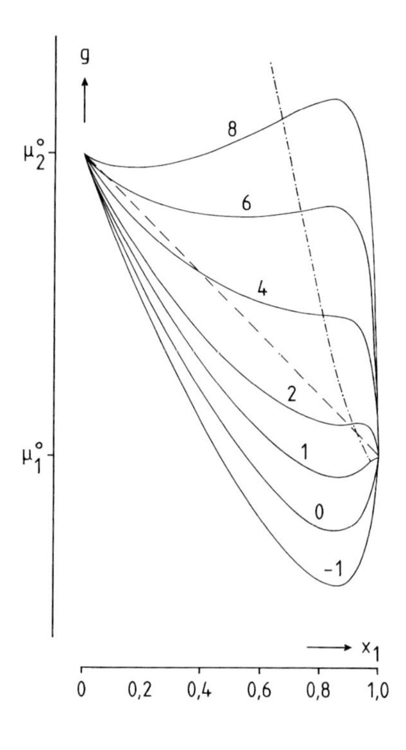

Fig. 3.1. Gibbs free energy g per mole of a polymer solution as a function of the mole fraction x_1 of the solvent for different values of the Huggins' interaction parameter χ according to Eq. (3.93). [Chemical potential of the pure polymer: μ_2^o; chemical potential of the pure solvent: μ_1^o. Assumed parameters: $\mu_2^o - \mu_1^o = 10$ kJ/mol, $T = 300$ K, $P_1 = 1$, and $P_2 = 30$. The dash-dotted line indicates where $\partial^2 g/\partial x_1^2 = 0$ holds. To the right of this curve, $\partial^2 g/\partial x_1^2 < 0$ is valid. Hence, the solution is unstable there. The dashed straight line would hold for g if the solution were to behave additively.]

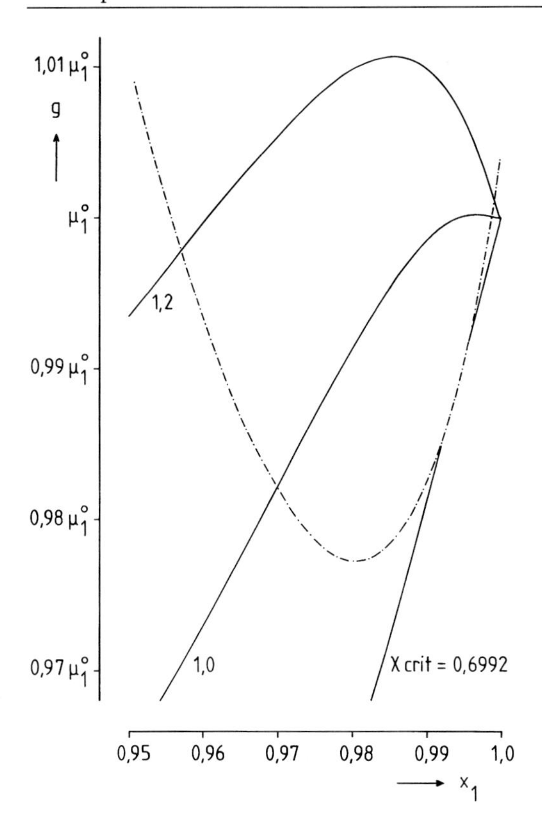

Fig. 3.2. Section of Fig. 3.1. The dash-dotted line $g'' = \partial^2 g/\partial x_1^2 = 0$ forms a parabola according to Eq. (3.102). The system is only unstable within the parabola. Miscibility is thus given for very small polymer concentrations $x_2 = 1 - x_1$, even if $\chi > \chi_{\text{crit}}$ holds

The fact that polymers have a certain molecular mass distribution can be approximatively taken into account by replacing the mole fraction x_2^* of the polymer in the interaction term (3.92) by

$$1 - x_1^* = x_2^* + x_3^* + \ldots + x_r^* \equiv \Sigma' x_i^* \,,$$

i.e., by using the Gibbs fundamental equation in the form [Flory (1953)]:

$$G = G_o + RTN\left[\sum_i x_i \ln x_1^* + \chi\langle P\rangle x_1^*(1 - x_1^*)\right]. \tag{3.106}$$

It makes sense in this regard to refer the mean degree of polymerization only to the components $i = 2 \ldots r$ of the polymer by excluding the low-molecular-mass substance No. 1:

$$\langle P\rangle^* \equiv \frac{\sum_i' P_i N_i}{\sum_i' N_i} = \frac{\sum_i' P_i x_i}{\sum_i' x_i} \,.$$

The relations

$$\langle P\rangle^* \, (1 - x_1) = \langle P\rangle - x_1$$

and

$$\left(1 - \frac{1}{\langle P\rangle}\right) = \left(1 - \frac{1}{\langle P\rangle^*}\right)(1 - x_1^*)$$

exist between the mean degree of polymerization $\langle P\rangle$, which includes the low-molecular-mass substance and the true mean degree of polymerization $\langle P\rangle^*$ of the polymer. However, formally simpler equations can be obtained by retaining $\langle P\rangle$. From (3.106), for example, follow for the chemical potentials:

$$\mu_1 = \mu_1^o + RT\left[\ln x_1^* + \left(1 - \frac{1}{\langle P\rangle}\right) + \chi(1 - x_1^*)^2\right], \tag{3.107a}$$

$$\mu_i = \mu_i^o + RT\left[\ln x_i^* + \left(1 - \frac{P_i}{\langle P\rangle}\right) + \chi P_i x_1^{*2}\right], \quad (i \neq 1). \tag{3.107b}$$

On the basis of the ideal-athermal mixture (3.76), one can generally use for a polymer homologous mixture [Koningsveld, Kleintjens (1977), Kurata (1982)]:

$$G = G_o + RTN \sum_i x_i \ln x_1^* + \Gamma(T, p, x_1^* \ldots x_r^*). \tag{3.108}$$

The more or less empirically determinable "excess quantity" Γ contains the intermolecular interactions and possibly correction factors, which are due to the fact that the polymer units have different shapes and the molecules are not statistically distributed. However, since the statistical distribution of the molecules in (3.76) [and, thus, also in (3.93), (3.106), and (3.108)] has, as already mentioned, not been absolutely correctly taken into account, Γ and χ may still contain a factor which corrects the approximations present in (3.76). In this case, for example, Huggins' interaction parameter can no longer be unequivocally associated with the intermolecular binding energy Δw_{12} [Sariban *et al.* (1987)].

CHAPTER 4

Heterogeneous Systems

Every region of space in a thermodynamic system which is homogeneous with respect to its macroscopic physical properties is referred to as a phase. Heterogeneous systems are composed of several phases. It is generally assumed here that the individual phases are mutually independent (autonomous) and that the volumes of the individual phases are large enough so that the effects produced by the singular inhomogeneities at the phase boundaries can be neglected in contrast to the volume effects [for a consideration of surface effects and non-autonomous phases, see Prigogine, Defay (1951), with respect to polymers also Sect. 8.2].

4.1 Equilibrium and Stability With Respect to the Macroscopic Degrees of Freedom

We will specifically examine the equilibrium and stability of the states of a fluid, one-component system with respect to its macroscopic degrees of freedom. We will only consider changes of state in which the internal molecular degrees of freedom are not macroscopically relevant [*cf.* Sect. 2.4]. Let us, in addition, assume that the system does not have any exchange with its surroundings. No interaction whatsoever with the surroundings means that the system is energetically closed, with $U, V, N = \text{const}$. Under this boundary condition, the easiest way to describe the system is by means of the S-representation [Sect. 2.1].

Consider a heterogeneous system consisting of two homogeneous parts "1" and "2" which differ in the values of the thermodynamic variables. Nothing separates the parts from one another, permitting a free exchange of internal energy, volume, and mass. The values of the extensive variables of state of the total system are additively composed of the values of the individual phases. Hence, the Gibbs fundamental equation of the total system in the S-representation is

$$S = S_1(U_1, V_1, N_1) + S_2(U_2, V_2, N_2) , \tag{4.1}$$

or in a differential form, according to the equations (2.32–2.36),

$$\mathrm{d}S = \frac{1}{T_1}\,\mathrm{d}U_1 + \frac{p_1}{T_1}\,\mathrm{d}V_1 + \frac{\mu_1}{T_1}\,\mathrm{d}N_1 + \frac{1}{T_2}\,\mathrm{d}U_2 + \frac{p_2}{T_2}\,\mathrm{d}V_2 + \frac{\mu_2}{T_2}\,\mathrm{d}N_2 .$$

As there is no interaction with the surroundings, it must hold for the internal exchange that:

$$\mathrm{d}U_1 = -\,\mathrm{d}U_2\,; \quad \mathrm{d}V_1 = -\,\mathrm{d}V_2\,; \quad \mathrm{d}N_1 = -\,\mathrm{d}N_2\,.$$

Thus,

$$\mathrm{d}S = \left(\frac{1}{T_1} - \frac{1}{T_2}\right)\mathrm{d}U_1 + \left(\frac{p_1}{T_1} - \frac{p_2}{T_2}\right)\mathrm{d}V_1 + \left(\frac{\mu_1}{T_1} - \frac{\mu_2}{T_2}\right)\mathrm{d}N_1$$

is obtained for the change in entropy of the total system. According to the second law [compare the Eqs. (2.43) and (2.91)], $\mathrm{d}S = 0$ is characteristic for all changes in the thermodynamic equilibrium under the condition $U, V, N =$ const. The changes $\mathrm{d}U_1$, $\mathrm{d}V_1$ and $\mathrm{d}N_1$, on the other hand, can be selected completely arbitrarily and independently of each other. Therefore, equilibrium can only exist if

$$\frac{1}{T_1} - \frac{1}{T_2} = \frac{p_1}{T_1} - \frac{p_2}{T_2} = \frac{\mu_1}{T_1} - \frac{\mu_2}{T_2} = 0\,.$$

One obtains the following conditions for

thermal equilibrium: $T_1 = T_2 = T\,,$ (4.2)

mechanical equilibrium: $p_1 = p_2 = p\,,$ (4.3)

material equilibrium: $\mu_1 = \mu_2 = \mu\,.$ (4.4)

Thermodynamic equilibrium only exists in a heterogeneous system if the system is homogeneous with respect to its intensive variables. In the case of a mixture, the equilibrium condition

$$\mu_i^{(1)} = \mu_i^{(2)} = \mu_i. \tag{4.5}$$

can be obtained in the same way in place of (4.4). A heterogeneous mixture with a free mass exchange inside the system is homogeneous with respect to the chemical potentials of each component [*cf.* Eq. (3.41)].

These statements can safely be generalized for arbitrary heterogeneous, as well as inhomogeneous systems: at equilibrium, thermodynamic systems are homogeneous with respect to their intensive variables. This requires, however, that the systems do not contain any impermeable or semi-permeable walls and are not exposed to external force fields [*e.g.*, the gravitational field of the earth; for a consideration of external fields, see *e.g.*, Guggenheim (1959), De Groot and Mazur (1962), Baur (1984)]. Hence, heterogeneity or inhomogeneity with respect to the macroscopic intensive variables indicates thermal, mechanical, or material non-equilibrium. Systems in such non-equilibrium states always tend to compensate the heterogeneities or inhomogeneities through energy transport (heat conduction), transport of momentum, and transport of mass (diffusion). The dissipative transport of momentum leads to viscosity, a very pronounced phenomenon in polymer liquids [*cf.* Sect. 7.6].

As to the question about the stability of the states of equilibrium, one must first inquire about the thermal and mechanical stability of a homogeneous system, which was not considered in Sect. 2.4 [see Sect. 3.2 regarding the question about the stability of a homogeneous system with respect to material changes]. If a homogeneous system is in a stable equilibrium, each of its parts must, of course, also be in a stable equilibrium, irrespective of the (macroscopic) size of the parts. We will now divide a homogeneous system into two subsystems and separate both subsystems by a movable diathermal wall. A diathermal wall is permeable to heat, but not to mass. We introduce the separation by a diathermal wall in order to avoid material non-equilibrium states, which are not of interest for the present. A single prime denotes the extensive quantities of the total system and a double prime denotes the extensive quantities of one of the subsystems. The Gibbs fundamental equation of the total system is then in the S-representation

$$S' = Ns(u, v) + N''s''(u'', v'') \tag{4.6}$$

(small letters denote the molar quantities; *cf.* Sect. 2.2., footnote on p. 13). The subsidiary conditions

$$Nu + N''u'' = U' = \text{const.},$$
$$Nv + N''v'' = V' = \text{const.},$$

apply if the total system is energetically closed (see above), whereby individually, due to the diathermal wall:

$$N = \text{const.} \quad \text{and} \quad N'' = \text{const.}.$$

It then follows for the possible changes in the internal energy and the volume of the subsystems that:

$$\mathrm{d}u'' = -\frac{N}{N''}\,\mathrm{d}u\,; \quad \mathrm{d}v'' = -\frac{N}{N''}\,\mathrm{d}v\,. \tag{4.7}$$

Expansion of the fundamental equation (4.6) in a Taylor series about the equilibrium state "e", whose thermal and mechanical stability is to be examined, results in:

$$S' = S'_e + N\left[(\mathrm{d}s)_e + \frac{1}{2}(\mathrm{d}^2s)_e + \ldots\right] + N''\left[(\mathrm{d}s'')_e + \frac{1}{2}(\mathrm{d}^2s'')_e + \ldots\right], \tag{4.8}$$

whereby, according to (2.91),

$$(\mathrm{d}S')_e = N(\mathrm{d}s)_e + N''(\mathrm{d}s'')_e = 0\,,$$

holds for the deviations from the equilibrium, *i.e.*:

$$N(\mathrm{d}s)_e = -N''(\mathrm{d}s'')_e\,. \tag{4.9}$$

Without restrictions of generality, we can now assume that one subsystem is considerably smaller than the other one, so that $N \ll N''$. With respect to (4.7), all differentials $N''(\mathrm{d}^n s'')_e$ with $n \geq 2$ are negligible as compared to the differentials $N(\mathrm{d}^n s)_e$. Hence with (4.9), we obtain from (4.8)

$$S' = S'_e + \frac{1}{2} N(\mathrm{d}^2 s)_e + \ldots \tag{4.10}$$

Thus, the homogeneous total system is stable or metastable [see (2.91)] with respect to a deviation from the equilibrium "e" if

$$(\mathrm{d}^2 s)_e = \left(\frac{\partial^2 s}{\partial u^2}\right)_e (\mathrm{d}u)^2 + 2\left(\frac{\partial^2 s}{\partial u\,\partial v}\right)_e \mathrm{d}u\,\mathrm{d}v + \left(\frac{\partial^2 s}{\partial v^2}\right)_e (\mathrm{d}v)^2 < 0\,, \tag{4.11}$$

i.e.,

$$\left(\frac{\partial^2 s}{\partial u^2}\right)_e < 0\,, \tag{4.12a}$$

$$\left(\frac{\partial^2 s}{\partial v^2}\right)_e < 0\,, \tag{4.13a}$$

$$\left(\frac{\partial^2 s}{\partial u^2}\right)_e \left(\frac{\partial^2 s}{\partial v^2}\right)_e - \left(\frac{\partial^2 s}{\partial u\,\partial v}\right)_e^2 > 0 \tag{4.14a}$$

are valid [the matrix of the coefficients in (4.11) is necessarily negative definite].

Some transformations are necessary in order to draw further conclusions: First, from (2.33–35) and (2.52) it follows with $N = $ const. that:

$$\left(\frac{\partial^2 s}{\partial u^2}\right)_e = \frac{\partial}{\partial u}\left(\frac{1}{T}\right) = -\frac{1}{T^2}\left(\frac{\partial T}{\partial u}\right)_v = -\frac{1}{T^2 c_v}\,, \tag{4.15}$$

$$\left(\frac{\partial^2 s}{\partial v^2}\right)_e = \frac{\partial}{\partial v}\left(\frac{p}{T}\right) = \frac{1}{T^2}\left[T\left(\frac{\partial p}{\partial v}\right)_u - p\left(\frac{\partial T}{\partial v}\right)_u\right], \tag{4.16}$$

$$\left(\frac{\partial^2 s}{\partial u\,\partial v}\right)_e = \frac{\partial}{\partial v}\left(\frac{1}{T}\right) = -\frac{1}{T^2}\left(\frac{\partial T}{\partial v}\right)_u\,. \tag{4.17}$$

Corresponding to (2.57b) and (2.61b), we now have

$$\left(\frac{\partial T}{\partial v}\right)_u \left(\frac{\partial v}{\partial u}\right)_T \left(\frac{\partial u}{\partial T}\right)_v = -1\,,$$

i.e., with (2.52)

$$\left(\frac{\partial T}{\partial v}\right)_u = -\frac{1}{c_v}\left(\frac{\partial u}{\partial v}\right)_T . \tag{4.18}$$

From the differential form of the fundamental equation

$$\mathrm{d}s = \frac{1}{T}\,\mathrm{d}u + \frac{p}{T}\,\mathrm{d}v$$

it results further that:

$$\left(\frac{\partial s}{\partial v}\right)_T = \frac{1}{T}\left(\frac{\partial u}{\partial v}\right)_T + \frac{p}{T} ,$$

and with (2.45)

$$\left(\frac{\partial u}{\partial v}\right)_T = p\,(T\beta - 1) ,$$

so that from (4.18) one finds:

$$\left(\frac{\partial T}{\partial v}\right)_u = -\frac{p}{c_v}\,(T\beta - 1) . \tag{4.19}$$

Finally, the differential mechanical equation of state in the F-representation

$$\mathrm{d}p = \left(\frac{\partial p}{\partial T}\right)_v \mathrm{d}T + \left(\frac{\partial p}{\partial v}\right)_T \mathrm{d}v$$

results in

$$\left(\frac{\partial p}{\partial v}\right)_u = \left(\frac{\partial p}{\partial T}\right)_v \left(\frac{\partial T}{\partial v}\right)_u + \left(\frac{\partial p}{\partial v}\right)_T$$

and, according to (2.45), (2.47), and (4.19), in

$$\left(\frac{\partial p}{\partial v}\right)_u = -\frac{\beta p^2}{c_v}\,(T\beta - 1) - \frac{1}{v\kappa_T} . \tag{4.20}$$

Substitution of (4.19) and (4.20) into (4.16) and (4.17) yields the expressions

$$\left(\frac{\partial^2 s}{\partial v^2}\right)_e = -\frac{1}{T^2}\left[\frac{p^2}{c_v}\,(T\beta - 1)^2 + \frac{T}{v\kappa_T}\right]$$

and

$$\left(\frac{\partial^2 s}{\partial u\,\partial v}\right)_e = \frac{1}{T^2}\frac{p}{c_v}(T\beta - 1)\,.$$

Hence, the stability conditions (4.12a–4.14a) are also

$$\frac{1}{T^2 c_v} > 0\,, \tag{4.12b}$$

$$\frac{1}{T^2}\left[\frac{p^2}{c_v}(T\beta - 1)^2 + \frac{T}{v\kappa_T}\right] > 0\,, \tag{4.13b}$$

$$\frac{T}{v\kappa_T} > 0\,. \tag{4.14b}$$

These conditions are fulfilled if $\kappa_T > 0$ and $c_v > 0$. Using the relations (2.60) and (2.62), we arrive at the conclusion: the states of equilibrium of a simple, homogeneous fluid system are thermally and mechanically stable or metastable if

$$C_p \geq C_V > 0\,, \tag{4.21}$$

$$\kappa_T \geq \kappa_S > 0\,. \tag{4.22}$$

The equality sign only applies here if the thermal expansivity α vanishes. In addition, it was assumed that $T \neq 0$. The states of equilibrium are stable when the system reacts with an increase in temperature on heating and with a decrease in pressure in response to an increase in volume.

Let us now return to the heterogeneous system (4.1) and inquire about its thermal and mechanical stability. If both phases are separated by a movable diathermal wall, we obtain with

$$N_1 = \text{const.}\,; \quad N_2 = \text{const.}\,, \quad \text{that}$$

$$S = S_e + N_1\left[(\mathrm{d}s_1)_e + \frac{1}{2}(\mathrm{d}^2 s_1)_e + \ldots\right] + N_2\left[(\mathrm{d}^2 s_2)_e + \frac{1}{2}(\mathrm{d}^2 s_2)_c + \ldots\right],$$

or, due to the equilibrium conditions,

$$(\mathrm{d}S)_e = N_1(\mathrm{d}s_1)_e + N_2(\mathrm{d}s_2)_e = 0\,,$$

$$S = S_e + \frac{N_1}{2}(\mathrm{d}^2 s_1)_e + \frac{N_2}{2}(\mathrm{d}^2 s_2)_e + \ldots$$

According to (2.91), the heterogeneous total system is stable or metastable with respect to a deviation from the equilibrium "*e*" if

$$(\mathrm{d}^2 S)_e = N_1(\mathrm{d}^2 s_1)_e + N_2(\mathrm{d}^2 s_2)_e < 0\,.$$

With the subsidiary conditions

$$N_1 \mathrm{d}u_1 + N_2 \mathrm{d}u_2 = \mathrm{d}U = 0\,,$$
$$N_1 \mathrm{d}v_1 + N_2 \mathrm{d}v_2 = \mathrm{d}V = 0\,,$$

we explicitly obtain the stability conditions

$$\frac{1}{N_1}\frac{\partial^2 s_1}{\partial u_1^2} + \frac{1}{N_2}\frac{\partial^2 s_2}{\partial u_2^2} < 0\,, \tag{4.23}$$

$$\frac{1}{N_1}\frac{\partial^2 s_1}{\partial v_1^2} + \frac{1}{N_2}\frac{\partial^2 s_2}{\partial v_2^2} < 0\,, \tag{4.24}$$

$$\left[\frac{1}{N_1}\frac{\partial^2 s_1}{\partial u_1^2} + \frac{1}{N_2}\frac{\partial^2 s_2}{\partial u_2^2}\right]\left[\frac{1}{N_1}\frac{\partial^2 s_1}{\partial v_1^2} + \frac{1}{N_2}\frac{\partial^2 s_2}{\partial v_2^2}\right] - \left[\frac{1}{N_1}\frac{\partial^2 s_1}{\partial u_1 \partial v_1} + \frac{1}{N_2}\frac{\partial^2 s_2}{\partial u_2 \partial v_2}\right]^2 > 0. \tag{4.25}$$

The heterogeneous total system is certainly only stable if the phases of the total system are individually stable, *i.e.*, if (4.12–4.14) are valid for each phase. However, if the inequalities (4.12–4.14) are fulfilled, the inequalities (4.23–4.25) are also always fulfilled. This conclusion is obvious for the inequalities (4.12, 4.13) and (4.23, 4.24). A few algebraic transformations show that (4.12–4.14) necessarily also imply (4.25). Hence, heterogeneous systems are stable or metastable if the individual (autonomous!) phases, of which they are composed, are stable or metastable. In particular, the inequalities (4.21) and (4.22) hold for each phase individually in a stable or metastable heterogeneous system. Ultimately, this is a result of the assumed autonomy of the phases.

4.2 Gibbs' Fundamental Equation, Gibbs' Phase Rule and Response Functions

The Gibbs fundamental equation of a heterogeneous fluid system is given by the sum of the Gibbs potentials of the individual phases:

$$G = \sum_k G_k\,. \tag{4.26}$$

If the phases are autonomous, the G_k in the G-representation are dependent on the (homogeneous) temperature, the (homogeneous) pressure, and only on the mole numbers $N_i^{(k)}$ of the substances $i = 1 \ldots r$ contained in the phase k:

$$G_k = G_k(T, p, N_1^{(k)} \ldots N_r^{(k)})\,. \tag{4.27}$$

Corresponding to (2.12) and (3.28), the fundamental equation can always be put into the form

$$G = \sum_k \sum_i \mu_i N_i^{(k)}. \tag{4.28}$$

Due to the autonomy, the Gibbs–Duhem relation holds individually for each phase:

$$S_k \mathrm{d}T - V_k \mathrm{d}p + \sum_i N_i^{(k)} \mathrm{d}\mu_i = 0 . \tag{4.29}$$

Thus, if the heterogeneous system is comprised of a total of q phases ($k = 1 \ldots q$), only $r + 2 - q$ variables of the $r + 2$ intensive variables

$$T, p, \mu_1 \ldots \mu_r$$

can be mutually independent.

With a constant total mass

$$N = \sum_k \sum_i N_i^{(k)} = \text{const.}$$

of the system, the number of mutually independent intensive variables is equal to the number of macroscopic degrees of freedom of the system. Thus a fluid heterogeneous mixture in equilibrium with constant mass containing r components and q phases, possesses $r + 2 - q$ macroscopic degrees of freedom (Gibbs' phase rule). The case $r = 1, q = 1$ leads, of course, back to the simple fluid system [Chap. 2]. With constant mass, a simple homogeneous fluid system possesses two macroscopic degrees of freedom. It is divariant. With constant mass, a fluid two-phase system composed of a pure substance ($r = 1$, $q = 2$), on the other hand, only has a single macroscopic degree of freedom. It is monovariant and can, therefore, at a given pressure, only exist in equilibrium at one specific temperature. Based on this, the partially crystalline state of a polymer composed of chain molecules of uniform length necessarily appears to be a non-equilibrium state if regarded as a non-arrested two-phase system at constant pressure and variable temperature [*cf.* Sect. 8.2]. In the case of elastic systems with more than one mechanical degree of freedom [Chap. 5] or electromagnetic systems which include the electric field strength and magnetic induction in addition as intensive variables, the Gibbs phase rule must be changed with respect to the number of additional degrees of freedom. The same applies to a system containing semi-permeable, impermeable, or rigid walls, *e.g.*, in the case of an osmotic equilibrium in which the pressure is not homogeneous [*e.g.*, see Moelwyn-Hughes (1961)].

Like the free enthalpy (4.26), all extensive quantities of a heterogeneous system of autonomous phases are additively composed of the contributions of the individual phases. Specifically in the G-representation we obtain, for example, for the volume

$$V = \sum_k V_k = \sum_k \sum_i \bar{v}_i N_i^{(k)} \tag{4.30}$$

and for the heat capacity

$$C_p = \sum_k C_p^{(k)} = \sum_k \sum_i \bar{c}_{pi} N_i^{(k)} . \tag{4.31}$$

The homogeneity of a heterogeneous mixture with respect to the partial molar quantities (3.17) results directly from the homogeneity with respect to the intensive variables T, p, $\mu_1 \ldots \mu_r$. For the partial molar volumes $\bar{v}_i$ and the partial molar heat capacities $\bar{c}_{pi}$, the homogeneity results directly from

$$\bar{v}_i = \frac{\partial \mu_i}{\partial p} \quad \text{and} \quad \bar{c}_{pi} = -T \frac{\partial^2 \mu_i}{\partial T^2}$$

[*cf.* Eqs. (3.29) and (3.60)].

Specific mixing rules follow from the additivity of the extensive variables for those functions of state and response functions which can be derived from extensive variables. From the additivity of the volume (4.30), for example,

$$\alpha = \frac{1}{V} \frac{\partial V}{\partial T} = \sum_k \left(\frac{\partial V_k}{\partial T} \right)_{p,\, N_k} \Big/ \sum_k V_k$$

is valid for the coefficient of expansion of a heterogeneous one-component system in equilibrium [*cf.* Eq. (2.48)]. The coefficients of expansion of the individual phases are given by

$$\alpha_k = \frac{1}{V_k} \left(\frac{\partial V_k}{\partial T} \right)_{p,\, N_k} .$$

If

$$v_k = V_k \,/\, \sum_k V_k .$$

denotes the fraction of the volume of the phase k, we obtain

$$\alpha = \sum_k \alpha_k v_k . \tag{4.32}$$

Correspondingly, we obtain for the compressibility of the heterogeneous system [*cf.* Eq. (2.47)]:

$$\kappa_T = \sum_k \kappa_T^{(k)} v_k , \tag{4.33}$$

whereby

$$\kappa_T^{(k)} = -\frac{1}{V_k} \left(\frac{\partial V_k}{\partial p} \right)_{T,\, N_k} = \frac{1}{k_T^{(k)}} \tag{4.34}$$

denotes the compressibility and $k_T^{(k)}$ the bulk modulus of the phase k. For the bulk modulus of a two-phase system, for example, it follows from (4.33, 4.34):

$$k_T = \frac{k_T^{(1)} k_T^{(2)}}{k_T^{(2)} v_1 + k_T^{(1)} v_2} = \frac{k_T^{(1)} k_T^{(2)} V}{k_T^{(2)} V_1 + k_T^{(1)} V_2} . \tag{4.35}$$

The conditions are more complicated when the individual phases are so small that surface effects and geometry of the phases are of importance, or when the systems are anisotropic [*e.g.*, see Holliday (1966)]. In most cases, filled or reinforced polymers are heterogeneous systems in which the individual phases are not in an internal material equilibrium. If, however, no mass exchange takes place between the phases, the states of these systems can be regarded as materially arrested equilibrium states (*cf.* Fig. 2.1), which can also be described by the formalism of equilibrium thermodynamics.

CHAPTER 5

Homogeneous Elastic Systems

Polymers in a crystalline, partially crystalline, glass-like or rubber-elastic state, but also frequently polymer liquids with mechanically relevant internal degrees of freedom (Chap. 7), are classified as elastic systems. As opposed to fluid systems, elastic systems not only resist a change in their volume but also a change in their shape by means of a restoring force. In the following, we will confine ourselves to homogeneous, rigid elastic systems in equilibrium, *i.e.*, to elastic systems whose characteristic field quantities neither depend on position coordinates nor on time.

5.1 Fundamental Equations and Equations of State

For an unequivocal description of the mechanical states of elastic systems, the stress tensor

$$\boldsymbol{\sigma} = \begin{pmatrix} \sigma_{11} & \sigma_{12} & \sigma_{13} \\ \sigma_{21} & \sigma_{22} & \sigma_{23} \\ \sigma_{31} & \sigma_{32} & \sigma_{33} \end{pmatrix}$$

is introduced in place of the pressure, and the strain tensor

$$\boldsymbol{\varepsilon} = \begin{pmatrix} \varepsilon_{11} & \varepsilon_{12} & \varepsilon_{13} \\ \varepsilon_{21} & \varepsilon_{22} & \varepsilon_{23} \\ \varepsilon_{31} & \varepsilon_{32} & \varepsilon_{33} \end{pmatrix}$$

in place of the volume [*e.g.*, see Landau, Lifshitz (1986); Ward (1971)]. The components of the stress and strain tensors refer to volume elements $\mathrm{d}V$ of the considered body which are infinitesimally small from the macroscopic point of view. Generally, they represent position- and time-dependent field quantities. The appropriate extensive thermodynamic quantities E for these field quantities are the equally position- and time-dependent densities ϱe ($\varrho = 1/v$: density of mass; $e = E/M$: specific value of the quantity E). One assumes that each individual volume element $\mathrm{d}V$ is in thermodynamic equilibrium [*e.g.*, see De Groot, Mazur (1962); Baur (1984)]. The quantities related to the total system or to macroscopic parts of the total system can be obtained *via* integration over the volume of the system or a subsystem. If the elastic system is homogeneous, all the densities and the components σ_{ik}, ε_{ik} of the stress and strain tensors are independent of the position. The integration is then reduced to a multiplication with the volume V_o of the non-deformed state:

$$E = \int_{V_o} \varrho e \, dV = V_o \varrho e \, .$$

In equilibrium, the stress and strain tensors are time-independent and symmetric, *i.e.*,

$$\sigma_{ik} = \sigma_{ki} \, ; \quad \varepsilon_{ik} = \varepsilon_{ki} \tag{5.1}$$

is valid. [The strain tensor is symmetric by definition; the stress tensor is symmetric if the angular momentum of the system is conserved; *e.g.*, see Gyarmati (1970)]. Using this symmetry, the following variables are introduced into the thermodynamics of homogeneous elastic media [Tsuan, Li (1957)].

$$\begin{aligned}
V_o\varepsilon_1 &\equiv V_o\varepsilon_{11} \, , & \sigma_1 &\equiv \sigma_{11} \, , \\
V_o\varepsilon_2 &\equiv V_o\varepsilon_{22} \, , & \sigma_2 &\equiv \sigma_{22} \, , \\
V_o\varepsilon_3 &\equiv V_o\varepsilon_{33} \, , & \sigma_3 &\equiv \sigma_{33} \, , \\
V_o\varepsilon_4 &\equiv 2V_o\varepsilon_{23} = 2V_o\varepsilon_{32} \, , & \sigma_4 &\equiv \sigma_{23} = \sigma_{32} \, , \\
V_o\varepsilon_5 &\equiv 2V_o\varepsilon_{13} = 2V_o\varepsilon_{31} \, , & \sigma_5 &\equiv \sigma_{13} = \sigma_{31} \, , \\
V_o\varepsilon_6 &\equiv 2V_o\varepsilon_{12} = 2V_o\varepsilon_{21} \, , & \sigma_6 &\equiv \sigma_{12} = \sigma_{21} \, .
\end{aligned}$$

$(\varepsilon_1, \varepsilon_2, \varepsilon_3)$ describe the stretchings in the directions 1, 2, 3 in a rectangular Cartesian system of coordinates and $(\varepsilon_4, \varepsilon_5, \varepsilon_6)$ the shearings. (The shearings $2\varepsilon_{ik}$ are often referred to as γ_{ik} in practice). $(\sigma_1, \sigma_2, \sigma_3)$ denote the tensile stresses in the directions 1, 2, 3 and $(\sigma_4, \sigma_5, \sigma_6)$ the shear stresses. During a pure dilatation, the current volume V is given by

$$V = V_o(1 + \varepsilon_1 + \varepsilon_2 + \varepsilon_3) \, . \tag{5.2}$$

The mean normal pressure

$$p = -\frac{1}{3}(\sigma_1 + \sigma_2 + \sigma_3) \, . \tag{5.3}$$

corresponds to the hydrostatic pressure. Elastic systems are always in a condensed state and, therefore, assume a finite volume V_o in a stress-free (non-deformed) state. This should be taken into account when comparing with the thermodynamics of gases, since the volume of gases increases to infinitely large values in a stress-free state.

If the homogeneous elastic system only consists of a single chemically inert substance, and if a constant mass is maintained (N = const.), the Gibbs fundamental equation in the F-representation is

$$F = F(T, V_o\varepsilon_1 \dots V_o\varepsilon_6) \, . \tag{5.4}$$

This results in seven equations of state

$$\left(\frac{\partial F}{\partial T}\right)_{\varepsilon_1 \ldots \varepsilon_6} \equiv - S(T, V_o\varepsilon_1 \ldots V_o\varepsilon_6) , \tag{5.5}$$

$$\frac{1}{V_o}\left(\frac{\partial F}{\partial \varepsilon_i}\right)_{T, \varepsilon_{k \neq i}} \equiv \sigma_i(T, V_o\varepsilon_1 \ldots V_o\varepsilon_6) . \tag{5.6}$$

The differential form of Gibbs fundamental equation is

$$\mathrm{d}F = - S\mathrm{d}T + V_o \sum_i \sigma_i \mathrm{d}\varepsilon_i . \tag{5.7}$$

If we expand the mechanical equations of state (5.6) in a Taylor series about the non-deformed state $Z(T_o, V_o)$ and break off the series after the linear terms, we obtain, since $\sigma_i^o = 0$ and $\varepsilon_i^o = 0$ in the non-deformed state, the linear equations of state:

$$\sigma_i = \sum_k c_{ik}^o \varepsilon_k + \beta_i^o(T - T_o) , \quad i = 1 \ldots 6 . \tag{5.8}$$

The $c_{ik}^o = (\partial\sigma_i/\partial\varepsilon_k)_o$ denote the elastic moduli of the system relative to the initial state. The $\beta_i^o = (\partial\sigma_i/\partial T)_o$ denote the coefficients of thermal stress relative to the initial state [*cf.* Eqs. (5.17) and (5.18) further below]. Hence, the deformation-induced mechanical stresses are generally superimposed by thermal stresses. If the system is isothermally stressed ($T = T_o = \text{const.}$), (5.8) is reduced to Hooke's law of the theory of elasticity:

$$\sigma_i = \sum_k c_{ik}^o \varepsilon_k . \tag{5.9}$$

The fundamental and state equations in the G-representation are:

$$G = G(T, \sigma_1 \ldots \sigma_6) , \tag{5.10}$$

$$\left(\frac{\partial G}{\partial T}\right)_{\sigma_1 \ldots \sigma_6} \equiv - S(T, \sigma_1 \ldots \sigma_6) . \tag{5.11}$$

$$\left(\frac{\partial G}{\partial \sigma_i}\right)_{T, \sigma_{k \neq i}} \equiv - V_o\varepsilon_i(T, \sigma_1 \ldots \sigma_6) . \tag{5.12}$$

$$\mathrm{d}G = - S\mathrm{d}T - V_o \sum_k \varepsilon_i \mathrm{d}\sigma_i . \tag{5.13}$$

In this case, linearization of the mechanical equations of state (5.12) leads to

$$\varepsilon_i = \sum_k \kappa_{ik}^o \sigma_k + \alpha_i^o(T - T_o) , \quad i = 1 \ldots 6 \tag{5.14}$$

and in the case $T = T_o$ = const. to Hooke's law of the theory of elasticity in the form

$$\varepsilon_i = \sum_k \kappa_{ik}^o \sigma_k \,. \tag{5.15}$$

The expressions $\kappa_{ik}^o = (\partial\varepsilon_i/\partial\sigma_k)_o$ and $\alpha_i^o = (\partial\varepsilon_i/\partial T)_o$ denote the coefficients of elastic compliance and linear thermal expansion relative to the initial state $Z(T_o, V_o)$ [*cf.* Eqs. (5.20) and (5.21) below; about the consistency of the equations (5.9) and (5.15), see Eq. (5.25)].

The Gibbs formalism of thermodynamics presented in Chap. 2 is completely maintained for homogeneous elastic systems if the product

$$pV \quad \text{is replaced there by} \quad -V_o \sum_k \sigma_k \varepsilon_k \,.$$

As a finite volume V_o has to be assumed for elastic bodies in a stress-free state, the product $p(V - V_o)$ appears in place of the product pV in the case of a hydrostatic deformation with $\sigma_1 = \sigma_2 = \sigma_3 = -p$, $\sigma_4 = \sigma_5 = \sigma_6 = 0$. Instead of the energy conservation law in the form (2.42) (the first law of thermodynamics), we now specifically have with N = const.:

$$\mathrm{d}U = \mathrm{d}_a U = \mathrm{d}Q + V_o \sum_k \sigma_k \mathrm{d}\varepsilon_k \,; \quad \mathrm{d}_i U = 0 \,. \tag{5.16}$$

5.2 Response Functions

The increased number of mechanical degrees of freedom leads to a considerable increase in the number of possible modes of stress or strain (boundary conditions) and responses in an elastic system in comparison to fluid systems.

We obtain from the F-representation (5.4–5.7) as mechanical response functions, the 36 coefficients of isothermal elastic stiffness (elastic moduli)

$$c_{ik} \equiv \frac{1}{V_o} \frac{\partial^2 F}{\partial\varepsilon_i \partial\varepsilon_k} = \left(\frac{\partial\sigma_i}{\partial\varepsilon_k}\right)_{T,\varepsilon_{j\neq k}} , \tag{5.17}$$

as thermomechanical response functions, the 6 coefficients of thermal stress

$$\beta_i \equiv \frac{1}{V_o} \frac{\partial^2 F}{\partial T \partial\varepsilon_i} = \left(\frac{\partial\sigma_i}{\partial T}\right)_{\varepsilon_1 \ldots \varepsilon_6} = \frac{1}{V_o} \left(\frac{\partial S}{\partial\varepsilon_i}\right)_{T,\varepsilon_{k\neq i}} , \tag{5.18}$$

and as the thermal response function, the heat capacity at constant strain

$$C_\varepsilon \equiv -T \frac{\partial^2 F}{\partial T^2} = T \left(\frac{\partial S}{\partial T}\right)_{\varepsilon_1 \ldots \varepsilon_6} . \tag{5.19}$$

Likewise, the G-representation (5.10–5.13) results in the 36 coefficients of isothermal elastic compliance

$$\kappa_{ik} \equiv -\frac{1}{V_o}\frac{\partial^2 G}{\partial\sigma_i\partial\sigma_k} = \left(\frac{\partial\varepsilon_i}{\partial\sigma_k}\right)_{T,\sigma_{j\neq k}}, \tag{5.20}$$

the 6 coefficients of linear thermal expansion

$$\alpha_i \equiv -\frac{1}{V_o}\frac{\partial^2 G}{\partial T\partial\sigma_i} = \left(\frac{\partial\varepsilon_i}{\partial T}\right)_{\sigma_1\ldots\sigma_6} = \frac{1}{V_o}\left(\frac{\partial S}{\partial\sigma_i}\right)_{T,\sigma_{j\neq i}}, \tag{5.21}$$

and the heat capacity at constant stress

$$C_\sigma \equiv -T\frac{\partial^2 G}{\partial T^2} = T\left(\frac{\partial S}{\partial T}\right)_{\sigma_1\ldots\sigma_6}. \tag{5.22}$$

From the U- and H-representations, we obtain the 36 coefficients of adiabatic elastic stiffness

$$c_{ik}^s \equiv \frac{1}{V_o}\frac{\partial^2 U}{\partial\varepsilon_i\partial\varepsilon_k} = \left(\frac{\partial\sigma_i}{\partial\varepsilon_k}\right)_{S,\varepsilon_{j\neq k}}, \tag{5.23}$$

and the 36 coefficients of adiabatic elastic compliance

$$\kappa_{ik}^s \equiv \frac{1}{V_o}\frac{\partial^2 H}{\partial\sigma_i\partial\sigma_k} = \left(\frac{\partial\varepsilon_i}{\partial\sigma_k}\right)_{S,\sigma_{j\neq k}}. \tag{5.24}$$

Corresponding to the relations (2.54, 2.55), (2.58, 2.59) and (2.62), the purely mathematical relations

$$(c_{ik}) = (\kappa_{ik})^{-1}, \tag{5.25}$$

$$\alpha_i = -\sum_k \kappa_{ik}\beta_k \quad \text{or} \quad \beta_i = -\sum_k c_{ik}\alpha_k, \tag{5.26}$$

$$C_\sigma = C_\varepsilon - TV_o\sum_k \alpha_k\beta_k, \tag{5.27}$$

$$|(\kappa_{ik}^s)|\, C_\sigma = |(\kappa_{ik})|\, C_\varepsilon, \tag{5.28}$$

exist between the response functions [Tsuan, Li (1957)]. In Eq. (5.25) the parentheses denote the 6,6-matrix of the elastic coefficients, and in Eq. (5.28) $|(\ldots)|$ represent their determinant. From (5.25–5.27), for example, one obtains in place of (2.60)

$$C_\sigma = C_\varepsilon + TV_o\sum_i\sum_k c_{ik}\alpha_i\alpha_k, \tag{5.29}$$

and in place of (2.63)

$$\kappa_{ik} = \kappa_{ik}^{s} + TV_o \alpha_i \alpha_k / C_\sigma . \tag{5.30}$$

From the equilibrium and stability conditions it follows further, corresponding to the relations (4.21) and (4.22), that the matrix of the coefficients c_{ik} must be positive definite, and that $C_\sigma \geq C_\varepsilon > 0$ must be valid.

As $\boldsymbol{\sigma}$ and $\boldsymbol{\varepsilon}$ are tensors of the second rank (with 9 components), the moduli and compliances connecting $\boldsymbol{\sigma}$ and $\boldsymbol{\varepsilon}$ are actually tensors of the fourth rank with $9 \times 9 = 81$ components each. The coefficients of thermal stress and expansion, which connect a tensor of the second rank with a scalar quantity, form tensors of the second rank with $9 \times 1 = 9$ components. The symmetry relations (5.1) reduce these tensors to the matrices (5.17, 5.18) and (5.20, 5.21) with only 36 or 6 components, respectively. It should be noted that the matrices (5.17, 5.18) and (5.20, 5.21) do not possess tensor or vector properties, so that one must always return to the original tensors when inquiring about the properties with respect to spatial transformations [*e.g.*, see Nye (1975)].

The defining Eqs. (5.17) and (5.18) show that the matrices of the c_{ik} and κ_{ik} are also symmetric, so that

$$c_{ik} = c_{ki} \,; \quad \kappa_{ik} = \kappa_{ki} \tag{5.31}$$

is valid. Out of the original 81 coefficients of isothermal elastic stiffness and the original 81 coefficients of isothermal elastic compliance, only 21 coefficients, respectively, can actually be mutually independent. Structural symmetry and manner of stress of a medium often cause a further considerable reduction in the number of relevant response functions [*cf.* Nye (1975)]. The full set of 21 mechanical and 6 thermomechanical response functions only come into play in media without any symmetry and in media with the symmetry properties of a triclinic lattice.

The macroscopic mechanical and thermal properties of polymers often prove to be isotropic, even if the polymers are partially crystallized. In the case of isotropic media (with the highest structural symmetry), the number of mutually independent mechanical response functions is reduced to only two. In the F-representation, one obtains

$$\left.\begin{aligned} &c_{12} = c_{13} = c_{23} = \lambda \,; \qquad && c_{44} = c_{55} = c_{66} = \mu \\ &c_{11} = c_{22} = c_{33} = \lambda + 2\mu \,; \qquad && c_{ik} = 0, \quad \text{otherwise} \,. \end{aligned}\right\} \tag{5.32}$$

μ, λ are referred to as the Lamé constants of an isotropic medium. The meaning of the constants becomes clear when one explicitly formulates Hooke's law (5.9) with the coefficients (5.32):

$$\sigma_i = 2\mu\varepsilon_i + \lambda(\varepsilon_1 + \varepsilon_2 + \varepsilon_3) \qquad \text{for } i = 1, 2, 3 \,, \tag{5.33a}$$

$$\sigma_i = \mu\varepsilon_i \qquad \text{for } i = 4, 5, 6 \,. \tag{5.33b}$$

The constant μ is identical with the so-called shear modulus, which is generally defined by the ratio of shear stress σ_{ik} to shear strain $2\varepsilon_{ik}$ $(i \neq k)$. Summation of the Eqs. (5.33 a) leads to

$$\sigma_1 + \sigma_2 + \sigma_3 = (2\mu + 3\lambda)\,(\varepsilon_1 + \varepsilon_2 + \varepsilon_3)\,,$$

hence, with (5.2) and (5.3) one finds:

$$p = -\left(\lambda + \frac{2}{3}\,\mu\right)\frac{V - V_o}{V_o}\,. \tag{5.34}$$

The bulk modulus [*cf.* Eq. (2.44)] is thus given by

$$k_T = \lambda + \frac{2}{3}\,\mu = \frac{1}{3}\,(c_{11} + 2\,c_{12})\,. \tag{5.35}$$

By inversion of Eq. (5.33) [transformation to the G-representation; *cf.* Eq. (5.25)], we obtain

$$\varepsilon_i = \frac{1}{2\mu}\,\sigma_i - \frac{\lambda}{2\mu\,(2\mu + 3\lambda)}\,(\sigma_1 + \sigma_2 + \sigma_3) \quad \text{for } i = 1, 2, 3\,, \tag{5.36a}$$

$$\varepsilon_i = \frac{1}{\mu}\,\sigma_i \quad \text{for } i = 4, 5, 6\,. \tag{5.36b}$$

If the system is subjected to a pure tensile stress $\sigma_1 \neq 0, \sigma_2 \ldots \sigma_6 = 0$, we specifically have

$$\varepsilon_1 = \frac{\mu + \lambda}{\mu\,(2\mu + 3\lambda)}\,\sigma_1\,, \tag{5.37a}$$

$$\varepsilon_2 = -\frac{\lambda}{2\,\mu\,(2\mu + 3\lambda)}\,\sigma_1\,, \tag{5.37b}$$

$$\varepsilon_3 = \frac{\lambda}{2\,\mu\,(2\mu + 3\lambda)}\,\sigma_1\,. \tag{5.37c}$$

The ratio of tensile stress to stretch:

$$\frac{\sigma_1}{\varepsilon_1} = \frac{\mu\,(2\mu + 3\lambda)}{\mu + \lambda} \equiv E \tag{5.38}$$

is referred to as elasticity or Young's modulus. The ratio of lateral contraction to longitudinal stretch

$$-\frac{\varepsilon_2}{\varepsilon_1} = -\frac{\varepsilon_3}{\varepsilon_1} = \frac{\lambda}{2(\mu + \lambda)} \equiv \nu \tag{5.39}$$

is the so-called Poisson's ratio. From (5.38) and (5.39) follows inversely

$$\lambda = \frac{\nu E}{(1 + \nu)(1 - 2\nu)} \quad \text{and} \quad \mu = \frac{E}{2(1 + \nu)} \tag{5.40}$$

and thus also from (5.35)

$$k_T = \frac{E}{3(1 - 2\nu)} \; . \tag{5.41}$$

The matrix of the coefficients (5.32) is only positive definite if

$$\mu > 0 \; ; \quad \lambda > -2\mu/3 \tag{5.42a}$$

is valid. Hence, the mechanical stability of an isotropic medium requires

$$k_T > 0 \; ; \quad -1 < \nu \leq \frac{1}{2} \; ; \quad 0 < E \leq 3\mu \; . \tag{5.42b}$$

The equality sign in (5.42b) only applies to incompressible media, for which $\kappa_T = 1/k_T = 0$ and (with a finite μ) $\lambda = \infty$.

The coefficients α_i of linear thermal expansion (5.21) are usually positive quantities, but can definitely also assume negative values. The coefficient α of thermal expansion of the volume (2.48) is given by

$$\alpha = \alpha_1 + \alpha_2 + \alpha_3 \; . \tag{5.43}$$

As already mentioned, the 6 coefficients α_i are only mutually independent in the case of media without any symmetry and in the case of triclinic lattices [Nye (1975)]. In the case of orthorhombic lattices, the coefficients (5.21) are reduced to

$$\alpha_1 \neq \alpha_2 \neq \alpha_3 \neq 0 \; ; \quad \alpha_4 = \alpha_5 = \alpha_6 = 0 \; . \tag{5.44}$$

In the case of polyethylene crystallized in an orthorhombic lattice, for example, the linear expansion coefficient in the chain direction is slightly negative [Baughman (1973)]. In the case of cubic lattices or isotropic media, we have

$$\alpha_1 = \alpha_2 = \alpha_3 = \frac{1}{3}\alpha \; ; \quad \alpha_4 = \alpha_5 = \alpha_6 = 0 \; . \tag{5.45}$$

With respect to (5.26), (5.32), and (5.45), the coefficients of thermal stress of an isotropic medium are given by

$$\beta_i = -(2\mu + 3\lambda)\frac{\alpha}{3} = -k_T\alpha \qquad \text{for } i = 1, 2, 3\,, \tag{5.46a}$$

$$\beta_i = 0 \qquad \text{for } i = 4, 5, 6\,. \tag{5.46b}$$

The thermal stress in an isotropic medium is actually a thermal pressure if $\alpha > 0$. When comparing with the coefficient of thermal pressure (2.45), one should note that β and β_i have different dimensions. From (5.3) one obtains

$$\beta = -\frac{1}{3p}(\beta_1 + \beta_2 + \beta_3)\,. \tag{5.47}$$

Thus, (5.46) and (2.58) become identical.

The boundary conditions $\boldsymbol{\varepsilon}$ = const. or $\boldsymbol{\sigma}$ = const., under which the heat capacities (5.19) and (5.22) should be measured, can, depending on the values ε_i = const. or σ_i = const. of the components, respectively, correspond to entirely different physical states of the system. Hence, a different heat capacity is measured under uniaxial tensile stress with the boundary condition σ_1 = const. $\neq 0$, $\sigma_{i\neq 1} = 0$ than under hydrostatic pressure with the boundary condition $\sigma_1 = \sigma_2 = \sigma_3 = -p$ = const. $\neq 0$, $\sigma_4 = \sigma_5 = \sigma_6 = 0$. It is also possible that the given physical situation requires a set of variables mixed from ε_i and σ_i. Mixed sets of variables correspond neither to the *F*- nor to the *G*-representation. In order to define the heat capacity or other response functions, a Gibbs function corresponding to the mixed set of variables must first be derived from the free energy *F* or the free enthalpy *G* using the Legendre transformations. For reversible changes of state, one obtains *via* (5.16), corresponding to the relations (2.53) and (2.54),

$$C_{\boldsymbol{\varepsilon}} = \left(\frac{\partial U}{\partial T}\right)_{\varepsilon_1 \dots \varepsilon_6} = \left(\frac{\mathrm{d}Q}{\mathrm{d}T}\right)_{\varepsilon_1 \dots \varepsilon_6}, \tag{5.48}$$

$$C_{\boldsymbol{\sigma}} = \left(\frac{\partial H}{\partial T}\right)_{\sigma_1 \dots \sigma_6} = \left(\frac{\mathrm{d}Q}{\mathrm{d}T}\right)_{\sigma_1 \dots \sigma_6}. \tag{5.49}$$

For the heat capacities of isotropic media, (5.29) with (5.32), (5.35), and (5.45) results in the relation

$$C_{\boldsymbol{\sigma}} = C_{\boldsymbol{\varepsilon}} + TV_o k_T \alpha^2\,, \tag{5.50}$$

which is identical with (2.60). If the bulk modulus and the expansion coefficient are known, (5.50) specifically allows a conversion of C_V into C_p. As already mentioned in Sect. 2.3, we are often dependent on such a conversion in the case of condensed media as C_V is theoretically and C_p experimentally more accessible for these media. In solids at not too high temperatures, the bulk modulus is usually practically constant, and the bulk expansion coefficient approximately proportional to the heat capacity C_p (*cf.* Sect. 6.5.4). Equation (5.50) reduces then to the approximate formula

$$c_p = c_v + ATc_p^2 \,. \tag{5.51}$$

c_p, c_v denote the specific heat capacities (relative to the mass unit). The constant $A \equiv \varrho k_T$ (ϱ: density of mass) is substance-specific. In an additional assessment, whose universal validity, however, is rather questionable, Nernst and Lindemann (1911) set $A = A_o/T_M$, so that one obtains

$$c_p = c_v + A_o \left(\frac{T}{T_M}\right) c_p^2 \,. \tag{5.52}$$

$A_o = 5.12 \cdot 10^{-3}$ Kmol/J is a universal constant, T_M is the melting temperature of the substance [for critical comments on the applicability of the Nernst–Lindemann formula for polymers, see Choy (1975) and Pan et al. (1989); see also Sect. 6.5.4].

5.3 Energy and Entropy Elasticity

Substitution of relation (2.18) into the mechanical equations of state (5.6) results in

$$\sigma_i = \frac{1}{V_o}\left(\frac{\partial U}{\partial \varepsilon_i}\right)_{T,\varepsilon_{k\neq i}} - \frac{T}{V_o}\left(\frac{\partial S}{\partial \varepsilon_i}\right)_{T,\varepsilon_{k\neq i}} \,. \tag{5.53}$$

Hence, the stress components are composed of two parts: an energy-elastic part due to the change in internal energy upon deformation and an entropy-elastic part induced by the change in entropy upon deformation. According to (5.18), we can also replace (5.53) by

$$\sigma_i = \frac{1}{V_o}\left(\frac{\partial U}{\partial \varepsilon_i}\right)_{T,\varepsilon_{k\neq i}} + T\left(\frac{\partial \sigma_i}{\partial T}\right)_{\varepsilon_1 \ldots \varepsilon_6} \,. \tag{5.54}$$

This form of equation of state indicates how energy-elastic and entropy-elastic contributions can be distinguished experimentally: If one measures (reversible changes of state assumed!) the stress components σ_i at different temperatures T in a specified state of deformation $\varepsilon_1 \ldots \varepsilon_6 =$ const., the tangent at any point of the curve of $\sigma_i(T)$ thus obtained gives information on the entropy-elastic part. The intercept of the tangent with the σ_i-axis gives information on the energy-elastic part at T. There is no energy-elastic part if the tangent passes through the origin (0, 0). There is no entropy-elastic part if the tangent runs parallel to the T-axis.

According to (5.18) and (5.26), every system whose coefficients of linear thermal expansion vanish, acts purely energy-elastically, with $\partial S/\partial \varepsilon_i = 0$ for all *i*. These are primarily solids which fulfil the so-called harmonic approximation (Sect. 6.4.5). The intermolecular bonds are particularly perturbed during mechanical deformation of an energy-elastic medium. No temperature or heat effects whatsoever are observed during a reversible mechanical perturbation of a purely energy-elastic system. This fact alone allows the proposal of a theory of elasticity without any reference to thermodynamics. The differential form

$$\mathrm{d}S = \left(\frac{\partial S}{\partial T}\right)_{\varepsilon_1 \ldots \varepsilon_6} \mathrm{d}T + \sum_i \left(\frac{\partial S}{\partial \varepsilon_i}\right)_{T, \varepsilon_{k \neq i}} \mathrm{d}\varepsilon_i \tag{5.55a}$$

of the thermal equation of state (5.5), or with (5.18) and (5.19)

$$\mathrm{d}S = \left(\frac{C_\varepsilon}{T}\right) \mathrm{d}T - V_o \sum_i \beta_i \, \mathrm{d}\varepsilon_i \,, \tag{5.55b}$$

shows directly that no heat is produced during an isothermal reversible equilibrium deformation of purely energy-elastic systems, and that the temperature remains constant during an adiabatic reversible equilibrium deformation. Because of $\partial S/\partial \varepsilon_i = 0$, (5.55) (with $\mathrm{d}T = 0$) results in $\mathrm{d}S = 0$ for isothermal processes, and thus according to (2.43 a, b) in $\mathrm{d}Q = 0$. The adiabatic deformation is characterized by $\mathrm{d}Q = 0$, $\mathrm{d}N = 0$. Furthermore, $\mathrm{d}S = 0$ holds for adiabatic reversible deformations according to (2.43 a, b), *i. e.*, $\mathrm{d}T = 0$ according to (5.55).

A reversible mechanical perturbation of entropy-elastic systems, on the other hand, always involves heat effects or changes in temperature, so that a separation of the theory of elasticity and thermodynamics is not possible. A system in which the entropy-elastic part is non-negligible can, according to (5.55), only undergo a reversible isothermal deformation if the heat

$$\mathrm{d}Q = -\, V_o T \sum_i \beta_i \, \mathrm{d}\varepsilon_i$$

consumed or produced during deformation is simultaneously (depending on the signs of the β_i) absorbed from or released to the surroundings. According to (5.55), the adiabatic reversible deformation is linked to the change in temperature

$$\mathrm{d}T = \left(\frac{V_o T}{C_\varepsilon}\right) \sum_i \beta_i \, \mathrm{d}\varepsilon_i \,.$$

Entropy-elastic effects are found in every medium whose coefficients of thermal expansion do not vanish. In solids, however, the entropy elasticity is usually much less pronounced than the energy elasticity. Substances or mixtures in a gas-like state and media in a so-called rubber-elastic state, on the other hand, behave almost opposite. The mechanical properties of gases and rubber-elastic media are decisively determined by a change in entropy upon deformation. In certain temperature ranges at least, some of these systems can even be regarded as being purely entropy-elastic. However, the applicability of the infinitesimal linear theory of elasticity used in Sect. 5.1/2 is very limited in exactly these media. The definition of the strain tensor $\boldsymbol{\varepsilon}$ is tailored to media which only respond with very small deformations to even relatively intense stress, as is the case in energy-elastic systems. Gases and rubber-elastic media, on the other hand, already respond to small changes in stress or pressure with a relatively large deformation, and behave non-linearly with regard to the relations between stress and strain variables [*e.g.*, see Eq. (5.61) further below]. For isotropic, purely entropy-elastic systems, with $\partial U/\partial \varepsilon_i = 0$ for all i, we obtain from (5.55) with (5.46) and (5.2)

$$\mathrm{d}S = \frac{C_V}{T}\,\mathrm{d}T + k_T \alpha \,\mathrm{d}V\,. \tag{5.56a}$$

From (5.46) follows further with (5.3), (5.6), and (5.18)

$$p = T k_T \alpha\,,$$

i.e., from (5.56a)

$$\mathrm{d}S = \frac{C_V}{T}\,\mathrm{d}T + \frac{p}{T}\,\mathrm{d}V\,. \tag{5.56b}$$

This equation definitely applies to gases. For isotropic rubbers, however, it only applies to very small deformations. Equation (5.56) would imply that the reversible deformation of an incompressible isotropic rubber-elastic medium (with $\mathrm{d}V = 0$) does not involve any heat or temperature effects. This is, however, not true for larger deformations.

Rubber elasticity is a specific effect which only occurs in polymers. The cause of this effect is to be found in the conformational isomerism of flexible chain molecules. The chemical repeating units of a polymer chain can generally be distorted with respect to each other. If the valence angles between the units deviate from 180°, a distortion leads to coiling of the chains. From a statistic point of view (Sect. 6.2), all chains in an ensemble of identical chain molecules in equilibrium now assume in a time and ensemble average a most probable coil form with a specific radius of gyration or a specific end-to-end distance of the chains. Maximum entropy is attributed to this coiled state. A change in the radius of gyration or in the end-to-end distance of the coils by an external mechanical action leads to a necessary decrease in the entropy of the ensemble. If the external mechanical perturbation is removed again, the ensemble strives to return to the original state with maximum entropy. Hence, the ensemble responds elastically, *i.e.*, entropy-elastically.

The rubber-elastic effect, however, can only be achieved as long as the chain molecules do not slip past each other during the external perturbation, so that the external force is transmitted to the ends of the chain molecules or at least to the ends of longer chain segments. This can be achieved by intercoupling (cross-linking) the chain molecules *via* covalent bonds. In natural rubber, for example, the polyisoprene molecules are cross-linked by sulphur bridges

–S–S–...–S– (so-called vulcanization). An at least temporary cross-linking in sufficiently long chain molecules, however, is already induced by the interpenetration and the resulting entanglement or snagging of the molecule coils (*cf.* Sect. 7.7).

Let us specifically consider a homogeneous and isotropic rectangular parallelepiped, composed of a rubber-elastic substance with the edge lengths ℓ_{01}, ℓ_{02}, ℓ_{03}, which is exposed to pure stretching (without shearing). The edge lengths in a stretched state are written as ℓ_1, ℓ_2, ℓ_3. The ratios $\lambda_i = \ell_i/\ell_{0i} > 0$ $(i = 1, 2, 3)$ can be introduced as a measure of the stretching. The volume in a stretched state is then given by

$$V = V_o \lambda_1 \lambda_2 \lambda_3 \,. \tag{5.57}$$

If the internal energy of the rectangular parallelepiped (*i.e.*, especially the inter- and intramolecular interaction of the chain molecules) does not undergo a significant change during the stretching, *i.e.*, if the material behaves purely entropy-elastically, we obtain from the statistics of chain coiling for the free energy of the rectangular parallelepiped in a stretched state [Sect. 6.2, Eq. (6.90)]

$$F = F_o + \frac{1}{2} RNT \left(\lambda_1^2 + \lambda_2^2 + \lambda_3^2 - 3\right), \tag{5.58}$$

where F_o is the free energy in the unstretched state, N the mole number of the chain molecules (or more precisely, the mole number of the chain segments between two cross-links of the network or rather the mole number of the effective chain segments) and R the gas constant. If the material is incompressible, *i.e.*, if according to (5.57)

$$\lambda_1 \lambda_2 \lambda_3 = 1 \,,$$

then an elongation in the 1-direction in the ratio $\lambda_1 \equiv \lambda$ leads to a contraction in the 2- and 3-directions in the ratio $\lambda_2 = \lambda_3 = 1/\sqrt{\lambda}$ (λ must, of course, not be confused with the equally labelled Lamé constant in Sect. 5.2). In this case, we obtain for the free energy

$$F = F_o + \frac{1}{2} RNT \left(\lambda^2 + \frac{2}{\lambda} - 3\right). \tag{5.59}$$

Corresponding to (5.5), the thermal equation of state of the system is

$$S = -\left(\frac{\partial F}{\partial T}\right)_\lambda = S_o - \frac{1}{2} RN \left(\lambda^2 + \frac{2}{\lambda} - 3\right), \tag{5.60}$$

and corresponding to (5.6), the mechanical equation of state

$$\sigma = \frac{1}{V_o}\left(\frac{\partial F}{\partial \lambda}\right)_T = \frac{RNT}{V_o}\left(\lambda - \frac{1}{\lambda^2}\right). \tag{5.61}$$

Hence, the relation between stress σ and strain λ is non-linear. The quantity

$$\left(\frac{\partial \sigma}{\partial \lambda}\right)_T = \frac{RNT}{V_o}\left(1 + \frac{2}{\lambda^3}\right), \tag{5.62}$$

appropriate for the modulus of elasticity (5.38), is not a constant of the material but depends on the temperature and strain. For very small elongations

$$\Delta \ell \equiv \ell_1 - \ell_{01} \ll \ell_{01},$$

we obtain with the strain component

$$\varepsilon_1 = \frac{\Delta \ell}{\ell_{01}} \ll 1$$

from Sect. 5.1 the relation

$$\lambda = 1 + \varepsilon_1 .$$

For very small elongations, the incompressible rubber-elastic medium can be described by the modulus of elasticity

$$E = 3\mu = 3\, RNT/V_o . \tag{5.63}$$

It is typical for rubber-elastic materials that the "modulus of elasticity" increases proportionally to the temperature.

In place of the differential thermal equations of state, (5.55) or (5.56), we now have

$$\mathrm{d}S = \left(\frac{\partial S}{\partial T}\right)_\lambda \mathrm{d}T + \left(\frac{\partial S}{\partial \lambda}\right)_T \mathrm{d}\lambda$$

with

$$\left(\frac{\partial S}{\partial T}\right)_\lambda = \left(\frac{\partial S_o}{\partial T}\right)_\lambda = C_\lambda^o/T > 0$$

and

$$\left(\frac{\partial S}{\partial \lambda}\right)_T = -\, RN\left(\lambda - \frac{1}{\lambda^2}\right) = -\, V_o\sigma/T, \tag{5.64}$$

therefore,

$$\mathrm{d}S = \frac{1}{T}\left(C_\lambda^o\, \mathrm{d}T - V_o\sigma\, \mathrm{d}\lambda\right). \tag{5.65}$$

Thus, the reversible isothermal elongation of an incompressible rubber [with $\mathrm{d}T = 0$, $\mathrm{d}\lambda > 0$, and $\lambda > 1$, *i.e.*, $\sigma > 0$ according to (5.61)] always involves an evolution of heat

$$dQ = -V_o \sigma \, d\lambda < 0$$

($dQ = T d_a S = T dS < 0$ means that the heat needs to be emitted to the surroundings). In case of a reversible adiabatic elongation (with $T dS = dQ = 0$), the temperature is increased by the amount

$$dT = V_o \sigma \, d\lambda / C^o_\lambda > 0 \, .$$

Corresponding to (5.18), the coefficient of thermal stress is given by

$$\beta \equiv \frac{1}{V_o} \frac{\partial^2 F}{\partial T \partial \lambda} = \left(\frac{\partial \sigma}{\partial T} \right)_\lambda = -\frac{1}{V_o} \left(\frac{\partial S}{\partial \lambda} \right)_T . \tag{5.66}$$

Via (5.64), we thus obtain

$$\sigma = \left(\frac{\partial \sigma}{\partial T} \right)_\lambda T \, . \tag{5.67}$$

One can observe experimentally that the stress increases with the temperature during constant elongation ($\lambda = \text{const.} > 1$). The coefficient of thermal stress of a rubber is positive: $\beta > 0$. From the relation

$$\left(\frac{\partial \sigma}{\partial T} \right)_\lambda \left(\frac{\partial T}{\partial \lambda} \right)_\sigma \left(\frac{\partial \lambda}{\partial \sigma} \right)_T = -1 \, ,$$

which is equivalent to (2.57),

$$\alpha = \left(\frac{\partial \lambda}{\partial T} \right)_\sigma = -\beta \Big/ \left(\frac{\partial \sigma}{\partial \lambda} \right)_T < 0 \tag{5.68}$$

is obtained as the coefficient of thermal expansion [*cf.* Eq. (5.62)]. Hence, under constant stress, rubber exhibits a contractile behaviour with increasing temperature.

Ideal rubber-elastic, *i.e.*, purely entropy-elastic behaviour can, however, only be found – if at all – in a limited temperature and stretching range. Following a decrease in temperature, every rubber ultimately changes into a glass-like state. A rather abrupt transition from the entropy-elastic to a largely energy-elastic behaviour takes place in the glass transition region. With increasing stretch, the chain molecules, or rather the chain segments between the cross-links of the network, become more and more parallelized and possibly even partially crystallized, so that the intermolecular interaction becomes increasingly important. Therefore, the energy-elastic part can by no means be neglected in the case of high elongations. [For the theory of rubber elasticity, *e.g.*, see Treloar (1975) or Eirich (1978); for a theory of rubber elasticity established in analogy to the theory of real gases, see Ambacher, Kilian (1992)].

CHAPTER 6

Molecular Aspects of Some Equilibrium Properties of Polymer

6.1 Linkage of Thermodynamics with Statistical Mechanics

Derivation of Gibbs fundamental equation or the equations of state of a thermodynamic system from molecular data requires a linkage of thermodynamics with the mechanics or the quantum mechanics of the molecules contained in the system. Due to the multiplicity of molecular particles – the number of particles in a macroscopic body is of the order of magnitude $N_A \approx 10^{24}$ (see Sect. 3.1) – the accessible information on the molecular situation is insufficient for an exact determination of the current mechanical states. On the other hand, an exact determination of the macroscopic state of a thermodynamic system does not mean that the molecular situation is unequivocally determined. A macroscopic state with a specific internal energy value, for example, can be realized by a multitude of molecular configurations (microstates). An unequivocal linkage of the macroscopic thermodynamics with the mechanics of the molecules can only be achieved if the possible states of the molecular system are known, if these possible states can be associated with probability values, and if the resulting statements of the statistical mechanics can be identified with the corresponding statements of thermodynamics [*e.g.*, see, also with respect to the following, Huang (1963), Pathria (1972), or Toda, Kubo, Saitô (1983)].

One has to assume that the defining quantities of the molecular system which allow a determination of all possible states of the system are already known. Essentially, this means that the Hamiltonian function or the Hamiltonian operator of the molecular system must be known. One must assume that all of the possible mechanical states which are accessible under the macroscopic boundary conditions are passed through in times which are shorter or equal to the times of measurement. Furthermore, one must assume that each of the possible mechanical states can be associated with a quantity which indicates the probability of finding the molecular system in this state. The existence of such a probability distribution simultaneously guarantees the existence of a measure for the information we have on the molecular system [Shannon (1962); compare also Landsberg (1961)]. According to Boltzmann, this measure for the information can, with the exception of the sign and a universal constant of proportionality, be identified with the thermodynamic entropy of the macroscopic system [Eq. (6.1), below]. For a given macroscopic equilibrium state, the probability distribution of the molecular states can be determined on the basis of the maximum entropy principle. The probability distribution is essentially dependent on the macroscopic variables of state which can be varied independently.

The macroscopic (thermodynamic) equilibrium state must be completely known. In particular, the exact values of the mutually independent variables of the macroscopic system must be known. If the state is determined by the values of m intensive variables $t_1 \ldots t_m$ and n-m extensive variables $U_{m+1} \ldots U_n$, only mean values $\langle U_1 \rangle \ldots \langle U_m \rangle$ can be determined from the molecular data for the extensive variables $U_1 \ldots U_m$ which are conjugated to the intensive variables $t_1 \ldots t_m$. If the distribution of the values of $U_1 \ldots U_m$ is sufficiently narrow, the mean values can be identified with the values of the corresponding thermodynamic variables. In this case, neglect of the less probable states only leads to a minor error. Fixation of the m intensive variables $t_1 \ldots t_m$ to specific values requires contacting the considered system with a very large "reservoir". The specific values of the intensive variables can basically be considered as properties of this "reservoir".

The fundamental relationship between thermodynamics and statistical mechanics is Boltzmann's entropy principle

$$S = -k \sum_i P_i \ln P_i \tag{6.1}$$

[see Landsberg (1961)]. In Eq. (6.1) S is the entropy of the macroscopic system, k the Boltzmann constant ($kN_A = R$ is the gas constant), $P_i \equiv P(Z_i)$ are the probabilities of finding the states Z_i of the molecular system. The subsidiary condition

$$\sum_i P_i = 1 \tag{6.2}$$

holds for these probabilities. In (6.1) and (6.2), one must sum over all molecular states accessible under the given macroscopic conditions. Furthermore, the thermodynamic requirement that the entropy of the system at equilibrium is maximized is of fundamental importance. As S, according to Eq. (6.1), is only a function of the probabilities P_i, $S(P_1 \ldots P_\Omega)$ as function of the P_i must also be maximized at equilibrium. It is useful when observing this requirement to include the subsidiary conditions, which the probabilities P_i depending on the macroscopic boundary conditions are subject to, by means of Lagrange's method of multipliers [*e.g.*, see Courant, Hilbert (1965)].

In the following part of Sect. 6.1, N (in contrast to the previous sections) will denote the absolute number of the molecular particles. The mole number (Sect. 3.1) will then be given by N/N_A. If the macroscopic boundary condition is U, V, N = const., there will be no subsidiary conditions for the probabilities P_i with the exception of Eq. (6.2). Maximizing the entropy (6.1) is then equivalent to maximizing the function

$$L = -k \sum_i P_i [\alpha + \ln P_i] ,$$

where α is the (still undefined) Lagrange multiplier by means of which one takes into account the subsidiary condition (6.2). L has an extreme value with respect to the P_i if

$$dL = -k \sum_i [\alpha + \ln P_i + 1] \, dP_i = 0 \tag{6.3}$$

is valid. If

$$\Omega = \Omega(U, V, N)$$

designates the number of molecular states Z_i, accessible under the condition $U, V, N = \text{const.}$, then, according to Eq. (6.2), $\Omega - 1$ of the probabilities P_i can be varied independently of each other. If α is selected in such a way that one of the brackets [...] in (6.3) vanishes, all the other brackets [...] in (6.3), due to the independence of the remaining variations dP_i, must also vanish. Hence, (6.3) is equivalent to the requirement

$$\alpha + \ln P_i + 1 = 0$$

or

$$P_i = e^{-(1+\alpha)} = \text{const.} \tag{6.4}$$

Under the condition $U, V, N = \text{const.}$, the entropy of the macroscopic system only achieves an extreme value if all the molecular states Z_i are equally probable.

If we insert (6.4) into (6.2), we have

$$\sum_i P_i = e^{-(1+\alpha)} \, \Omega = 1$$

or

$$e^{-(1+\alpha)} = 1/\Omega \, ,$$

i.e., according to (6.4),

$$P_i = 1/\Omega \, . \tag{6.5}$$

The probability P_i of finding the molecular state Z_i under the condition $U, V, N = \text{const.}$ is reciprocal to the total number of molecular states that can be realized under the same condition. With (6.5), one obtains from (6.1) as the equilibrium entropy of the system

$$S(U, V, N) = k \ln \Omega(U, V, N) \, . \tag{6.6}$$

The molecular states Z_i which must be enumerated in (6.5) and (6.6) correspond to the mutually independent eigensolutions (wave functions)

$$\mathcal{H}\,\Psi_i = E\,\Psi_i$$

of the Hamiltonian operator $\mathcal{H}$ for the N particles contained in the volume V with the eigenvalue $E = U$. In the case of the continuous manifold of states in classical mechanics, Ω is proportional to the volume of the phase space of the N molecules:

$$\Omega^* = \int \dots \int \mathrm{d}q_1 \dots \mathrm{d}q_\mathrm{f}\, \mathrm{d}p_1 \dots \mathrm{d}p_\mathrm{f}\,, \tag{6.7}$$

where $q_1 \dots p_\mathrm{f}$ are the generalized coordinates and momenta of the molecular system, f is the number of molecular degrees of freedom. One must integrate over the "very thin" energy shell

$$U - \mathrm{d}U/2 < \mathcal{H} < U + \mathrm{d}U/2$$

of the phase space, whereby

$$\mathcal{H} = \mathcal{H}(q_1 \dots q_\mathrm{f}\,; p_1 \dots p_\mathrm{f})$$

is the Hamiltonian function of the system. As the volume of a high-dimensional structure (the dimension $2f$ is in the order of magnitude $> 6N_A$) lies practically under its surface area, one can also extend the integration in (6.7) over the domain $0 \le \mathcal{H} \le U$. The relation

$$\Omega = \Omega^*/h^\mathrm{f} N! \tag{6.8}$$

exists between Ω and Ω^*. The factor $1/N!$ is due to the fact that the permutation of the particle numbers (coordinates) leads to different points in the classical phase space, however, the states corresponding to these points cannot be distinguished for identical particles. The factor $1/h^\mathrm{f}$ is a pure dimensional factor from the viewpoint of classical mechanics. Quantum-mechanically, it can be explained by the fact that the Heisenberg uncertainty principle

$$\Delta q_i\, \Delta p_i = h$$

(h: Planck constant) exists between the coordinates and momenta of the particles. For this reason, the states surrounding each point $(q_1 \dots p_\mathrm{f})$ within the phase space with the "area" h^f cannot be distinguished.

In a mixture of r components (Chap. 3), only the N_j ($j = 1 \dots r$) molecules of the individual components, respectively, are not distinguishable. In a mixture, (6.8) is thus replaced by

$$\Omega\,(U, V, N_1 \dots N_r) = \Omega^* \Big/ \prod_{j=1}^{r} h^{\mathrm{f}_j} N_j!\,. \tag{6.9}$$

f_j is the number of molecular degrees of freedom of the component No. j. The integration in Ω^* must be performed in the phase space of the whole mixture.

Furthermore, the temperature of the macroscopic system, according to (6.6) and (2.33), is given by

$$\frac{1}{T} = \frac{k}{\Omega}\left(\frac{\partial \Omega}{\partial U}\right)_{V,N} . \tag{6.10}$$

The macroscopic boundary condition $T, V, N =$ const. requires an overall contact of the system with a heat reservoir (thermostat) *via* a fixed diathermal wall in order to maintain a constant temperature. In this case, the internal energy of the system is given by the mean value

$$U \equiv \langle U \rangle = \sum_i P_i E_i , \tag{6.11}$$

where E_i is the energy of the molecular state Z_i. Besides Eq. (6.2), Eq. (6.11) represents an additional subsidiary condition for the probabilities P_i. Maximizing the entropy (6.1) with the subsidiary conditions (6.2) and (6.11) is equivalent to maximizing the function

$$L = -k \sum_i P_i [\alpha + \beta E_i + \ln P_i] .$$

This function only has an extreme value if

$$\alpha + \beta E_i + \ln P_i + 1 = 0 ,$$

i.e.,

$$P_i = e^{-(1+\alpha)} e^{-\beta E_i} \tag{6.12}$$

holds for all i. Insertion of (6.12) into (6.2) yields the expression

$$e^{1+\alpha} = \sum_i e^{-\beta E_i} \tag{6.12}$$

for the multiplier α, and thus with (6.12), the distribution

$$P_i = \frac{e^{-\beta E_i}}{\sum_i e^{-\beta E_i}} \tag{6.13}$$

for the probabilities P_i. According to (6.1), (6.11), and (6.13), the equilibrium entropy of the macroscopic system, when $T, V, N =$ const., is given by

$$S = k\beta U + k \ln \sum_i e^{-\beta E_i} \tag{6.14}$$

If one identifies this equation with the thermodynamic relation (2.18), one obtains

$$\beta = 1/kT \tag{6.15}$$

for the multiplier β and the relation

$$F(T, V, N) = -kT \ln Q(T, V, N) \tag{6.16}$$

with

$$Q(T, V, N) = \sum_i e^{-E_i/kT} \tag{6.17}$$

for the free energy of the system. The function $Q(T, V, N)$ is designated as the partition function of the system. In (6.17), one must sum over all the states accessible under the condition $T, V, N = \text{const.}$ which satisfy the Schrödinger equation

$$\mathcal{H}\, \Psi_i = E_i\, \Psi_i\,. \tag{6.18}$$

If the energy levels E_i are degenerate, and if the number of states which belong to the eigenvalues E_n of the Hamiltonian operator $\mathcal{H}$ amount to g_n, we can write

$$Q = \sum_n g_n\, e^{-E_n/kT} \tag{6.19}$$

instead of (6.17). One must then only sum over the accessible states with different energies. In the case of a continuous spectrum of eigenvalues of $\mathcal{H}$, or if the eigenvalues E_n are at least sufficiently dense (which is to be expected considering the large number of particles), one can also replace (6.19) with

$$Q = \int g(E, V, N)\, e^{-E/kT}\, dE\,. \tag{6.20}$$

$g\,dE$ indicates the number of states in the energy interval $< E - dE/2\,; E + dE/2 >$. Hence, $g = \Omega$ holds for $E = U$. In the case of classical mechanics, the partition function, corresponding to (6.7) and (6.8), is given by

$$Q = \frac{1}{h^f N!} \int \ldots \int e^{-\mathcal{H}/kT}\, dq_1 \ldots dp_f\,, \tag{6.21}$$

where $\mathcal{H}$ denotes the Hamiltonian function of the molecular system.

Among the relevant statistical functions, the partition function Q is the most accessible to the theory. Theoretically, the F-representation of Gibbs fundamental equation in the form (6.16) is, therefore, of central importance. As equations of state of the system, we obtain from (6.16) according to (2.3–2.5)

$$S = k\left[\ln Q + T\left(\frac{\partial \ln Q}{\partial T}\right)_{V,N}\right], \tag{6.22}$$

$$p = kT\left(\frac{\partial \ln Q}{\partial V}\right)_{T,N}, \tag{6.23}$$

$$\mu = -\,kT\left(\frac{\partial \ln Q}{\partial N}\right)_{T,V}, \tag{6.24}$$

From the comparison of (6.22) with (6.14–6.17) follows in addition

$$U = kT^2\left(\frac{\partial \ln Q}{\partial T}\right)_{V,N}. \tag{6.25}$$

For the response functions (2.44–2.46), we obtain

$$k_T = -\,kTV\left(\frac{\partial^2 \ln Q}{\partial V^2}\right)_{T,N}, \tag{6.26}$$

$$\beta = \frac{k}{p}\left[\left(\frac{\partial \ln Q}{\partial V}\right)_{T,N} + T\frac{\partial^2 \ln Q}{\partial T\,\partial V}\right] = \frac{1}{T} + \frac{\partial^2 \ln Q}{\partial T\,\partial V}\Big/\left(\frac{\partial \ln Q}{\partial V}\right)_{T,N}, \tag{6.27}$$

$$C_V = kT\left[2\left(\frac{\partial \ln Q}{\partial T}\right)_{V,N} + T\left(\frac{\partial^2 \ln Q}{\partial T^2}\right)_{V,N}\right]. \tag{6.28}$$

When determining the quantum-mechanical partition function, the decisive question deals with the possibilities of occupying the elemental wave functions, which the wave functions Ψ_i of the total system are composed of. One must ask whether the wave functions of the particles are restricted to separate volumes (so-called localized particles, like, at least approximatively, the atomic or molecular cores in a crystal) or whether they overlap. If the wave functions overlap, one should ask further whether the particles are subject to Pauli's exclusion principle (fermions, such as the electrons) or not (bosons, such as the phonons in Sect. 6.4.1). These distinctive features lead to completely different results on statistical determination of the numbers Ω, g_n, or g in (6.6), (6.19), or (6.20), respectively. Bose particles obey Bose–Einstein statistics, fermions Fermi–Dirac statistics, and localized particles classical Maxwell–Boltzmann statistics. The results of these three statistical enumerations, however, become equal and correspond to the result of Maxwell–Boltzmann statistics if the probabilities of occupation of the states of the total system are considerably smaller than the ratio of Ω, g_n, or $g\,dE$ to the total particle number. The latter is usually the case when the temperature is not too low and the pressure is not too high.

Calculation of the partition function is almost always a difficult task, which can only be approximatively realized. This is mainly due to the fact that the mechanical many-body problem can only be solved approximatively. However, another reason is that even in relatively simple cases it is often not possible to obtain a mathematically closed expression for the partition function (as, *e.g.*, in the case of the discrete quantum-mechanical energy spectrum of the degree of freedom for a free rotation around a fixed axis). Considerable simplifications result particularly when the Hamiltonian operator (or the Hamiltonian function) can be separated into mutually independent, additive terms. This is possible if several degrees of freedom of the system are at least approximatively independent of each other. According to (6.17–6.20), the additive separation of the Hamiltonian operator means a multiplicative separation of the partition function and further an additive separation of the thermodynamic functions of state (6.16, 6.22–6.24) and response functions (6.26–6.28). The contributions of the translative centre of mass motion of the molecules (of the diffusive translational motion of the molecules in liquids and gases or the collective lattice vibrations in solids) and the contributions of the internal degrees of freedom of the molecules (of the oscillations or rotations of molecular parts and the degrees of freedom of the electrons and nucleons) can often be separated in this way. From

$$\mathcal{H} = \mathcal{H}_{\text{trans}} + \mathcal{H}_{\text{int}} \quad \text{or} \quad E_i = E_i^{\text{trans}} + E_i^{\text{int}}$$

follows

$$Q = Q_{\text{trans}}\, Q_{\text{int}}$$

and thus for the entropy

$$S = S_{\text{trans}} + S_{\text{int}}$$

or the heat capacity

$$C_V = C_V^{\text{trans}} + C_V^{\text{int}}$$

If the particles are independent of each other, *i. e.*, if the molecular interaction can be neglected and classical Maxwell–Boltzmann statistics are valid, the partition function collapses into a product

$$Q = q^N$$

of identical partition functions q which then only refers to the individual molecules. Since the Hamiltonian operator or the Hamiltonian function does not depend on the positional coordinates x, y, z of the individual molecules (or on their quantum-mechanical equivalent), a factor corresponding to the integral

$$\int\!\!\int\!\!\int \mathrm{d}x\,\mathrm{d}y\,\mathrm{d}z = V$$

can be split off in q in the classical formulation (6.21). The partition function of the total system can then be represented in the form

$$Q = V^N Q^*\,, \tag{6.29}$$

in which Q^* contains the coordinates of all degrees of freedom except for the positional coordinates of the centre of mass of the molecules.

Experimentally, T, p, N = const. is usually given. This boundary condition requires an overall contact of the considered system with a heat and pressure reservoir *via* a movable diathermal wall. In addition to the internal energy, the volume is also only determined by a mean value

$$V \equiv \langle V \rangle = \sum_i P_i V_i \tag{6.30}$$

due to this condition. When maximizing the entropy (6.1), the subsidiary condition (6.30) must in addition to (6.2) and (6.11) also be taken into account for the probabilities P_i. This leads (see above) to the extremum condition

$$\alpha + \beta E_i + \gamma V_i + \ln P_i + 1 = 0 ,$$

i. e., to

$$P_i = \exp [-(1+\alpha) - \beta E_i - \gamma V_i] .$$

Insertion of this expression into (6.2) yields

$$e^{1+\alpha} = \sum_i e^{-\beta E_i - \gamma V_i} ,$$

that is the probability distribution

$$P_i = \frac{e^{-\beta E_i - \gamma V_i}}{\sum_i e^{-\beta E_i - \gamma V_i}} , \tag{6.31}$$

Furthermore:

$$S = k\beta U + k\gamma V + k \ln \sum_i e^{-\beta E_i - \gamma V_i} , \tag{6.22}$$

results from (6.1), (6.11), (6.30), and (6.31) for the equilibrium entropy. Identification of this equation with the thermodynamic relation

$$S = \frac{1}{T} U + \frac{p}{T} V - \frac{1}{T} G$$

following from (2.12) and (2.19) or (2.31) yields

$$\beta = 1/kT ; \quad \gamma = p/kT \tag{6.33}$$

for the Lagrange multipliers β and γ and the expression for the free enthalpy:

$$G(T, p, N) = -kT \ln Q'(T, p, N) \tag{6.34}$$

with

$$Q'(T, p, N) = \sum_i e^{-(E_i + pV_i)/kT} . \tag{6.35}$$

The classical analogue to (6.35) is, corresponding to (6.7, 6.8) and (6.20),

$$Q' = \frac{1}{h^{f} N!} \int \dots \int e^{-(\mathcal{H}+pV)/kT} \, dq_1 \dots dp_f \,. \tag{6.36}$$

Corresponding to (6.9),

$$Q'(T, p, N_1 \dots N_r) = \frac{1}{\prod_j h^{f_j} N_j!} \int \dots \int e^{-(\mathcal{H}+pV)/kT} \, dq_1 \dots dp_f \tag{6.37}$$

holds in the case of a mixture of r components, where $f = \Sigma f_j$ is the number of molecular degrees of freedom of the total system, and f_j $(j = 1 \dots r)$ denotes the number of molecular degrees of freedom of the individual components. For the individual components in a pure (unmixed) state, on the other hand,

$$Q'_{oj}(T, p, N_j) = \frac{1}{h^{f_j} N_j!} \int \dots \int e^{-(\mathcal{H}_j+pV_j)/kT} \, dq_1^{(j)} \dots dp_{f_j}^{(j)} \tag{6.38}$$

is valid under the same pressure and at the same temperature, where $V_j \neq V$ is the volume which the N_j molecules of the component No. j assume in a pure state at the temperature T and under the pressure p, and $\mathcal{H}_j$ is the Hamiltonian function of these molecules. If there is no interaction between the molecules, the integrands in (6.37) and (6.38) are independent of the coordinates of the centre of mass of the molecules. Volume factors can then, as in (6.29), be placed before the integrals, and with $N = \Sigma N_j$, we have

$$Q' = V^N R' = \prod_j V^{N_j} R' = \prod_j \left(\frac{V}{V_j}\right)^{N_j} V_j^{N_j} R'\,; \quad Q'_{oj} = V_j^{N_j} R'_j \,. \tag{6.39}$$

The Hamiltonian function $\mathcal{H}$ of the mixture can be split into a sum

$$\mathcal{H} = \mathcal{H}_1 + \dots + \mathcal{H}_r \tag{6.40}$$

of the Hamiltonian functions $\mathcal{H}_j$ of the pure substances. In addition, if all molecules are of the same shape and size (the molecules of the different components still differ in their masses and the internal degrees of freedom), we have further

$$\frac{V}{V_j} = \frac{N}{N_j}\,; \quad i.e., \quad V = V_1 + \dots + V_r \,. \tag{6.41}$$

With (6.40) and (6.41), we obtain from (6.37) and (6.38)

$$R' = \prod_j R'_j$$

and thus from (6.39)

$$Q' = \prod_j \left(\frac{V}{V_j}\right)^{N_j} V_j^{N_j} R'_j = \prod_j \left(\frac{N}{N_j}\right)^{N_j} Q'_{oj} \tag{6.42}$$

The free enthalpy (6.34) of the mixture is thus given by

$$G = -kT \sum_j \ln Q'_{oj} - kT \sum_j N_j \ln \frac{N}{N_j}. \tag{6.43a}$$

Here,

$$G_j^o = \mu_j^o N_j = -kT \ln Q'_{oj}$$

is the free enthalpy of the pure component No. j. Instead of (6.43a), one can also write

$$G = N \sum_j \mu_j^o \frac{N_j}{N} + kTN \sum_j \frac{N_j}{N} \ln \frac{N_j}{N}. \tag{6.43b}$$

If we now, as in the previous sections, introduce the mole numbers N/N_A in place of the particle numbers N and consider the relations $kN_A = R$ and $N_j/N = x_j$, (6.43b) leads to Gibbs fundamental equation (3.62) of the ideal mixture given in Sect. 3.4. A precondition for the validity of this equation is, as already mentioned, that the molecular interaction can be neglected and that all the molecules have about the same shape and size. The mixing term in (6.43) is a purely entropic effect, which is obviously due to the fact that, at the same temperature and under the same pressure, a larger volume is available for the molecules of the individual components in a mixture than in the pure form.

6.2 The Polymer Coil

The molecules of a polymer are characterized by the linkage of a large number of atoms or atomic groups by covalent bonds. In the so-called linear polymers, which we will consider exclusively in the following, this leads to an enormous length for the molecules. In linear (not or only moderately branched) polyethylene,

$$R\!-\!\left[\begin{array}{c} H \\ | \\ C \\ | \\ H \end{array}\right]_n\!\!-\!R'$$

for example, the number n of the repeating structural units – (CH_2) – lies between ca. 1000 and 100000. The distance between the C-atoms in the C–C-single bond is approximately 0,15 nm. Thus, the length of the extended polyethylene molecule may be of an order of magnitude of 10 μm. If one constructs a model of such a molecule on the scale of $1:2 \cdot 10^8$ [*e.g.*, by means of the Stuart–Brigleb spacefilling model; Stuart (1967)], the effective radius of the CH_2-group is about 5 cm and the length of the model can reach up to 2 km (1.24 miles).

The C-atoms in a fully extended polyethylene molecule are arranged in a zigzag pattern in a plane, whereby the angle enclosed by two neighbouring bonds (valence angles) is about 110°. As the single C–C-bond (in contrast to the C=C-double bond) is symmetrical with respect to rotation and the hydrogen atoms hardly lead to a stereometric hindrance and are also only linked with each other by weak van der Waals' forces, there is a certain possibility of rotation around each of the C–C-bonds. This rotation, however, is not completely free. The geometrical constitution of the structural units and the interaction of the units result in a threefold potential of revolution which possesses pronounced minima at the rotation angles $\varphi = 0°, 120°$ and $240°$ (see Fig. 6.1). The C–C-bond with $\varphi = 0°$ is designated as the trans-bond t, the C–C-bonds with $\varphi = 120°$ and $\varphi = 240°$ as gauche-bonds g or $\bar{g}$, respectively. These bond types enable the molecule to produce a multitude of relatively stable conformational isomers.

The fact that the gauche-bonds point out of the plane of the extended all-trans molecule is of great importance. Incorporation of a single gauche-bond into the all-trans skeleton has the effect that the molecular segments separated by the gauche-bond lie in two planes which are tilted against each other. The incorporation of many gauche-bonds results in the formation of a molecular coil. Due to the hindered rotation around the C–C-bonds, the molecules have a flexibility which far exceeds the flexibility that can be achieved by deformation of the valence angles (see Sect. 6.4.3).

If we disregard the mutual intramolecular interaction of the bond types (see the comments at the end of Sect. 6.3), the conformational isomerism (and thus also the coil-formation tendency and the flexibility) of an isolated polyethylene molecule is energetically characterized by three different parameters (Fig. 6.1): by the energy differences $\Delta\varepsilon_{d_1}$ and $\Delta\varepsilon_{d_2}$ between the maxima and minima of the rotational potential and by the energy difference $\Delta\varepsilon_g$ between the gauche-bonds and the trans-bond.

The differences $\Delta\varepsilon_{d_1}$ and $\Delta\varepsilon_{d_2}$ act as activation energies with regard to the transitions $t \rightarrow g, \bar{g}$, $g \rightarrow \bar{g}$ and $g, \bar{g} \rightarrow t$. They determine the dynamic behaviour of the conformational isomerism. They give information regarding the question on how the long chain-like molecules react to a more or less rapid external perturbation and at which rate a dynamically disturbed molecule returns to the internal equilibrium. If one puts a disturbed molecule into a temperature bath and if $\Delta\varepsilon_{d_i} \gg kT$ $(i = 1, 2)$ is valid, the potential barriers between the trans- and gauche-bonds are only overcome at those few "hot" points where the thermal fluctuations considerably exceed the mean thermal energy kT. One must wait a long time until the internal equilibrium between trans- and gauche-bonds is reached. If, on the other hand, $\Delta\varepsilon_{d_i} \leq kT$ is valid, the potential barriers in the whole molecule are overcome without restraints. The internal equilibrium between trans- and gauche-bonds can be reached practically instantaneously. For the frequency ν with which the potential barriers are overcome, we can set according to Arrhenius

$$\nu_i = \nu_{\infty i}\, e^{-\Delta\varepsilon_{di}/kT} \quad (i = 1, 2)$$

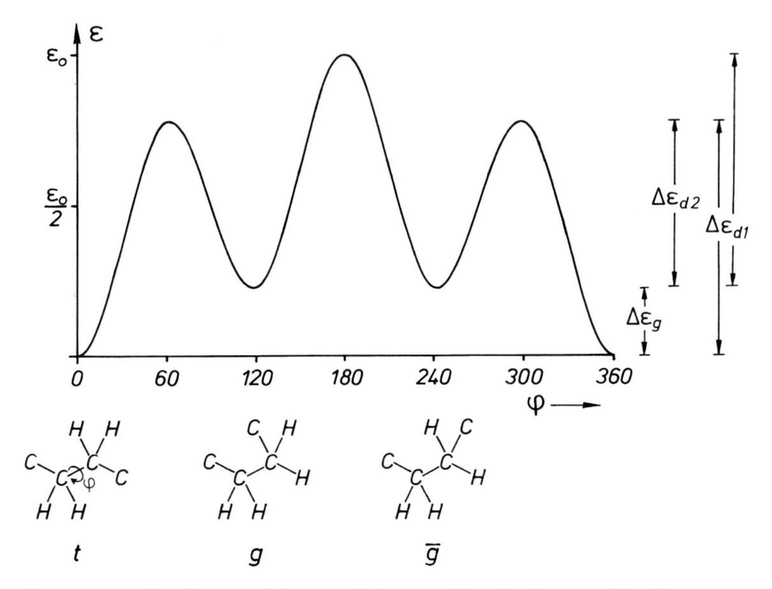

Fig. 6.1. Rotational potential around the C–C-bond of a paraffin chain according to Taylor (1947/1948):

$$\varepsilon(\varphi) = \frac{1}{2}\,\varepsilon_o\,[x(1-\cos\varphi) + (1-x)(1-\cos 3\varphi)]\,.$$

The parameter x allows a continuous transition from the onefold rotation axis ($x = 1$) to the threefold rotation axis ($x = 0$). In the figure $x = 0.3$ was assumed

[*e. g.*, see Glasstone, Laidler, Eyring (1941)]. The relaxation times

$$\tau_i = \tau_{\infty i}\,e^{\Delta\varepsilon_{di}/kT} \sim \frac{1}{\nu_i} \tag{6.44}$$

are a measure of the time a molecule, which has been disturbed and then left to itself, needs in order to reach internal equilibrium with regard to its conformation (see Sect. 7.4). If the molecule is stressed within the times $\Delta t \ll \tau_i$, it does not have time to change its conformation in any way during the perturbation. The molecule appears stiff when responding to this perturbation. However, if the perturbation is sufficiently slow so that the molecule is always able to adjust its equilibrium distribution between trans- and gauche-bonds, it appears quite flexible in response to this perturbation (see Fig. 7.17).

The internal equilibrium of the linear polyethylene molecule with respect to the conformational isomerism is determined by the energy difference $\Delta\varepsilon_g$. The equilibrium distribution between trans- and gauche-bonds can be determined on the basis of Eqs. (6.13) and (6.15). If the events of finding a trans- or a gauche-bond are independent of each other, and if T = const. designates the temperature of the temperature bath, V = const., the natural volume of the molecule, and $n = n_t + n_g$ = const., the number of bonds in the molecule (n_t: number of trans-bonds, n_g: number of gauche-bonds), the probability of finding a trans-bond is given by

$$P_t = \frac{n_t}{n} = \frac{e^{-\varepsilon_t/kT}}{e^{-\varepsilon_t/kT} + 2\,e^{-\varepsilon_g/kT}} = \frac{e^{\Delta\varepsilon_g/kT}}{e^{\Delta\varepsilon_g/kT} + 2} \tag{6.45}$$

(ε_t: energy of the trans-bond; ε_g: energy of the gauche-bond; $\Delta\varepsilon_g \equiv \varepsilon_g - \varepsilon_t$). For the probability of finding a gauche-bond, one obtains

$$P_g = \frac{n_g}{n} = 1 - P_t = \frac{2}{e^{\Delta\varepsilon_g/kT} + 2}\,. \tag{6.46}$$

If there is no energetic difference between the trans- and gauche-bonds, *i.e.*, $\Delta\varepsilon_g = 0$, (6.45, 6.46) result in: $P_t = 1/3$, $P_g = 2/3$. The binding states are evenly distributed independent of the temperature over the minima of the rotational potential. However, if $\Delta\varepsilon_g \neq 0$, we have for $T = 0$: $P_t = 1$, $P_g = 0$. There are no gauche-bonds for the case of $T = 0$. The molecule is in an extended form. Increasing the temperature then leads to an increase in the number of gauche-bonds as well as in the coiling tendency until, with $T \to \infty$, the equipartition $P_t = 1/3$, $P_g = 2/3$ is reached. Hence, the equipartition characterizes a saturation value which is usually not exceeded in the internal equilibrium. In a completely isolated molecule, however, negative absolute temperature values are also conceivable, whose scale starting from $T = \pm\infty$ goes from the negative side as far as $T = 0$ [see Ramsey (1956)]. The number of gauche-bonds increases from $n_g = 2n/3$ up to $n_g = n$ when traversing the scale of negative absolute temperatures.

The knowledge of the ratio n_g/n_t, however, does not say anything about the shape and size of the conformational isomers. With a given ratio of n_g/n_t, a very large variety of conformational isomers is still possible. Their shapes range from almost extended to highly coiled and are subject to continuous changes due to thermal fluctuations. Hence, even statements about the shape and size of the individual polymer molecules can only be made statistically. The so-called segment model by W. Kuhn [Kuhn (1934), (1943)] provides, without great difficulties, the first qualitative evaluations.

Kuhn proceeded from the fact that the fixed valence angles solely determine the mutual orientation of directly neighbouring chain links. The orientation of an arbitrary chain link is even freer with respect to another, the more chain links there are between the two selected chain links. Kuhn assumed, therefore, that every flexible linear polymer molecule can be divided into rigid segments of the same length which can be rotated completely freely with respect to each other (without any restrictions by valence angles or rotational potentials). As every segment comprises several chain links, the segment length itself is, of course, already a statistical (mean) quantity. Determination of the shapes and sizes of the molecules in this segment model corresponds completely to the solution of the random walk problem, which is already known from the theory of diffusion and Brownian motion [*e.g.*, see Feller (1957), Isihara (1971),

Pathria (1972)]. Two important characteristic statistical quantities can easily be determined in this model: the mean distance of the molec-ule ends and the mean radius of gyration of the molecular coil (whereby the mean values can either be interpreted as a time average or as an ensemble average).

In the following, we will assume that the linear polymer molecule can be divided into n rigid segments of length $|\vec{a}_i| = a$ and mass m which can be rotated completely freely with respect to each other. Let us suppose that $n \gg 1$ is valid. $\vec{a}_i$ represents the vector whose position, orientation, and length correspond to the segment No. $i\,(i = 1 \ldots n)$. We can visualize the masses m of the segments as being concentrated in the endpoints of the vectors $\vec{a}_i$. Because of this, no mass is found at the beginning of the molecule, however, this is not important when determining the centre of mass as $n \gg 1$. One denotes by $\vec{r}_{i-1}$ the vector which points from the centre of mass C of the molecule towards the origin of the segment vector $\vec{a}_i$, and $\vec{h}$ represents the vector from the beginning A of the molecule to the end B of the molecule (Fig. 6.2).

The end-to-end vector $\vec{h}$ is given by

$$\vec{h} = \sum_{i=1}^{n} \vec{a}_i \,. \tag{6.47}$$

Hence, we have

$$\vec{h}^2 = h^2 = \sum_{i=1}^{n} \sum_{k=1}^{n} \vec{a}_i \vec{a}_k = a^2 \sum_{i=1}^{n} \sum_{k=1}^{n} \cos(\vec{a}_i; \vec{a}_k)$$

$\vec{a}_i \vec{a}_k$ designates the scalar product of the vectors, $(\vec{a}_i; \vec{a}_k)$ the angle between the directions of the vectors $\vec{a}_i$ and $\vec{a}_k$, and $h = |\vec{h}|$ the magnitude of the vector $\vec{h}$, *i.e.*, the end-to-end distance. As the $\vec{a}_i$ should be completely freely rotatable with respect to each other,

$$\langle \cos(\vec{a}_i; \vec{a}_k) \rangle = \delta_{ik} \tag{6.48}$$

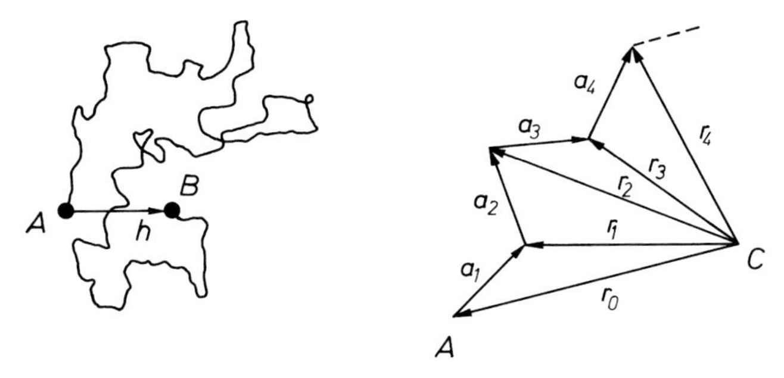

Fig. 6.2. Definition of the characteristic parameters of a polymer coil. On the left: a polymer coil with the end-to-end vector $\vec{h}$ which leads from the beginning A to the end B of the chain. On the right: Kuhn's segment vectors $\vec{a}_i$ and position vectors $\vec{r}_i$, referenced to the centre of mass C of the chain molecule

(δ_{ik}: Kronecker delta, see p. 39) holds for the mean value. This leads to

$$\langle h^2 \rangle = a^2 \sum_{i=1}^{n} \delta_{ii} = n a^2 \,. \tag{6.49}$$

In the segment model, the mean end-to-end distance of a linear polymer molecule is given by

$$\bar{h} \equiv \sqrt{\langle h^2 \rangle} = n^{1/2} a \,. \tag{6.50}$$

It increases proportionally to the square root of the "length" of the molecule na (one should bear in mind that na does not give the exact length of the molecule as there already is a certain tendency for coiling within the segments).

In an assembly of $n+1$ mass points with the mass m_i, the position vector of the centre of mass $\vec{r}_C$ referring to an arbitrary origin O is given by

$$\vec{r}_C = \sum_{i=0}^{n} m_i \vec{r}_i \,.$$

Here, the position vectors $\vec{r}_i$ of the mass points m_i must also be referred to the origin O. If the centre of mass itself is chosen as the origin of the position vectors, we have

$$\sum_{i=0}^{n} m_i \vec{r}_i = 0 \,.$$

$$\vec{r}_o = - \sum_{i=1}^{n} \vec{r}_i \tag{6.51}$$

follows if all the mass points are the same ($m_i = m$ for all i). The relationship between the position vectors $\vec{r}_i \, (i \neq 0)$ referring to the centre of mass and the segment vectors $\vec{a}_i$ is given by

$$\vec{r}_i = \vec{a}_i + \vec{r}_{i-1}$$

(see Fig. 6.2). If one successively inserts the values for the preceding vectors $\vec{r}_{i-1}, \vec{r}_{i-2}, \ldots$, respectively, one finally obtains

$$\vec{r}_i = \sum_{k=1}^{i} \vec{a}_k + \vec{r}_o \qquad (i \neq 0) \tag{6.52}$$

and according to (6.51),

$$\vec{r}_o = - \sum_{i=1}^{n} \sum_{k=1}^{i} \vec{a}_k - n\vec{r}_o = - \sum_{i=1}^{n} (n-i+1)\, \vec{a}_i - n\vec{r}_o \,,$$

i.e.,

$$\vec{r}_o = -\frac{1}{n+1}\sum_{i=1}^{n}(n-i+1)\,\vec{a}_i\,. \tag{6.53}$$

For the scalar square of the starting vector $\vec{r}_o$

$$\vec{r}_o^{\,2} = r_o^2 = \frac{1}{(n+1)^2}\sum_{i=1}^{n}\sum_{k=1}^{n}(n-i+1)\,(n-k+1)\,\vec{a}_i\,\vec{a}_k \tag{6.54}$$

then follows and from that with (6.48) the mean value

$$\langle \vec{r}_o^{\,2}\rangle = \left(\frac{a}{n+1}\right)^2\sum_{i=1}^{n}(n-i+1)^2 = \left(\frac{a}{n+1}\right)^2\sum_{k=1}^{n}k^2\,.$$

With

$$\sum_{k=1}^{n}k^2 = \frac{1}{6}\,n\,(n+1)\,(2n+1)\,, \tag{6.55}$$

one obtains

$$\langle r_o^2\rangle = \frac{a^2}{6}\,\frac{n}{n+1}\,(2n+1)$$

and with $n \gg 1$ in good approximation

$$\langle r_o^2\rangle = \frac{1}{3}\,na^2 = \frac{1}{3}\,\langle h^2\rangle\,. \tag{6.56}$$

If one inserts (6.53) into (6.52), one gets further for $i \neq 0$

$$\vec{r}_i = \sum_{k=1}^{i}\vec{a}_k - \frac{1}{n+1}\sum_{k=1}^{n}(n-i+1)\,\vec{a}_k\,,$$

thus, when taking into account (6.54)

$$\vec{r}_i^{\,2} = \sum_{k=1}^{i}\sum_{\ell=1}^{i}\vec{a}_k\,\vec{a}_\ell - \frac{2}{n+1}\sum_{k=1}^{i}\sum_{\ell=1}^{n}(n-\ell+1)\,\vec{a}_k\,\vec{a}_\ell + r_o^2$$

and from this with (6.48) the mean square of the distances

$$\langle r_i^2\rangle = -\,ia^2 + \frac{2a^2}{n+1}\sum_{\ell=1}^{i}\ell + \langle r_o^2\rangle\,.$$

With

$$\sum_{\ell=1}^{i}\ell = \frac{1}{2}\,i\,(i+1)\,, \tag{6.57}$$

one obtains

$$\langle r_i^2 \rangle = \langle r_o^2 \rangle - \frac{n}{n+1}\, i a^2 + \frac{i^2 a^2}{n+1}$$

and with $n \gg 1$ and (6.49), (6.56) in good approximation

$$\langle r_i^2 \rangle = \langle h^2 \rangle \left[\frac{1}{3} - \frac{i}{n} + \left(\frac{i}{n} \right)^2 \right]. \tag{6.58}$$

A comparison with (6.56) shows that this equation is also valid for $i = 0$. For $i = n$, one gets

$$\langle r_n^2 \rangle = \frac{1}{3} \langle h^2 \rangle = \langle r_o^2 \rangle ,$$

which expresses that it is arbitrary which of the two chain ends is defined as the beginning and which as the end of the chain.

For the segment coil, one can define a mean radius

$$\bar{r} \equiv \sqrt{\langle r^2 \rangle}$$

via

$$\langle r^2 \rangle = \frac{1}{n} \sum_{i=0}^{n} \langle r_i^2 \rangle , \tag{6.59}$$

When one inserts (6.58), one obtains

$$\langle r^2 \rangle = \frac{1}{n} \langle h^2 \rangle \left[\frac{n}{3} - \frac{1}{n} \sum_{i=0}^{n} i + \frac{1}{n^2} \sum_{i=0}^{n} i^2 \right]$$

or with (6.55) and (6.57)

$$\langle r^2 \rangle = \frac{1}{n} \langle h^2 \rangle \left[\frac{n}{3} - \frac{1}{2} (n+1) + \frac{1}{6n} (n+1)(2n+1) \right]$$

and because of $n \gg 1$

$$\langle r^2 \rangle = \frac{1}{6} \langle h^2 \rangle = \frac{1}{6} n a^2 . \tag{6.60}$$

Furthermore, a mean degree of coiling can be defined by the ratio

$$Q = \frac{n}{\bar{r}} \sim n^{1/2} . \tag{6.61}$$

The degree of coiling also increases with the square root of the "length" of the molecule.

As there is no preferred orientation in the segment model, the mass distribution of the molecular coil is characterized by spherical symmetry regarding the spatial and temporal average. The moment of inertia I of a spherically symmetric rigid mass distribution is given by

$$I = \frac{1}{3}\,(I_{xx} + I_{yy} + I_{zz})\,, \tag{6.61}$$

where the principal moments of inertia around the coordinate axes are designated by:

$$I_{xx} = \sum_i m_i\,(r_i^2 - x_i^2)\,,$$

$$I_{yy} = \sum_i m_i\,(r_i^2 - y_i^2)\,,$$

$$I_{zz} = \sum_i m_i\,(r_i^2 - z_i^2)\,,$$

and the scalar squares of the position vectors of the mass points m_i are

$$r_i^2 = x_i^2 + y_i^2 + z_i^2$$

[*e.g.*, see Goldstein (1962)]. It follows when all of the masses are equal that:

$$I = \frac{2}{3}\,m \sum_i r_i^2\,,$$

Hence, the mean moment of inertia is, according to (6.59):

$$\langle I \rangle = \frac{2}{3}\,M\,\langle r^2 \rangle\,. \tag{6.62}$$

$M = nm$ is the mass of the molecule. When one defines the so-called radius of gyration *via*

$$I = M\,R_o^2\,,$$

(6.62) and (6.60) lead to the relation

$$\langle R_o^2 \rangle = \frac{2}{3}\,\langle r^2 \rangle = \frac{1}{9}\,\langle h^2 \rangle\,. \tag{6.63}$$

Thus, the mean distance $\bar{h}$ between the ends of the molecule is ultimately the defining quantity for the shape and size of the linear polymer molecules [Eq. (6.50)].

The mean radius of gyration, the mean radius of the coil, and the mean end-to-end distance are quantities which are accessible to macroscopic experiments (especially light and neutron scattering). From the standpoint of thermodynamics, the question about the number of conformational isomers (microstates) by means of which the macroscopic equilibrium states with given values of these mean quantities can be realized is, therefore, of particular importance. To answer this question, we will first proceed from an even simpler model: the one-dimensional segment model (without doubt somewhat hypothetical but still very evident and definitely also powerful).

In the strictly one-dimensional segment model (Fig. 6.3), the segment vectors $\vec{a}_i$ can only assume two positions. They either point to the right or to the left. We write $\vec{a}_i = +a$ if the vector points to the right and $\vec{a}_i = -a$ if the vector points to the left. To simplify matters, we assume that the total segment number n is even. This does not mean a major restriction for the case $n \gg 1$, which we are mainly interested in.

The total number of conformational isomers which can be produced in the one-dimensional segment chain is equal to the number of combinations of two distinct, unrestrictedly repeatable elements in n positions, *i.e.*, equal to 2^n (in Fig. 6.3 with $n = 6$ equal to $2^6 = 64$). If k segments point to the left, $n-k$ segments necessarily point to the right. The vector pointing from the beginning to the end of the chain is thus given by

$$\vec{h} = \sum_{i=1}^{n} \vec{a}_i = a\,[(n-k) - k] = 2a\left(\frac{n}{2} - k\right). \tag{6.64}$$

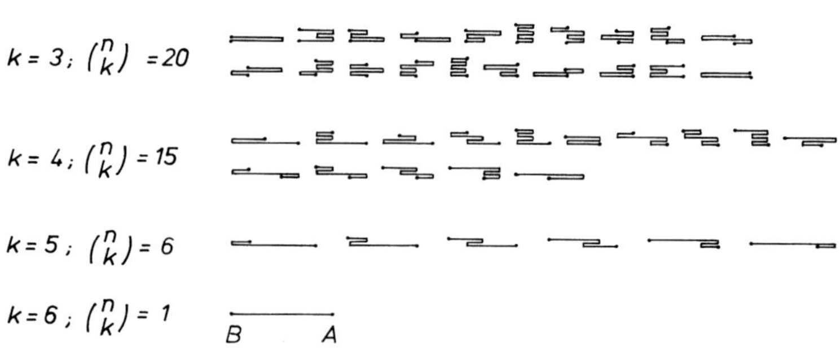

$k = 0;\ \binom{n}{k} = 1$ — a; A B

$k = 1;\ \binom{n}{k} = 6$ — h

$k = 2;\ \binom{n}{k} = 15$

$k = 3;\ \binom{n}{k} = 20$

$k = 4;\ \binom{n}{k} = 15$

$k = 5;\ \binom{n}{k} = 6$

$k = 6;\ \binom{n}{k} = 1$ — B A

Fig. 6.3. Conformational isomers in the one-dimensional segment model. $n = 6$: number of segments of the length a; k: number of segments pointing to the left; h: end-to-end distance

$\vec{h} = \pm h$ is valid depending on whether the beginning of the chain lies on the left (+) or on the right (–) of the chain end. The number $\omega(n, k)$ of conformational isomers which is possible for a fixed $\vec{h}$ is equal to the number of combinations of k elements of the one kind ($\vec{a}_i = -a$) with $n-k$ elements of the other kind ($\vec{a}_i = +a$), *i.e.*, equal to the binomial coefficient

$$\omega(n, k) = \binom{n}{k} \equiv \frac{n}{(n-k)!\,k!} \,. \tag{6.65}$$

The probability of finding the end-to-end vector (6.64) amounts to

$$P(\vec{h}) = \frac{1}{2^n}\,\omega(n, k) = \frac{n!}{2^n (n-k)!\,k!} \,. \tag{6.66}$$

The most probable end-to-end vector is the one that can be realized by the largest number of conformational isomers, *i.e.*, for which the binomial coefficient (6.65) reaches a maximum. As is generally known, this is the case for $k = n/2$. Hence, according to (6.64), the state with $\vec{h} = 0$, for which the beginning and the end of the chain are at the same position, is the most probable one (in Fig. 6.3 with $k = 3$ and $\omega = 20$). However, if one inquires about the number of conformational isomers and about the probability of states with a constant magnitude $|\vec{h}| = h$ of the end-to-end distance, one obtains a completely different result. With (6.64), (6.66) can be transformed into

$$P(\vec{h}) = \frac{n!}{2^n \left(\frac{n}{2} + \frac{\vec{h}}{2a}\right)! \left(\frac{n}{2} - \frac{\vec{h}}{2a}\right)!} \,. \tag{6.67}$$

The probability of finding the distance h is equal to finding the vector $\vec{h} = +h$ or the vector $\vec{h} = -h$. According to (6.67), for $h \neq 0$ one obtains:

$$P(h \neq 0) = \frac{2n!}{2^n \left(\frac{n}{2} + \frac{h}{2a}\right)! \left(\frac{n}{2} - \frac{h}{2a}\right)!} \,. \tag{6.68}$$

This probability has a maximum value at $h = 2a$ which exceeds the probability value

$$P(h = 0) = \frac{2n!}{2^n \left(\frac{n}{2}\right)! \left(\frac{n}{2}\right)!} \,.$$

Thus, judged on the basis of the magnitude h of the end-to-end vector, the state with $h = 2a$ and

$$\omega = \frac{2n!}{\left(\frac{n}{2}+1\right)!\left(\frac{n}{2}-1\right)!}$$

possibilities of realization is the most probable one (in Fig. 6.3 altogether $2 \cdot 15 = 30$ conformational isomers).

According to (6.64), the mean end-to-end vector with respect to all 2^n conformational isomers is:

$$\langle \vec{h} \rangle = \sum_{k=0}^{n} \frac{1}{2^n} \binom{n}{k} \vec{h} = \frac{a}{2^n} \sum_{k=0}^{n} \binom{n}{k} (n-2k) ,$$

and because of

$$\sum_{k=0}^{n} \binom{n}{k} = 2^n \quad \text{and} \quad \sum_{k=0}^{n} \binom{n}{k} k = \frac{1}{2} 2^n n , \tag{6.69}$$

$$\langle \vec{h} \rangle = 0 . \tag{6.70}$$

This result, which may be somewhat irritating at first glance, only expresses that an equally probable mirror-inverted conformational isomer exists for each of the 2^n conformational isomers (see Fig. 6.3). The average of the square $\vec{h}^2 = h^2$ (the so-called second moment), however, does not vanish. According to (6.64), it is

$$\langle h^2 \rangle = \sum_{k=0}^{n} \frac{1}{2^n} \binom{n}{k} \vec{h}^2 = \frac{a^2}{2^n} \sum_{k=0}^{n} \binom{n}{k} (n-2k)^2 .$$

$$= \frac{a^2}{2^n} \left[n^2 \sum_{k=0}^{n} \binom{n}{k} - 4n \sum_{k=0}^{n} \binom{n}{k} k + 4 \sum_{k=0}^{n} \binom{n}{k} k^2 \right] ,$$

i. e., because of (6.69) and

$$\sum_{k=0}^{n} \binom{n}{k} k^2 = \frac{1}{4} n(n+1) 2^n , \tag{6.71}$$

$$\langle h^2 \rangle = n a^2 . \tag{6.72}$$

In addition, (6.70) and (6.72) lead to the so-called standard deviation

$$\langle (\vec{h} - \langle \vec{h} \rangle)^2 \rangle^{1/2} = \sqrt{\langle h^2 \rangle} = n^{1/2} a = \bar{h} . \tag{6.73}$$

According to the De Moivre-Laplace limit theorem [*e.g.*, see Feller (1957)], a binomial probability distribution of the form (6.67) in the limit

$$n \to \infty\,, \quad \frac{h}{2na} \to 0$$

is identical to a Gaussian normal distribution. Hence, if the polymer molecules are very long flexible molecules in which the predominant portion is relatively strongly coiled so that $(h/2a) \ll n$ can be assumed, the expression (6.67) can in good approximation be replaced by

$$P(\vec{h}) = \left(\frac{2}{\pi n}\right)^{1/2} \mathrm{e}^{-h^2/2na^2}, \tag{6.74}$$

whereby $\vec{h}$ can now be understood as a continuous variable. When averaging $\sum\limits_{k=0}^{n} \ldots$ must then be replaced by $\frac{1}{2a}\int\limits_{-\infty}^{+\infty} \ldots \, \mathrm{d}h$, as the length $2a$ corresponds to the discrete unit step $\Delta k = 1$. If one uses the probability densities

$$R(\vec{h}) \equiv P(\vec{h})/2a = \left(\frac{1}{2na^2\pi}\right)^{1/2} \mathrm{e}^{-h^2/2na^2}, \tag{6.75}$$

instead of the probabilities $P(\vec{h})$ when averaging, the summation must simply be replaced by an integral over dh. In addition, the following values of the integrals must be known when averaging:

$$\int\limits_{-\infty}^{+\infty} \mathrm{e}^{-b^2x^2} \mathrm{d}x = \sqrt{\pi}/b\,, \tag{6.76a}$$

$$\int\limits_{-\infty}^{+\infty} x\, \mathrm{e}^{-b^2x^2} \mathrm{d}x = 0\,, \tag{6.76b}$$

$$\int\limits_{-\infty}^{+\infty} x^2\, \mathrm{e}^{-b^2x^2} \mathrm{d}x = \sqrt{\pi}/2b^2\,. \tag{6.76c}$$

Equations (6.74) or (6.75) lead, of course, only back to the mean values (6.70) and (6.72).

Let us now return to the three-dimensional problem and assume a cubic space lattice whose unit cells possess the edge length $a/\sqrt{3}$. The segment vectors $\vec{a}_i$ should, starting from the lattice point A_i and arbitrarily pointing in the direction of one of the eight diagonals, lead to the next lattice point (Fig. 6.4). The (directed) components of the segment vectors parallel to the rectangular lattice axes x, y, z are then given by

$$\vec{a}_{ix} = \pm\frac{a}{\sqrt{3}}\,, \quad \vec{a}_{iy} = \pm\frac{a}{\sqrt{3}}\,, \quad \vec{a}_{iz} = \pm\frac{a}{\sqrt{3}}\,. \tag{6.77}$$

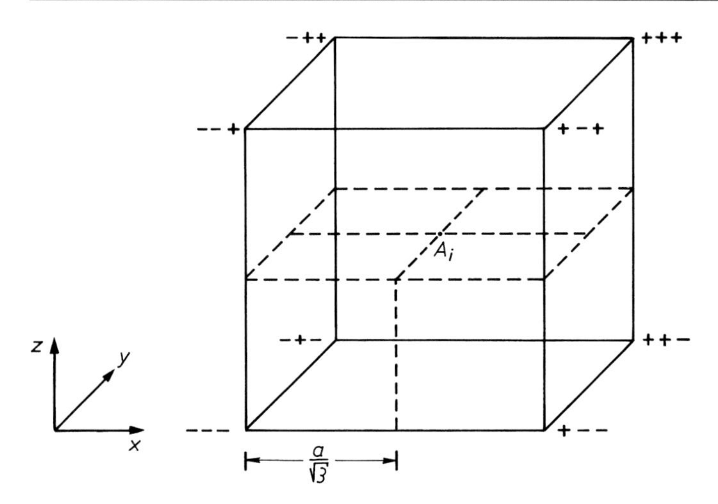

Fig. 6.4. Section of a cubic space lattice. The Kuhn segment vectors $\vec{a}_i$ with length a should, starting from a lattice point A_i and going in the direction of one of the eight spatial diagonals, arbitrarily lead to the next lattice point. The components of the segment vectors in the direction of the lattice axes then have the length $a/\sqrt{3}$. The signs of the components (6.77) are given at the eight possible end points

They possess a positive sign if they point in the direction of the positive axis and a negative sign if they point in the direction of the negative axis. Provided that the different directions in the space are not correlated, the one-dimensional segment model with the segmental length $a/\sqrt{3}$ is obviously valid along each of the axes x, y, z. In particular, the probability of finding the component $\vec{h}_x$ of the end-to-end vector $\vec{h}$ in the x-direction is, according to (6.74), given by

$$P(\vec{h}_x) = \left(\frac{2}{\pi n}\right)^{1/2} \mathrm{e}^{-3h_x^2/2na^2} \,. \tag{6.78}$$

Corresponding probabilities hold for the components $\vec{h}_y$ and $\vec{h}_z$. The probability of finding the end-to-end vector $\vec{h}$ in the three-dimensional space is equal to the product

$$P(\vec{h}) = P(\vec{h}_x)\, P(\vec{h}_y)\, P(\vec{h}_z)\,,$$

i.e., with

$$\vec{h}^2 = \vec{h}_x^2 + \vec{h}_y^2 + \vec{h}_z^2 = h^2\,, \tag{6.79}$$

equal to

$$P(\vec{h}) = \left(\frac{2}{\pi n}\right)^{3/2} \mathrm{e}^{-3h^2/2na^2} \,. \tag{6.80}$$

If $\vec{h}$ is regarded as a continuous variable, the threefold sum over the lattice points

$$\sum_k \sum_\ell \sum_m \ldots \quad \text{must be replaced by} \quad \left(\frac{\sqrt{3}}{2a}\right)^3 \iiint_{-\infty}^{+\infty} \ldots \quad dh_x\, dh_y\, dh_z$$

when calculating the mean values. The volume of the cube in Fig. 6.4 is $(2a/\sqrt{3})^3$. The probability density is given by

$$R(\vec{h}) \equiv \left(\frac{\sqrt{3}}{2a}\right)^3 P(\vec{h}) = \left(\frac{b^2}{\pi}\right)^{3/2} e^{-b^2h^2} \tag{6.81a}$$

if one writes

$$b^2 \equiv 3/(2na^2) \tag{6.81b}$$

as an abbreviation. It, of course, follows *via* (6.76a) that:

$$\iiint_{-\infty}^{+\infty} R(\vec{h})\, dh_x\, dh_y\, dh_z = 1\,. \tag{6.82}$$

One obtains:

$$R(\vec{h}_x) \equiv \frac{\sqrt{3}}{2a} P(\vec{h}_x) = \left(\frac{b^2}{\pi}\right)^{1/2} e^{-b^2h_x^2} \tag{6.83}$$

for the one-dimensional probability density corresponding to (6.78). With (6.80) and (6.81), we have already found our way to the three-dimensional random walk problem. (6.80) and (6.81) are identical with the normally used probabilities which result as approximative solutions from the random walk problem [*e.g.*, see Isihara (1971), Treloar (1975)].

Corresponding to (6.70), we get with (6.83) and (6.76b)

$$\langle\vec{h}_x\rangle = \int_{-\infty}^{+\infty} h_x R(\vec{h}_x)\, dh_x = 0\,,$$

and equally $\langle\vec{h}_y\rangle = \langle\vec{h}_z\rangle = 0$, *i.e.*, also

$$\langle\vec{h}\rangle = 0\,. \tag{6.84}$$

With (6.81) and (6.76c) follows further considering (6.79) and in agreement with (6.49):

$$\langle h^2\rangle = \iiint_{-\infty}^{+\infty} h^2 R(\vec{h})\, dh_x\, dh_y\, dh_z = na^2\,. \tag{6.85}$$

When inquiring about the probability of finding the distance $|\vec{h}| = h$, the probability (6.80) must be multiplied with the surface $4\pi h^2$ of that sphere which is swept by the vector $\vec{h}$ with constant magnitude h. This corresponds completely to the transition from (6.67) to (6.68), as the factor 2 in (6.68) can also be interpreted as the surface of a one-dimensional "sphere" [*e.g.*, see Pathria (1972), appendix C]. One thus obtains

$$P(h) = \left(\frac{2^7}{\pi n^3}\right)^{1/2} h^2 \, e^{-b^2 h^2} . \tag{6.86}$$

This function has a maximum at

$$h^2 = \frac{2}{3} n a^2 = \frac{2}{3} \langle h^2 \rangle = 1/b^2 .$$

Hence, the state with the end-to-end distance $1/b$ is the most probable one. It can be realized with the largest number of conformational isomers.

Let us now consider an ensemble of non-interacting polymer molecules under the thermodynamic boundary condition $U, V, N =$ const. Because of the multitude of conformational isomers with which a molecule with the end-to-end vector $\vec{h}$ can be realized, an entropy can already be attributed to each of the molecular states $\vec{h} =$ const. *via* (6.6). If $\Omega_t(n)$ is the total number of conformational isomers of the molecule, the number $\Omega(\vec{h})$ of the conformational isomers with which the state $\vec{h}$ can be realized is equal to

$$\Omega(\vec{h}) = \Omega_t(n) \, P(\vec{h}) .$$

According to (6.6), the conformational isomerism contributes to the entropy of the molecule in the state $\vec{h} =$ const. by

$$s(\vec{h}) = k \ln \Omega(\vec{h}) = k \left[\ln \Omega_t(n) + \ln P(\vec{h})\right] .$$

Within the scope of the segment model, one obtains with (6.80) and (6.81 b)

$$s(\vec{h}) = k \left[\ln \Omega_t'(n) - b^2 h^2\right] \tag{6.87}$$

for it, whereby

$$\Omega_t'(n) = \left(\frac{2}{\pi n}\right)^{3/2} \Omega_t(n) .$$

is used as an abbreviation. This requires, of course, that the degrees of freedom of conformation are completely independent of the remaining degrees of freedom of the molecule. If the possible states $\vec{h}$ are traversed due to thermal agitation by each molecule in a time average and, in addition, by the molecules of the ensemble in an ensemble average, the mean contribution per molecule of the conformational isomerism to the entropy is

$$\langle s(\vec{h}) \rangle = \iiint_{-\infty}^{+\infty} s(\vec{h}) \, R(\vec{h}) \, dh_x \, dh_y \, dh_z ,$$

i. e., with (6.85) and (6.82)

$$\langle s(\vec{h})\rangle = k\,[\ln \Omega_t'(n) - b^2 \langle h^2\rangle]\,,$$

or with (6.85) and (6.81 b)

$$\langle s(\vec{h})\rangle = k\left[\ln \Omega_t'(n) - \frac{3}{2}\right]. \tag{6.88}$$

Let us assume further that due to the action of an external force each end-to-end vector is subject to an affine deformation

$$\vec{h}_{\lambda x} = \lambda_x \vec{h}_x\,,\quad \vec{h}_{\lambda y} = \lambda_y \vec{h}_y\,,\quad \vec{h}_{\lambda z} = \lambda_z \vec{h}_z$$

(while the framework of the statistics, the spacial lattice, the segment vectors $\vec{a}_i$, and their number n remain untouched). Corresponding to (6.87), the contribution of the deformed molecules to the entropy is then:

$$s(\vec{h}_\lambda) = k\,[\ln \Omega_t'(n) - b^2(\lambda_x^2 h_x^2 + \lambda_y^2 h_y^2 + \lambda_z^2 h_z^2)]\,.$$

For the mean value it follows that:

$$\langle s(\vec{h}_\lambda)\rangle = k\left[\ln \Omega_t'(n) - \frac{1}{2}(\lambda_x^2 + \lambda_y^2 + \lambda_z^2)\right]. \tag{6.89}$$

If the remaining degrees of freedom of the molecules are not affected by the deformation, the difference between the entropies of the molecular ensemble in the deformed and non-deformed states, according to (6.88) and (6.89), is

$$\Delta S = N[\langle s(\vec{h}_\lambda)\rangle - \langle s(\vec{h})\rangle] = -\frac{1}{2} RN(\lambda_x^2 + \lambda_y^2 + \lambda_z^2 - 3)$$

($R = kN_A$). As the different conformational isomers are not energetically different in the segment model *($\Delta U = 0$)*, one obtains for the free energy difference

$$\Delta F = -T\Delta S = \frac{1}{2} RNT(\lambda_x^2 + \lambda_y^2 + \lambda_z^2 - 3)\,. \tag{6.90}$$

This relationship is the simplest basis for the theory of rubber elasticity [Sect. 5.3, particularly Eq. (5.58)].

Without doubt, the segment model can only provide a first qualitative insight into the specific physical behaviour of very long linear polymer molecules. The segment model also has certain drawbacks because the magnitude $|\vec{a}_i| = a$ of the segment vectors, which is a numerical component of almost all results, remains completely open and, therefore, it only serves as a fitting parameter when comparing with the experiment. The stiffer the polymer chain becomes, the larger must be the value of a that is be chosen. However, the prerequisites and results of the three-dimensional segment model are actually conclusive in the case of melts comprised of long homologous polymer molecules. Nevertheless, due to the neglect of the intramolecularly effective interaction potentials, the segment model does not answer the thermodynamically obvious question about the temperature dependence of the coiling tendency. (If one adheres to W. Kuhn's basic idea that every linear polymer molecule can be modeled by segments which freely rotate with respect to each other, consideration of the rotational potential within the segments would lead to a temperature dependence of the segmental length $a(T)$. That is, one should expect that the segmental length decreases with increasing temperature.) Moreover, the intermolecular interaction forces certainly also affect the coiling. Hence, it is easily understandable that solvation in a good low-molecular-mass solvent causes a reduction in the flexibility of the polymer molecules and thus in the coiling tendency. It is further obvious that, for example, the formation of a single gauche-bond in an ensemble of densely packed polymer molecules basically requires a cooperative process which can lead to noticeable temporal effects. The segment model does not give information about this either. For more complicated problems of the conformational isomerism see, for example, Volkenstein (1963), Flory (1969), Hopfinger (1973), Treloar (1975), de Gennes (1979), des Cloizeaux, Jannink (1990), and Sect. 6.3 and Chap. 7.

6.3 The Two-level System

The conformational isomerism of the chain molecules is without doubt the most prominent physical feature of a polymer liquid. If one places this characteristic at the centre of discussion, the polymer liquid comprised of very long chain molecules can be regarded to a first approximation as a mixture of trans- and gauche-bonds. If trans- and gauche-bonds are separated by a given energy gap $\Delta\varepsilon_g$, this leads to the problem of the so-called two-level system. Hence, the conformational isomerism causes characteristics in the physical properties of the polymer liquids that are comparable with the properties of a system of magnetic dipoles in an external field, which is also a two-level system [*e.g.*, see Huang (1963), Pathria (1972)]. In particular, taking the interaction into account leads to the so-called Ising problem [see specifically Green, Hurst (1964)]. However, all the problems which are connected with the fact that trans- and gauche-bonds are lined up on linear chains as carriers (for example, the stereometric problem and the problem of possible restraints) are not considered in such a comparison.

From the standpoint of phenomenological thermodynamics, the two-level system represents a system with a macroscopically relevant internal degree of freedom (Sect. 2.4 and Chap. 7). The concentration x_t of the trans-bonds or the concentration x_g of the gauche-bonds can be attributed to the internal degree of freedom as a macroscopic (internal) variable, whereby

$$x_t + x_g = 1 \tag{6.91}$$

is valid.

In the following, we will normalize the concentrations x_t and x_g over the constant number N of the possible bonds in the whole liquid and assume that the trans- and gauche-bonds are statistically distributed over the system. The concentrations x_t, x_g are then also significant as the probabilities of finding a trans- or a gauche-bond, respectively, in the system. We assume further that the gauche-bonds, such as in the case of polyethylene (see Fig. 6.1), are energetically doubly degenerate. For the concentrations x_{g1} and x_{g2} of the two gauche-bond types, we then have

$$x_{g1} + x_{g2} = x_g \qquad \text{and} \qquad x_{g1} = x_{g2} \,.$$

With this, it follows from (6.1) for the entropy of the system that:

$$S = -kN(x_t \ln x_t + 2x_{g1} \ln x_{g1}) = -kN[x_t \ln x_t + x_g \ln (x_g/2)] \,. \tag{6.92}$$

If we disregard for the present the mutual interaction of the statistical individuals, the internal energy of the system is given by

$$U = N(\varepsilon_t x_t + \varepsilon_g x_g) = N[\varepsilon_t + \Delta\varepsilon_g (1 - x_t)] \tag{6.93}$$

(for a definition of the energy parameters, see Sect. 6.2). As the free energy $F = U - TS$ of the system, we thus obtain

$$F = N\left[\varepsilon_t + \Delta\varepsilon_g (1 - x_t) + kT\left(x_t \ln x_t + (1 - x_t) \ln \frac{1 - x_t}{2}\right)\right] . \tag{6.94}$$

In order to maintain consistency with (6.92) it is necessary to assume that ε_t and $\Delta\varepsilon_g$ are constant with respect to temperature. However, ε_t and $\Delta\varepsilon_g$ can depend on the volume V of the system. The system is in internal equilibrium when $T, V, N =$ const. if, according to (2.77) with $f = F/N$, it is valid that:

$$\left(\frac{\partial f}{\partial x_t}\right)_e = -\Delta\varepsilon_g + kT \ln \frac{2x_t^e}{1 - x_t^e} = 0 \,. \tag{6.95}$$

According to (2.78), the equilibrium is stable or metastable if

$$\left(\frac{\partial^2 f}{\partial x_t^2}\right)_e = kT / x_t^e (1 - x_t^e) > 0 \tag{6.96}$$

holds. Because of $0 \le x_t^e \le 1$, the two-level system without interaction is, like the ideal mixture (Sect. 3.4), always stable. The expression

$$x_t^e = \frac{e^{\Delta\varepsilon_g/kT}}{2 + e^{\Delta\varepsilon_g/kT}} = \frac{1}{1 + 2e^{-\Delta\varepsilon_g/kT}} \tag{6.97}$$

results from (6.95) as the equilibrium value of the internal variable and thus, according to (6.91):

$$x_g^e = \frac{2}{2 + e^{\Delta\varepsilon_g/kT}} = \frac{2\,e^{-\Delta\varepsilon_g/kT}}{1 + 2\,e^{-\Delta\varepsilon_g/kT}} \tag{6.97}$$

These equations (except for the different normalization) are identical to (6.45) and (6.46). The internal variable in (6.94) can be eliminated by means of (6.97), so that only T, V, N appear as variables in the Gibbs fundamental equation in the equilibrium thermodynamics of the two-level system. Insertion of (6.97) into (6.94) leads to

$$F_e = N(\varepsilon_t + \Delta\varepsilon_g) - kNT \ln(2 + e^{\Delta\varepsilon_g/kT}) . \tag{6.99a}$$

Nevertheless, it is often also useful to keep the internal variable as a (although admittedly dependent) variable in the internal equilibrium. With (6.97), we can then write instead of (6.99a)

$$F_e = N\varepsilon_t + kNT \ln x_t^e . \tag{6.99b}$$

From (6.99) we obtain for the entropy $S_e = -(\partial F_e/\partial T)$, the internal energy $U_e = F_e + TS_e$, and the heat capacity $C_V^e = T(\partial S_e/\partial T)$ in the internal equilibrium

$$S_e = kN\left[\ln(2 + e^{\Delta\varepsilon_g/kT}) - \frac{\Delta\varepsilon_g\, e^{\Delta\varepsilon_g/kT}}{kT(2 + e^{\Delta\varepsilon_g/kT})}\right], \tag{6.100a}$$

$$U_e = N\left[\varepsilon_t + \frac{2\Delta\varepsilon_g}{2 + e^{\Delta\varepsilon_g/kT}}\right], \tag{6.101a}$$

$$C_V^e = kN\left(\frac{\Delta\varepsilon_g}{kT}\right)^2 \frac{2\,e^{\Delta\varepsilon_g/kT}}{(2 + e^{\Delta\varepsilon_g/kT})^2}\Bigg], \tag{6.102a}$$

or when keeping the internal variable (6.97)

$$S_e = kN\left[\frac{\Delta\varepsilon_g}{kT}(1 - x_t^e) - \ln x_t^e\right], \tag{6.100b}$$

$$U_e = N\,[\varepsilon_t + \Delta\varepsilon_g(1 - x_t^e)] , \tag{6.101b}$$

$$C_V^e = kN\left(\frac{\Delta\varepsilon_g}{kT}\right)^2 x_t^e(1 - x_t^e) . \tag{6.102b}$$

The two-level system with a doubly degenerate excited state is characterized by the fact that in the limit $T \to \infty$ with

$$x_t^e(\infty) = 1/3 ; \quad x_g^e(\infty) = 2/3$$

the energy levels are equally occupied. Correspondingly, when $T \to \infty$ the entropy and the internal energy reach a finite saturation value

$$S_e(\infty) = kN \ln 3 \,, \tag{6.100c}$$

$$U_e(\infty) = N\left(\varepsilon_t + \frac{2}{3}\,\Delta\varepsilon_g\right). \tag{6.101c}$$

The contribution of the trans-gauche isomerism to the entropy has the upper limit $R \ln 3 = 9.14$ J/(mol K). [If the gauche-state is not degenerate, the upper limit is given by $R \ln 2 = 5.77$ J/(mol K)]. Another characteristic is that the heat capacity first increases with increasing temperature, but then traverses a maximum (so-called Schottky anomaly) in order to finally vanish again when $T \to \infty$. The heat capacity reaches its extreme value if $\partial C_V^e/\partial T = 0$ is valid, that is if, according to (6.102a) with $\eta \equiv \Delta\varepsilon_g/kT$, we get

$$\frac{1}{2}\,e^{\eta} = \frac{\eta+2}{\eta-2}\,,$$

i.e., $\eta \approx 2.655\ldots$

The Schottky anomaly of the two-level system with a doubly degenerate excited state occurs at the temperature

$$T_{\mathrm{SA}} = \Delta\varepsilon_g/k\eta\,. \tag{6.103}$$

With $\Delta\varepsilon_g = 2000$ J/mol, the Schottky anomaly, for example, is around $T_{\mathrm{SA}} \approx 90$ K. (If there is no degeneration of the excited state, we have $\eta \approx 2.400$, and with this $T_{\mathrm{SA}} \approx 100$ K.) The equilibrium concentration of the trans-bonds, the internal energy, and the heat capacity according to (6.97), (6.101), and (6.102) are represented as a function of $kT/\Delta\varepsilon_g$ in Fig. 6.5.

When enumerating the possible distinguishable states of the bond system, one should consider that half of the gauche-bonds can be distinguished geometrically though not energetically. If $N_t \equiv Nx_t^e$ and $N_g \equiv Nx_g^e$ indicate the number of trans- or gauche-bonds in equilibrium, the number of distinguishable states is given by

$$\Omega = \frac{N!}{N_t!\left[\left(\frac{N_g}{2}\right)!\right]^2}\,. \tag{6.104}$$

If one normalizes the internal equilibrium energy of the system in such a way that it vanishes for $T \to \infty$, *i.e.*, sets

$$U_e^* \equiv U_e(T) - U_e(\infty)\,,$$

i.e., according to (6.91), (6.101b), and (6.101c),

$$U_e^* \equiv \frac{N}{3}\,\Delta\varepsilon_g\,(x_g^e - 2x_t^e)\,,$$

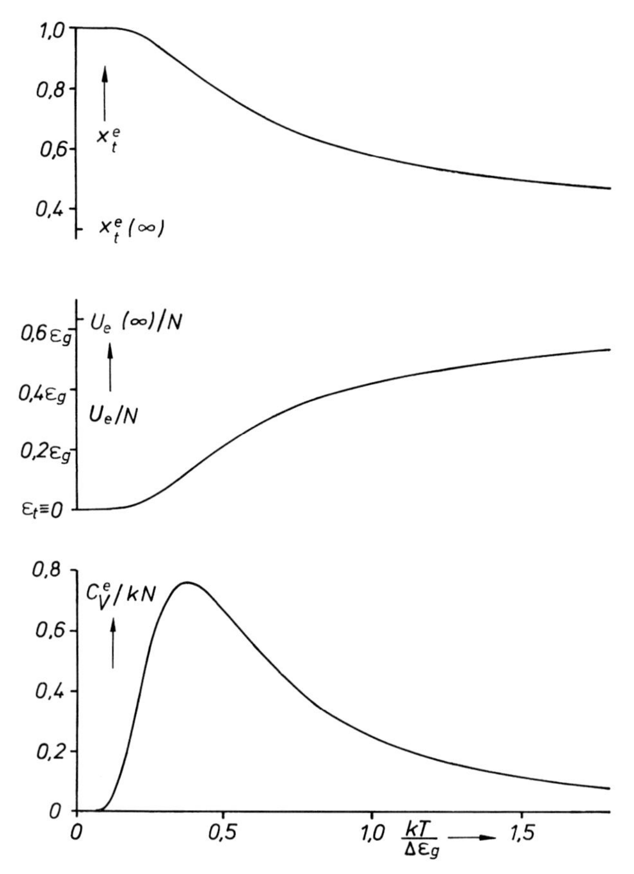

Fig. 6.5. Ideal two-level system. Top: equilibrium concentration x_t^e of the trans-bonds according to (6.97). Middle: internal energy U_e/N in internal equilibrium according (6.101). The energy of the ground level ε_t of the trans-bonds was set to zero. Bottom: specific heat capacity C_V^e/kN in internal equilibrium according to (6.102). All functions are dependent on $kT/\Delta\varepsilon_g$

one can also write for the number of bond types:

$$N_t = \frac{1}{3} N - \frac{U_e^*}{\Delta\varepsilon_g}\,; \quad N_g = \frac{2}{3} N + \frac{U_e^*}{\Delta\varepsilon_g}\,.$$

Thus, one obtains from (6.104)

$$\Omega\,(U, V, N) = \frac{N!}{\left(\dfrac{N}{3} - \dfrac{U_e^*}{\Delta\varepsilon_g}\right)!\,\left[\left(\dfrac{N}{3} + \dfrac{U_e^*}{2\Delta\varepsilon_g}\right)!\right]^2}\,. \tag{6.105}$$

for the number of distinguishable equilibrium states of the system under the condition $U, V, N = \text{const}$. By means of Stirling's formula

$$\ln N! \approx N \ln N - N \qquad (N \gg 1)$$

one obtains, according to (6.6), for the entropy

$$S_e(U, V, N) = k\left[N \ln N - \left(\frac{N}{3} - \frac{U_e^*}{\Delta\varepsilon_g}\right) \ln \left(\frac{N}{3} - \frac{U_e^*}{\Delta\varepsilon_g}\right)\right.$$
$$\left. - \left(\frac{2}{3} N + \frac{U_e^*}{\Delta\varepsilon_g}\right) \ln \left(\frac{N}{3} + \frac{U_e^*}{2\Delta\varepsilon_g}\right)\right] . \qquad (6.106)$$

This is the S-representation of Gibbs fundamental equation of the system. According to (6.10) and (6.105), the temperature of the system is given by

$$\frac{1}{T} = \frac{k}{\Delta\varepsilon_g} \ln \frac{\frac{N}{3} - \frac{U_e^*}{\Delta\varepsilon_g}}{\frac{N}{3} + \frac{U_e^*}{2\Delta\varepsilon_g}} . \qquad (6.107)$$

The (absolute) temperature is positive when $U_e^* \leq 0$ and negative when $U_e^* \geq 0$ (see Sect. 6.2). If one solves (6.107) with respect to U_e^*, one gets

$$U_e^* = \frac{2}{3} N\Delta\varepsilon_g \frac{1 - e^{\Delta\varepsilon_g/kT}}{2 + e^{\Delta\varepsilon_g/kT}} .$$

Insertion of this expression into (6.106) leads, of course, back to the F-representation (6.100) of the entropy.

As already mentioned, the two-level system without interaction is, according to (6.96), stable or metastable at all temperatures $0 \leq T \leq +\infty$. According to (6.97), the equilibrium concentration $x_t^e(T)$ of the trans-bonds is a unique function of the temperature which increases continuously with decreasing temperature from $x_t^e(\infty) = 1/3$ to $x_t^e(0) = 1$. This certainly does not agree with reality. With decreasing temperature, the polymer liquid either enters the vitreous state (Sect. 7.8) or it crystallizes (Sect. 8.2). In an ideal crystallization of polyethylene, in which all the chain molecules are completely extended, the equilibrium concentration of the trans-bonds discontinuously jumps to the value $x_t^e(T_M) = 1$ when one goes under a finite temperature $T_M > 0$. The question which should be treated further here is whether and under which conditions the two-level system of trans- and gauche-bonds is capable of such a discontinuous transformation without explicitly involving other internal degrees of freedom (*e.g.*, the degrees of freedom of diffusive translational motion).

A discontinuous change of the equilibrium concentration x_t^e of the trans-bonds is only conceivable as a cooperative process. Its explanation, therefore, requires consideration of the intermolecular interactions. To avoid the confusing complexity of this problem, we assume in the following that independent of the local relationships with immediate neighbours, a mean pair binding energy w_{tt} is given between the trans-bonds, a mean pair binding energy w_{gg} between the gauche-bonds, and a mean pair binding energy w_{tg} between the trans- and gauche-bonds. It is certainly useful to proceed from such mean binding quantities if we further assume a statistical distribution of the trans- and gauche-bonds [so-called Bragg–Williams approximation; *e.g.*, see Fowler, Guggenheim (1960)]. As for $T \geq 0$ the limiting values $x_t^e = 1$ correspond to the completely extended and $x_t^e = 1/3$ to the highly coiled chain molecules, x_t^e can in a certain sense also be regarded as a measure for the long-range order present in the system. In the same sense, the concentration x_{tt}^e of the trans-trans pairs then corresponds to a short-range order parameter. If we assume a statistical distribution of the trans- and gauche-bonds, the concentrations of the (t,t)-, (t,g)-, and (g,g)-pairs independent of the local conditions of interaction are fixed at

$$x_{tt} = x_t^2 \tag{6.108a}$$

$$x_{tg} = 2x_t(1 - x_t)\,, \tag{6.108b}$$

$$x_{gg} = (1 - x_t)^2 \tag{6.108c}$$

from the start (see Sect. 3.5, p. 44). The long-range-order parameter – this is characteristic for the statistical distribution of the individuals – also completely determines the pair distribution and thus the short-range order. Under these conditions, the quantities of interaction w_{tt}, w_{tg}, w_{gg} assigned to (6.108) can only be regarded as mean quantities which globally characterize the actual field of interaction.

If z is the mean number of the direct (intermolecular) neighbours of a trans- or gauche-bond, a total of $zN/2$ pairs exists in the system. The internal energy of the two-level system with interaction is then given by

$$U = N(\varepsilon_t x_t + \varepsilon_g x_g) + \frac{zN}{2}(w_{tt}x_{tt} + w_{tg}x_{tg} + w_{gg}x_{gg})$$

or, because of (6.91) and (6.108), by

$$U = N[\varepsilon_t + \Delta\varepsilon_g(1 - x_t)] + \frac{zN}{2}[w_{tt}x_t^2 + 2w_{tg}x_t(1 - x_t) + w_{gg}(1 - x_t)^2]\,.$$

With (6.92), the free energy of the system is

$$F = N[\varepsilon_t + \Delta\varepsilon_g(1 - x_t)] + \frac{zN}{2}[w_{tt}x_t^2 + 2w_{tg}x_t(1 - x_t) + w_{gg}(1 - x_t)^2] + kNT[x_t \ln x_t + (1 - x_t)\ln(1 - x_t)/2]\,. \tag{6.109}$$

From this,

$$\left(\frac{\partial f}{\partial x_t}\right)_e = -\Delta\varepsilon_g + z\left[(w_{tt} - w_{tg})x_t^e + (w_{tg} - w_{gg})(1 - x_t^e)\right] + kT\ln\frac{2x_t^e}{1 - x_t^e} = 0$$

results as the equilibrium condition, assuming that $T, V, N =$ const. With the abbreviation

$$\Delta w_{tg} \equiv 2w_{tg} - w_{tt} - w_{gg},$$

one obtains for the determination of the equilibrium concentration $x_t^e(T, V)$:

$$kT\ln\frac{2x_t^e}{1 - x_t^e} - z\Delta w_{tg}x_t^e = \Delta\varepsilon_g - z(w_{tg} - w_{gg}). \tag{6.110}$$

This is a transcendental equation which can only be solved numerically. In addition, the solutions are no longer necessarily unique. Several solutions $x_t^e(T)$ exist at a given temperature in certain ranges of the temperature and the interaction parameters. The system is stable or metastable if

$$\left(\frac{\partial^2 f}{\partial x_t^2}\right)_e = -z\Delta w_{tg} + \frac{kT}{x_t^e(1 - x_t^e)} > 0,$$

i.e., with (6.108b) if

$$kT > \frac{z}{2}\Delta w_{tg}x_{tg}^e \tag{6.111}$$

is valid. Hence, the system is stable for all concentrations if $\Delta w_{tg} < 0$ is valid, *i.e.*, if the trans- and gauche-bonds have a tendency towards solvation. However, this case must be excluded if only for stereometric reasons. We must rather assume that

$$0 > w_{tg} > w_{gg} > w_{tt}$$

holds, so that with $\Delta w_{tg} > 0$ the trans- and gauche-bonds tend more towards aggregation. (As binding energies the w_{ik} must, in addition, bear a negative sign.) If there is a tendency towards aggregation, the system, according to (6.111), only remains stable if the number of *(t, g)*-pairs is not too large, so that the product $z\Delta w_{tg}x_{tg}^e/2$ remains below the mean thermal energy kT. As the *(t, g)*-pairs can particularly be localized at the boundary layers between the t- and g-aggregates, (6.111) means that the total boundary layer between t- and g-aggregates in a stable equilibrium is subject to a certain minimum requirement, which is dependent on the aggregation tendency Δw_{tg}.

According to (6.110), we have

$$kT\ln\frac{1 - x_t^e}{2} = kT\ln x_t^e - z\Delta w_{tg}x_t^e + z(w_{tg} - w_{gg}) - \Delta\varepsilon_g. \tag{6.112}$$

If we refer (6.109) to $x_t = x_t^e$ and insert (6.112), we obtain

$$F_e = N\varepsilon_t + \frac{zN}{2} w_{tt} + kNT \ln x_t^e + \frac{zN}{2} \Delta w_{tg} (1 - x_t^e)^2 \tag{6.113}$$

as the free energy of the system in equilibrium (Gibbs fundamental equation in the F-representation). As compared to the ideal two-level system (6.99), consideration of the intermolecular interaction leads to a reduction in the ground state level by $zNw_{tt}/2 < 0$ and to an additive free energy of mixing

$$F_M = \frac{zN}{2} \Delta w_{tg} (1 - x_t^e)^2$$

(see Sect. 3.3).

With the transformation of coordinates

$$y \equiv 2x_t^e - 1 , \tag{6.114a}$$

i.e.,

$$x_t^e = \frac{1}{2}(1 + y) ; \qquad x_g^e = \frac{1}{2}(1 - y) , \tag{6.114b}$$

the equilibrium condition (6.110) can be applied in the form

$$B(T, y) = D(T) \tag{6.115a}$$

with

$$B(T, y) \equiv \ln \frac{1 + y}{1 - y} - \frac{z}{2kT} \Delta w_{tg} y , \tag{6.115b}$$

$$D(T) \equiv \frac{\Delta\varepsilon_g}{kT} + \frac{z}{2kT} (w_{gg} - w_{tt}) - \ln 2 . \tag{6.115c}$$

$B(T, y)$ is an antisymmetric function

$$B(T, -y) = -B(T, y) \tag{6.116}$$

with respect to y, which follows an s-shaped path (Fig. 6.6) in a certain range of values of the parameter

$$Q \equiv z\Delta w_{tg}/2kT .$$

The equilibrium condition (6.115) has three different solutions $y(T)$ in this range at certain values of $D(T)$. Accordingly, the equilibrium free energy F_e as a function of temperature takes a many-valued course (Fig. 6.7). Of the different values which $F_e(T)$ assumes at one and the same temperature T, however, only

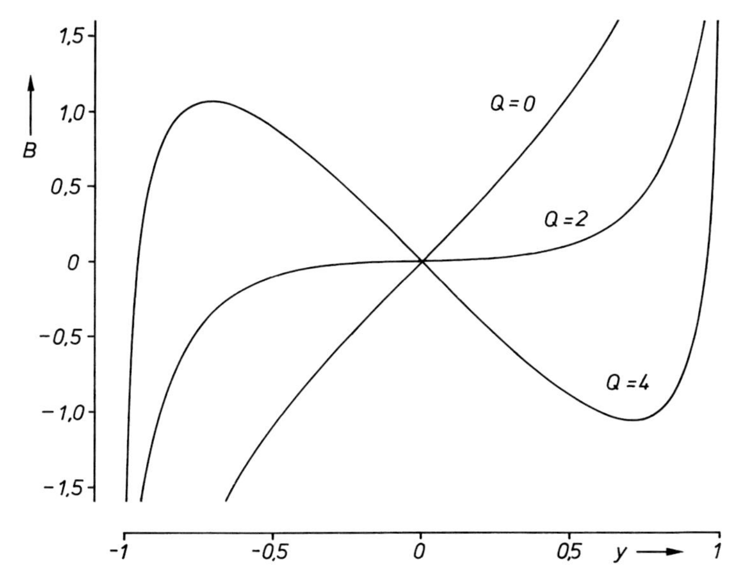

Fig. 6.6. The function $B(T, y)$ responsible for the ambiguity of the free energy F as a function of y according to (6.115b) with $Q \equiv z\Delta w_{tg}/2kT$ as the parameter

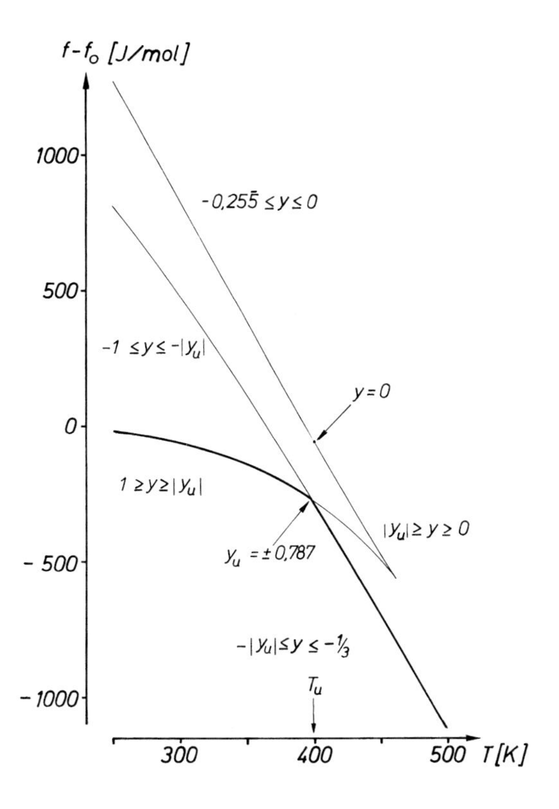

Fig. 6.7. Free energy $f_e \equiv F_e/N$ of the internal equilibrium according to (6.113–6.115) as a function of the temperature T. The respective ranges of the variable y are given on the different branches. Potential transition temperature according to (6.118): T_u; $z = 6$, $\Delta\varepsilon_g = 2$ kJ/mol, $\Delta w_{tg} = 3$ kJ/mol, $w_{gg} - w_{tt} = 100$ J/mol were the assumed parameters. The ground level of the free energy is $f_o = \varepsilon_t + z w_{tt}/2$. The thickly drawn branches correspond to the minimum free energy of the system. The irrelevant values of f_e form an open loop at $T = 0$; y reaches the values $-1, -0.25\bar{5}, +1$ in the limit $T \to 0$

the minimum one can be of physical significance. If we only consider these physically relevant values of F_e, the now unique function $F_e(T)$ is subjected to a break at a specific point T_u. The function $y(T)$ undergoes a discontinuous jump $\Delta y = 2|y_u|$ at this point (see Fig. 6.7). As $F_e(y_u) = F_e(-y_u)$ must hold at the break point,

$$B(T, y) = 0 \tag{6.117}$$

results from (6.113) and (6.114) as a defining equation for y_u with the triplet $(-y_u, 0, +y_u)$ as the solution. From (6.115) with (6.117), we obtain $D(T) = 0$ and from that for the critical temperature

$$T_u = \frac{1}{k \ln 2}\left[\Delta\varepsilon_g + \frac{z}{2}(w_{gg} - w_{tt})\right]. \tag{6.118}$$

$T_u \approx 399$ K, for example, is obtained for $\Delta\varepsilon_g = 2000$ J/mol, $z = 6$ and $w_{gg} - w_{tt} =$ 100 J/mol. However, there does not necessarily have to be discontinuity at this temperature. Independent of the energy parameters, (6.118) is rather always valid when y reaches the value $y = 0$, *i.e.*, if $x_t^e = x_g^e = 1/2$ according to (6.114) and $x_{tg}^e = 1/2$ according to (6.108b). Discontinuity is only observed at the temperature (6.118) if (6.117), in addition to the solution $y = 0$, also possesses, according to (6.116), the symmetric solutions $-y_u$ and y_u. This is the case (see Fig. 6.6) if the slope of the curve $B(y)$ is negative at the point $y = 0$, *i.e.*, if

$$\left(\frac{\partial B}{\partial y}\right)_{y=0} = 2 - \frac{z\Delta w_{tg}}{2kT} < 0\,, \tag{6.119}$$

or $Q > 2$ is valid. According to (6.111), this is the case when the system with the pair concentration $x_{tg}^e = 1/2$ given at $y = 0$ becomes unstable. Insertion of (6.118) into (6.119) leads to the condition

$$\frac{z}{2}\Delta w_{tg} > \frac{2}{\ln 2}\left[\Delta\varepsilon_g + \frac{z}{2}(w_{gg} - w_{tt})\right] \tag{6.120}$$

for the occurrence of a discontinuity.

If the tendency towards aggregation of the trans- and gauche-bonds, corresponding to (6.120), is sufficiently large, a sudden transition from gauche-bonds to trans-bonds occurs upon cooling of the system to the temperature T_u. The free energy $F_e(T)$ of the system undergoes a break at T_u, the internal energy $U_e(T)$ a jump, and the heat capacity $C_V^e(T)$ reaches a singular point of infinity. Hence, we are dealing with a so-called transition of the first order (Sect. 8.1). The cause of this discontinuous transition must be seen as follows: A large tendency towards aggregation means that the formation of *(t, g)*-pairs is energetically much more unfavourable than the formation of *(t, t)*- or *(g, g)*-pairs. We can thus observe the tendency that regions in which the segments of the chain molecules are extended separate from the regions in which the segments are highly coiled in order to keep the number of (t, g)-pairs as small as possible. As chain coiling prevails at higher temperatures, there are no pronounced t-aggregates. Cooling of the system in this state first suppresses the formation of trans-bonds as the formation of trans-bonds would necessarily be connected with the formation of energetically unfavourable (t, g)-pairs. With decreasing temperature, however, the "pressure" to form the energetically most favourable trans-bonds increases in such a way that finally, still avoiding the formation of new (t, g)-pairs, extended t-aggregates are suddenly produced (Fig. 6.8). The highest possible concentration $x^e_{tg} = 1/2$ of (t, g)-pairs thus remains a purely fictitious value.

Nevertheless, with a finite tendency towards aggregation Δw_{tg}, the transition at T_u is never complete. That is, a certain number of gauche-bonds always remains in the system even below T_u. The ideal value $x^e_t = 1$ can, provided there are no restraints, only be achieved in the limit $T \to 0$ (see Sect. 8.2). In the reverse case, the system reaches equipartition in the limit $T \to \infty$ with $x^e_t = 1/3$, as in the case of the ideal system without interaction. This value, however, no longer corresponds to the highest possible degree of coiling of the chain molecules. That is, with a higher tendency towards aggregation, $x^e_t < 1/3$ holds above the transition temperature (Fig. 6.8). Only when the mean thermal energy kT increases enough are some of the gauche-bonds produced at T_u able to change back into trans-bonds under the formation of (t, g)-pairs [see Eq. (6.111)], so

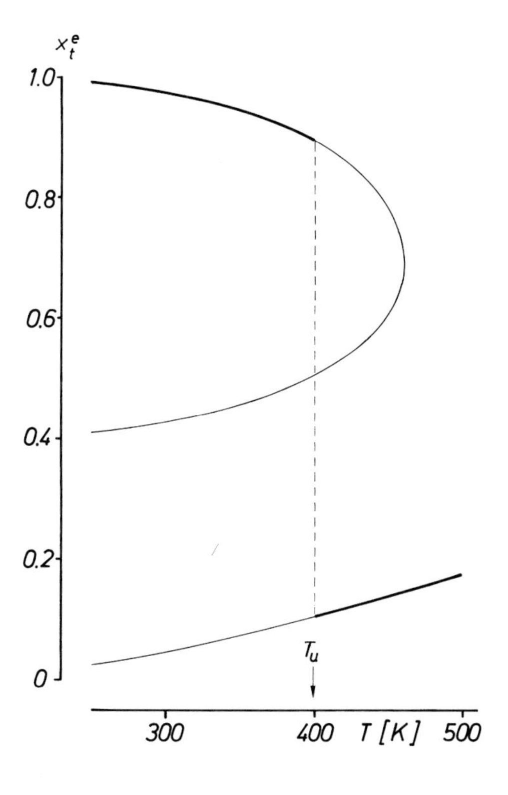

Fig. 6.8. Equilibrium concentration x^e_t of the trans-bonds as a function of the temperature T according to (6.110). Values of the parameters as in Fig. 6.7. If the system follows the minimum free energy f_e (thickly drawn curve in Fig. 6.7), x^e_t suffers a discontinuous jump at T_u (thickly drawn curve). It is remarkable that the equilibrium concentration of the trans-bonds above T_u increases with the temperature

that in the limit $T \to \infty$ $x_t^e = 1/3$. While the degree of coiling of the chain molecules always increases with increasing temperature in the case of a negligible intermolecular interaction, it can also decrease under the influence of intermolecular interactions with increasing temperature. With (6.108), the presumption of a statistical distribution of the trans- and gauche-bonds leads to

$$1 > x_t > 2/3 \; : \; x_{tt} > x_{tg} > x_{gg} \, ,$$

$$2/3 > x_t > 1/2 \; : \; x_{tg} > x_{tt} > x_{gg} \, ,$$

$$1/2 > x_t > 1/3 \; : \; x_{tg} > x_{gg} > x_{tt} \, ,$$

$$1/3 > x_t > 0 \quad : \; x_{gg} > x_{tg} > x_{tt} \, .$$

In the case of a high tendency towards aggregation, the ranges of values of the second and third lines are dropped for energetic reasons, so that the possible values of x_t^e for $0 < T < +\infty$ are restricted to the first and fourth lines. Moreover, the symmetry given by (6.116) around the value $x_t^e = x_g^e = 1/2$ is also a result of the assumed statistical distribution.

As already mentioned, we have neglected all the characteristics caused by the fact that the trans- and gauche-bonds are lined up on linear chains in the simple two-level model. In particular, one must understand that the intermolecular interaction of the bond types, which we referred to as Δw_{tg}, is considerably different from their intramolecular interaction. The sharp fold, for example, which develops when two gauche-bonds in a chain are located next to each other, is energetically much less favourable than the configuration where two gauche-bonds are separated by a trans-bond. Hence, a (g, g)-pair in the chain molecules is energetically less favourable than a (t, g)-pair. The intramolecular interaction of the different binding types is described in the so-called rotational isomeric state model [Flory (1969); Mattice, Suter (1994)].

6.4 The Extended Polymer Chain

In the following, we will also consider long, linear polymer molecules which are composed of a multitude of repeating identical units (prototype: linear polyethylene, Sect. 6.2). These chain molecules are completely extended in the ideal crystalline state. There is no conformational isomerism, and the degrees of freedom of the diffusive translational motion of the liquid state transform into the degrees of freedom of the so-called lattice vibrations of the crystal. If the polymer molecules are composed of identical repeating units, the unit cells of the crystal lattice do not comprise the whole molecules but only the units of the molecules or chain segments whose dimensions are of the order of magnitude of the dimensions of the chain units [*e.g.*, see Wunderlich (1973)]. Hence, the individual molecules traverse a multitude of unit cells. (In contrast, in the case of some irregularly structured biopolymers which are reinforced by intramolecular interaction forces, it is definitely possible that the huge molecules as a whole form building blocks of the unit cells.)

In view of the thermodynamic properties of the crystalline state, the possible motions of the crystal lattice units are in the centre of interest, especially the collective motions of the lattice units as a whole which are based on the translational degrees of freedom. If the internal degrees of freedom of the lattice units can be separated from the translational degrees of freedom, and if the shape and size of the lattice units are not of major importance, reduction of the real crystal to a lattice of structureless mass points is sufficient when examining the collective motions. The simplest model of such a point lattice is a strictly one-dimensional chain of periodically arranged mass points. This model can, of course, at the same time be regarded as the simplest model of an isolated, extended polymer molecule. The one-dimensional model already exhibits all the important characteristics of the collective motions of the lattice units of the three-dimensional crystal. In addition, compared to the three-dimensional case, it has the advantage that all results can be given in an analytically closed form. The one-dimensional model, however, cannot give useful information regarding the thermal fluctuations of the lattice units as well as the stability of the crystal lattice, as fluctuations and stability are decisively dependent on the three-dimensional field of interaction, in polymer crystals particularly on the field of intermolecular interaction (see Sect. 6.5.3).

6.4.1 The Linear Chain Composed of Equal Masses

First, we will consider the simplest mechanical model of a periodic lattice structure – the one-dimensional chain composed of equal structureless points with mass m, which are strictly arranged with a spacing a on a straight line (Fig. 6.9) – and inquire about the possible motions of the mass points (lattice units) in the field of their mutual interaction. To begin with, we will assume an infinitely long chain, so that there are no specific boundary conditions.

The equilibrium position of an arbitrary mass point can be selected as the origin of the coordinate axis z along the chain. The equilibrium positions of the mass points n are then characterized by the coordinates

$$z_n^o = an \quad \text{with} \quad n = 0, \pm 1, \pm 2, \ldots$$

The mass points undergo displacements s_n from these equilibrium positions, which occur only along the z-axis in the strictly one-dimensional case. The coordinates of the displaced mass points are given by

$$z_n = z_n^o + s_n\,.$$

According to Newton's *lex secunda*, the equations of motion of the mass points are

$$m\ddot{z}_n = m\ddot{s}_n = F_n\,. \tag{6.121}$$

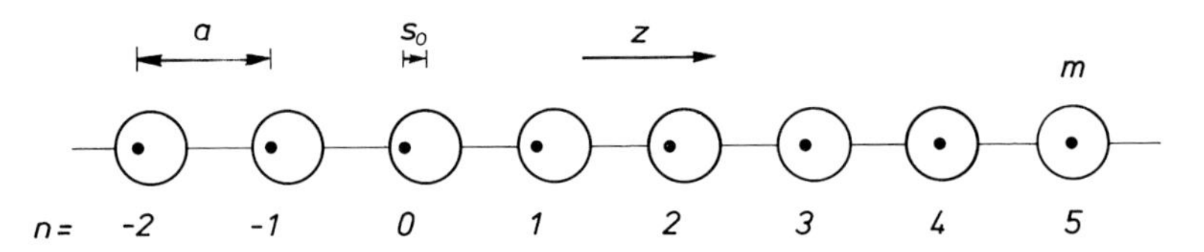

Fig. 6.9. One-dimensional lattice of equal masses m which are arranged strictly linearly at a distance a on a straight line z. Small filled circles: equilibrium positions $z_n^o = an$ of the centres of mass of the lattice units. Large circles: lattice units displaced from the equilibrium position in the z-direction. Displacement of the lattice unit $n = 0$: s_o

F_n denotes the total force acting on the nth mass point[5]. If we assume that the interaction forces between the mass points are conservative central forces, *i.e.*, forces which are not explicitly dependent on the time and only dependent on the distances between the mass points, the forces F_n can be derived *via*

$$F_n = -\frac{\partial \Phi}{\partial z_n} \tag{6.122}$$

from a potential

$$\Phi = \Phi(\ldots, z_{-2}, z_{-2}, z_0, z_1, z_2, \ldots)$$

The first integral of the equations of motion (6.121) [insert (6.122) into (6.121), multiply (6.121) by $\dot{z}_n$, sum over n and integrate over time] then yields the conservation law of energy

$$\mathcal{H} = \frac{1}{2} \sum_n m\dot{z}_n^2 + \Phi = \text{const.} \tag{6.123}$$

Here, $\mathcal{H}$ is the Hamiltonian function of the system of the mass points and $m\dot{z}_n$ the momentum of the nth mass point. The sum describes the kinetic energy and Φ the potential energy of the system.

The potential energy Φ can be expanded in a Taylor series about the equilibrium positions z_n^o:

$$\Phi = \Phi_0 + \Phi_1 + \Phi_2 + \Phi_3 + \ldots, \tag{6.124a}$$

with

$$\Phi_0 = \Phi(\ldots z_{-1}^o, z_0^o, z_1^o, \ldots), \tag{6.124b}$$

5 Here and in the following, a dot over a variable indicates the total derivative of this variable with respect to the time t: $\dot{x} \equiv \mathrm{d}x/\mathrm{d}t$.

$$\Phi_1 = \sum_n \left(\frac{\partial \Phi}{\partial z_n}\right)_0 (z_n - z_n^o) = \sum_n \left(\frac{\partial \Phi}{\partial s_n}\right)_0 s_n \,, \tag{6.124c}$$

$$\Phi_2 = \frac{1}{2}\sum_n \sum_\ell \left(\frac{\partial^2 \Phi}{\partial z_n \partial z_\ell}\right)_0 (z_n - z_n^o)(z_\ell - z_\ell^o) = \frac{1}{2}\sum_n \sum_\ell \left(\frac{\partial^2 \Phi}{\partial s_n \partial s_\ell}\right)_0 s_n s_\ell \,, \tag{6.124d}$$

$$\Phi_3 = \frac{1}{6}\sum_n \sum_\ell \sum_m \left(\frac{\partial^3 \Phi}{\partial z_n \partial z_\ell \partial z_m}\right)_0 (z_n - z_n^o)(z_\ell - z_\ell^o)(z_m - z_m^o)$$

$$= \frac{1}{6}\sum_n \sum_\ell \sum_m \left(\frac{\partial^3 \Phi}{\partial s_n \partial s_\ell \partial s_m}\right)_0 s_n s_\ell s_m \,, \tag{6.124e}$$

etc. One can select the potential energy Φ_0 of the equilibrium state, which, according to (6.123), is at the same time the total energy of the ground state, as the zero of the energy scale and thus set

$$\Phi_0 = 0 \,. \tag{6.124f}$$

According to the principles of classical mechanics, the equilibrium positions of the mass points are further characterized by a minimum of the potential energy, so that

$$\left(\frac{\partial \Phi}{\partial s_n}\right)_0 = 0 \tag{6.124g}$$

holds for every n. In addition, if the displacements s_n from the equilibrium positions are very small ($|s_n| \ll a$), one can terminate the series (6.124) after the quadratic terms. Thus,

$$\Phi = \Phi_2 = \frac{1}{2}\sum_n \sum_\ell \left(\frac{\partial^2 \Phi}{\partial s_n \partial s_\ell}\right)_0 s_n s_\ell \,. \tag{6.125}$$

For the forces (6.122), one thus obtains

$$F_n = -\sum_\ell \left(\frac{\partial^2 \Phi}{\partial s_n \partial s_\ell}\right)_0 s_\ell \,. \tag{6.126}$$

The formulations (6.125/6.126) are designated as harmonic approximation, as they lead to harmonic oscillations of the lattice units. This classical harmonic approximation forms the basis of the theory of lattice vibrations, even if certain corrections due to quantum mechanics and thermodynamics may be required. According to quantum mechanics, the equilibrium positions of the lattice units cannot be determined exactly (existence of the so-called zero-point oscillations). The ground state is characterized by a minimum of $\mathcal{H}$ with a non-zero kinetic energy. Within the scope of the harmonic approximation, this leads, however, only to an additional shift of the energy scale [see further below, Eq. (6.163)]. From the thermodynamic point of view, the ground state at $T \neq 0$ depends on the boundary conditions and is determined by the minimum of a Gibbs potential, at constant temperature and constant volume (fixed chain ends), for example, by a free energy minimum. Generally, we can expect that the harmonic approximation becomes invalid at higher temperatures as the displacements s_n increase with rising temperature, so that finally the higher terms can no longer be neglected in the Taylor series (6.124).

If we now confine ourselves to considering the interaction of directly neighbouring mass points and set

$$\left(\frac{\partial^2 \Phi}{\partial s_n^2}\right)_0 = 2\beta ,$$

$$\left(\frac{\partial^2 \Phi}{\partial s_n \partial s_{n-1}}\right)_0 = \left(\frac{\partial^2 \Phi}{\partial s_n \partial s_{n+1}}\right)_0 = -\beta ,$$

we finally obtain

$$F_n = -\beta(s_n - s_{n+1}) - \beta(s_n - s_{n-1}) . \tag{6.127}$$

This formulation corresponds to a coupling of neighbouring mass points by ideal-elastic, mass-less Hookean springs with the elastic spring constant β [see Hooke's law in the form (5.38)]. Insertion of (6.127) into (6.121) yields the equations of motion

$$m\ddot{s}_n = \beta(s_{n-1} - 2s_n + s_{n+1}) \tag{6.128}$$

and the Hamiltonian function of the system

$$\mathcal{H} = \frac{m}{2}\sum_n \dot{s}_n^2 + \frac{1}{2}\beta\sum_n (s_n - s_{n-1})^2 . \tag{6.129}$$

The solutions of the mutually coupled differential equations (6.128) in complex notation are (see annotation on p. 235)

$$s_n = A\,\mathrm{e}^{i(kan - \omega t)} . \tag{6.130}$$

In general, these are travelling sine waves which traverse the point lattice to the left or right, depending on the sign of k. k is the so-called propagation or wave vector whose magnitude, the wave number, is coupled *via*

$$|k| = 2\pi/\lambda \tag{6.131}$$

with the wavelength λ of the waves. The so-called angular frequency is $\omega \equiv 2\pi\nu$; ν is the oscillatory frequency of the waves, t the time. The magnitude of the velocity with which the phases of the waves propagate is given by

$$|w| = \lambda\nu = \omega/|k| \,. \tag{6.132}$$

The pre-exponential A of Eq. (6.130) is an amplitude factor common to all displacements s_n, which may still contain an arbitrary phase factor $e^{i\alpha}$. The ratio of amplitudes

$$\frac{s_{n+m}}{s_n} = e^{ikam}$$

for two arbitrary lattice units reflects the invariance of the chain with respect to translations by an arbitrary multiple of the lattice constant a.

In addition, we have

$$e^{i(kan-\omega t)} = \cos(kan - \omega t) + i\sin(kan - \omega t)$$

with $i \equiv (-1)^{1/2}$. Real solutions of the Eq. (6.128) are the real component as well as the imaginary component

$$\cos(kan - \omega t) = \cos kan \cos \omega t + \sin kan \sin \omega t \,,$$

$$\sin(kan - \omega t) = \sin kan \cos \omega t - \cos kan \sin \omega t \,.$$

The individual terms of the sums in these equations are also solutions of (6.128) and describe standing waves. The travelling waves are thus caused by the superimposition of standing waves. In the following, we will proceed from the specific solution

$$s_n = A \cos kan \cos \omega t \tag{6.130*}$$

for discussion and graphic representation of momentary images of the oscillatory phenomena.

It should be noted (and this is of particular importance considering the mutual interaction of the lattice waves; see further below and Sect. 6.5.4) that the medium in which the waves propagate only consists of discrete lattice points, *i.e.*, that it is extremely inhomogeneous but characterized by a certain periodicity. As a result, the waves whose wavelength is smaller than the double distance $2a$ between neighbouring lattice units lead to physical situations already described by the waves whose wavelengths are greater than or equal to this double distance. As shown by (6.130) or (6.130*), every wave vector

$$k' = k + 2\pi\ell/a \tag{6.133}$$

with an arbitrary integer ℓ leads back to the physical situation described by k. Restricting the wave vector to the so-called first Brillouin zone

$$-\frac{\pi}{a} < k \leq \frac{\pi}{a} \tag{6.134}$$

is, therefore, sufficient and leads to uniqueness. The wave number $|k|$ then remains, except for the factor π, entirely in the unit cell of the (one-dimensional) reciprocal lattice [Brillouin (1953)].

Insertion of (6.130) into (6.128) yields the equation

$$m\omega^2 = -\beta(e^{-ika} - 2 + e^{ika}) = 2\beta(1 - \cos ka) = 4\beta \sin^2 \frac{ka}{2}. \tag{6.135}$$

Hence, the relation

$$\omega_k \equiv \omega(k) = 2\left(\frac{\beta}{m}\right)^{1/2} \left|\sin \frac{ka}{2}\right| \tag{6.136}$$

exists necessarily between the frequency and the wave vector. Waves travelling to the left or right with $-k$ and $+k$ and the same wave number have the same frequency. For a representation of the relation (6.136), it is, therefore, sufficient to restrict oneself to the interval $0 \le k \le \pi/a$ (Fig. 6.10 A). According to (6.134), the frequency spectrum ranges from $\omega = 0$ to the finite maximum frequency

$$\omega_{\max} = 2(\beta/m)^{1/2}. \tag{6.137}$$

With $\beta = 230$ N/m and $m = 23.4 \cdot 10^{-27}$ kg (values more or less applicable to the CH_2-group and the C-C-bond in polyethylene), one finds $\omega_{\max} = 2 \cdot 10^{14}$ Hz or $\nu_{\max} = 3.2 \cdot 10^{13}$ Hz. According to (6.130), $\omega = 0$ corresponds to a rigid translation of the whole chain. For $\omega \neq 0$, however, the centre of mass $m \sum_n z_n$ of the chain stays at rest.

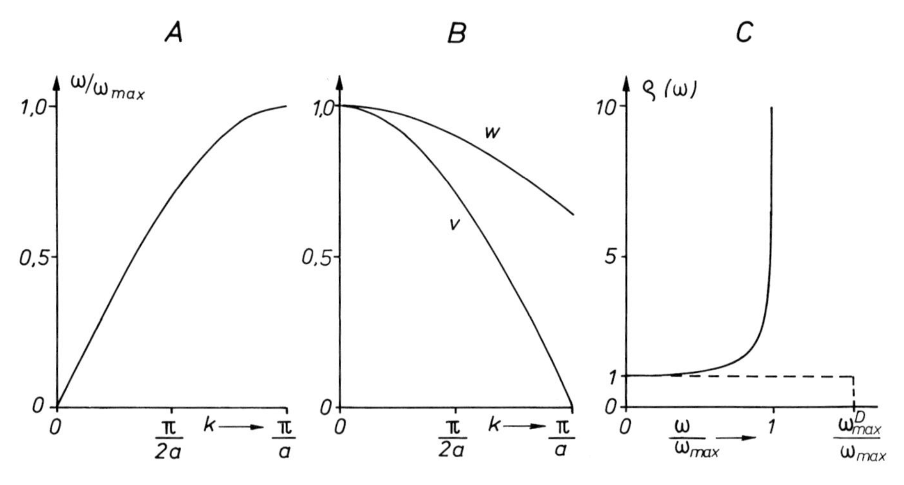

Fig. 6.10. Stretching modes of the linear chain of Fig. 6.9 in the harmonic approximation. A: dispersion relation $\omega(k)$ according to (6.136/6.137). B: phase velocity w and group velocity v as a function of the wave number k according to (6.132/6.136) as well as (6.139). The values of the velocities on the ordinate are multiples of $a\omega_{\max}/2$. C, solid curve: frequency distribution function $\varrho(\omega)$ according to (6.148). C, dashed line: continuum approximation according to (6.149). The values on the ordinate are multiples of $2N/\pi\omega_{\max}$

The individual lattice waves (6.130) are infinitely large with respect to time, *i.e.*, without beginning and end. They are, therefore, not capable of transporting energetic signals or disturbances along the lattice. Real, finitely large signals (or disturbances) are always composed of a certain number of waves of the form (6.130) with different wave numbers $|k|$ (Fourier's analysis of the signal). The velocity with which such signals propagate along the chain is given by the so-called group velocity

$$v = \frac{\partial \omega}{\partial k} . \tag{6.138}$$

From (6.136), one obtains for the magnitude of the group velocity

$$|v| \equiv a\,(\beta/m)^{1/2} \cos \frac{ka}{2} = \frac{a}{2}\,\omega_{\max} \cos \frac{ka}{2} . \tag{6.139}$$

Hence, the group velocity must, in general, be distinguished from the phase velocity (6.132) (Fig. 6.10 B). Due to the dependence of the group velocity on the wave number, a signal composed of waves with different wave numbers eventually changes its form and finally undergoes dispersion. The relation (6.136) is, therefore, also referred to as the dispersion relation.

Of the possible vibrational modes, three cases are of particular interest for the following:

1) $k = \pi/a$: The wavelength reaches the smallest value $\lambda = 2a$, which is physically significant in the discrete point lattice, at the boundary of the first Brillouin zone. The frequency ω reaches the maximum value (6.137). For arbitrary n we always have $\sin kan = \sin \pi n = 0$ and, therefore,

$$e^{i\,kan} = \cos \pi n = (-1)^n \tag{6.140}$$

is real. These are standing waves. Correspondingly, the group velocity (6.139) vanishes at the boundary of the first Brillouin zone (see Fig. 6.10 B). According to (6.140), neighbouring lattice units are displaced with the same amplitude in the opposite direction (Fig. 6.11). The point lattice disintegrates into two sublattices which rigidly oscillate with respect to each other. If the two sublattices are electrically charged with different signs, an oscillating dipole develops which is capable of interacting with an external electrical field. The oscillations can potentially be stimulated optically.

2) $k = \pi/2a$: The wavelength is $\lambda = 4a$, the frequency $\omega = \omega_{\max}/2$. One gets

$$\cos kan = \cos \frac{\pi n}{2} = \begin{cases} +1, \text{if } n = 0, \pm 4, \pm 8 \ldots \\ 0, \text{if } n = \pm 1, \pm 3, \pm 5 \ldots \\ -1, \text{if } n = \pm 2, \pm 6, \pm 10 \ldots . \end{cases}$$

In the case of the standing waves (6.130*), only the second neighbours, respectively, oscillate with an opposite amplitude while the lattice units in between remain at rest (Fig. 6.11). The point lattice disintegrates into two sublattices of which one vibrates whereas the other one remains at rest. This case is a limiting

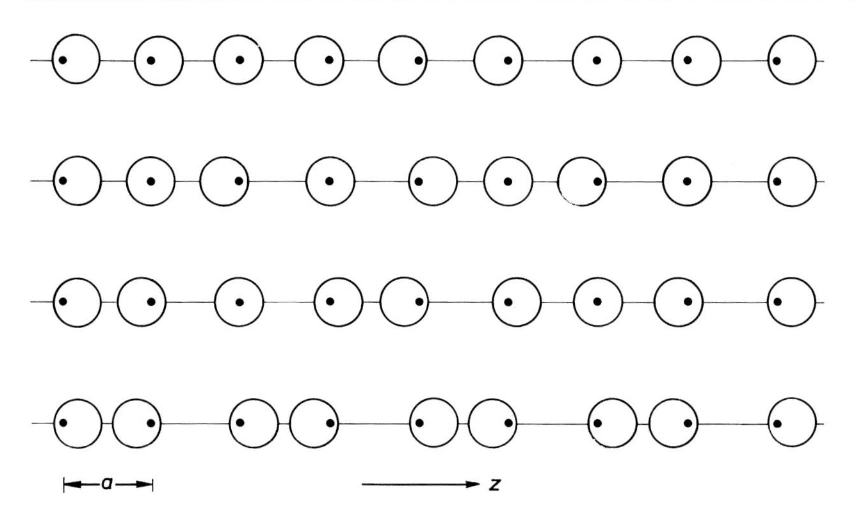

Fig. 6.11. Snapshots of the longitudinally polarized lattice waves (stretching modes = compression waves) of the linear chain according to (6.130*). From top to bottom: $k = \pi/4a, \pi/2a, 3\pi/4a, \pi/a$. According to (6.131), the wave lengths are $\lambda = 8a, 4a, 8a/3, 2a$, respectively. See text

case which separates two physically different regions:

For $\pi/2a < |k| \leq \pi/a$, $Re(s_{n+1}/s_n) \leq 0$ holds.

Neighbouring mass points oscillate with an oppositely directed amplitude.

For $0 \leq |k| < \pi/2a$, $Re(s_{n+1}/s_n) \geq 0$ holds.

Neighbouring mass points oscillate with an equally directed amplitude (see Fig. 6.12).

3) $|k| \ll \pi/a$: We have a very large wavelength and a small frequency with respect to ω_{max}. Half the wavelength comprises a multitude of lattice units which are all displaced in the same direction. As the difference between the displacements of neighbouring lattice units is so infinitely small compared with the wavelength, the lattice waves can be identified to a good approximation with the longitudinal elastic waves

$$s(z, t) = A\,e^{i(kz - \omega t)}$$

which propagate along a one-dimensional homogeneous continuum (a string) [*e.g.*, see Schaefer (1922)]. With $|k| \ll \pi/a$,

$$\omega = \omega_{max}\, a\, |k|/2 \tag{6.141}$$

follows from (6.136) and (6.137) and

$$|v| \equiv a\left(\frac{\beta}{m}\right)^{1/2} = \omega_{max}\, a/2 = \omega/|k| \tag{6.142}$$

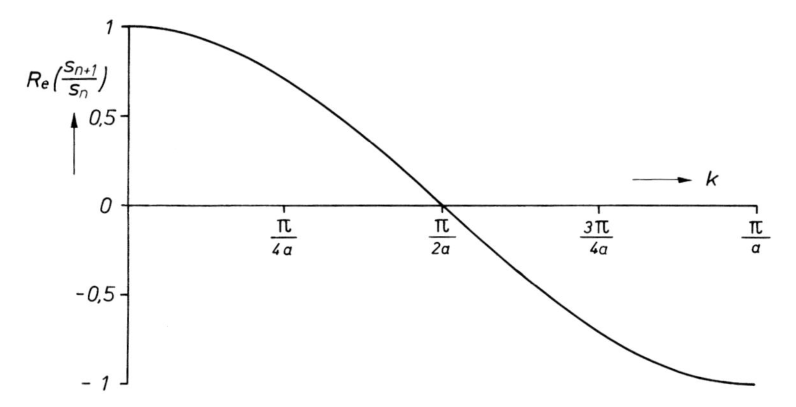

Fig. 6.12. Real part *Re* of the ratio of amplitudes s_{n+1}/s_n of two neighbouring lattice units as a function of the wave number k according to (6.130)

from (6.139) and (6.137). In the range of quasi-elastic waves $|k| \ll \pi/a$, the frequency is proportional to the wave number. As a result, the group velocity and the phase velocity become equal (Fig. 6.10B) and both do not depend on the wave number. In this range, wave signals are not subject to dispersion. We have

$$|v| = |w| = c_\ell \quad \text{or} \quad \lambda\nu = c_\ell \tag{6.143}$$

whereby the constant c_ℓ is identical with the sound velocity of the longitudinally polarized elastic compression waves which propagate along a string:

$$c_\ell = (E/\varrho_m)^{1/2}$$

[E: Young's modulus of the string, ϱ_m: mass density; the Poisson ratio for a one-dimensional string vanishes, so that the Young modulus becomes proportional to the bulk modulus; see Eq. (5.41)]. One thus obtains $E = \beta/a$ as the Young modulus of the linear chain.

The macroscopic properties of a real crystal, *e.g.*, the thermodynamic response functions also depend on the number of lattice units as well as on the volume occupied by the lattice units. In view of an explanation for the macroscopic properties, one has to consider a point lattice with a limited extension and its boundary conditions. This does not entail any difficulties in the case of the linear chain [see, for example, Zbinden (1964)], but considerably impedes the problem solution in the three-dimensional point lattice. In the following, we will also confine ourselves to a boundary condition in the case of the one-dimensional chain, which can easily be treated in the case of a three-dimensional lattice (Sect. 6.5): the so-called periodic boundary condition. One can show [Born, Huang (1954), Appendix IV] that this boundary condition, with a sufficiently large number N of lattice units, is completely equivalent to the boundary condition with a constant volume or fixed chain ends. The finitely long linear chain composed of $N+2$ mass points with fixed chain ends has (as one mechanical degree of freedom is attributed to every free mass point in the strictly one-dimensional case) a total of N degrees of freedom and is, therefore, capable of N mutually independent vibrational states. The periodic boundary condition now selects from the possible vibrational states of the infinitely long chain those for which

$$s_n = s_{n+N} \tag{6.144}$$

is valid. Arbitrary chain segments which contain $N+1$ lattice units and whose ends are subject to the same displacement from the equilibrium position are thus distinguished in the infinitely long chain. Each of these segments can be identified with the finitely long chain of $N+2$ lattice units with fixed ends. Insertion of (6.130) into (6.144) results in

$$e^{ikaN} = 1\,, \quad i.e., \quad \cos kaN = 1 \quad \text{and} \quad \sin kaN = 0$$

as a restricting condition with respect to the number of vibrational modes or

$$kaN = 2\pi\ell$$

with an arbitrary positive or negative integer ℓ. The wave vector is restricted to the values

$$k = 2\pi\ell/aN\,. \tag{6.145a}$$

If the wave vector is in the first Brillouin zone (6.134), and if we assume an even-numbered N, ℓ is restricted to the values

$$\ell = 0, \pm 1, \pm 2, \ldots \pm \left(\frac{N}{2} - 1\right), \frac{N}{2}\,. \tag{6.145b}$$

Corresponding to the number of degrees of freedom of the finitely long chain with N free links, N vibrational modes are then present in the first Brillouin zone. These states are identically repeated in the subsequent zones (with $|k| > \pi/a$).

The wave numbers are equally distributed over the first Brillouin zone. The number $\Delta\Omega$ of the wave numbers (number of states) is in every interval Δk

$$\Delta\Omega = \frac{aN}{2\pi}\,\mathrm{d}k\,.$$

If N is very large and if the k-values are practically densely distributed, one can also write

$$\mathrm{d}\Omega = \frac{aN}{2\pi}\,\mathrm{d}k\,.$$

The number of frequencies $\omega(k)$ in the interval $< \omega ; \omega + d\omega >$ then amounts to

$$\varrho(\omega)\,\mathrm{d}\omega = 2\,\frac{aN}{2\pi}\,\mathrm{d}k\,; \quad k \geq 0\,. \tag{6.146}$$

In this enumeration, one must confine oneself to positive k-values and multiply $\mathrm{d}\Omega$ by 2 since, according to (6.136), the states have the same frequency with $+k$ and $-k$. As the frequency distribution function on the frequency scale (frequency spectrum), one thus obtains

$$\varrho(\omega) = \frac{aN}{\pi}\,\frac{\mathrm{d}k}{\mathrm{d}\omega} = \frac{aN}{\pi}\,\frac{1}{|v|}\,; \quad k \geq 0\,. \tag{6.147}$$

and because of

$$\cos\frac{ka}{2} = +\left(1 - \sin^2\frac{ka}{2}\right)^{1/2} \quad \text{for} \quad 0 \leq k \leq \pi/a$$

with (6.139), (6.137) and (6.136)

$$\varrho(\omega) = \frac{2N}{\pi}\,\frac{1}{(\omega_{\max}^2 - \omega^2)^{1/2}}\,. \tag{6.148}$$

In the range of quasi-elastic waves ($|k| \ll \pi/a$, *i.e.*, $\omega \ll \omega_{\max}$), the frequency spectrum is practically constant:

$$\varrho_{el}(\omega) = \frac{2N}{\pi\omega_{\max}} = \text{const.} \tag{6.149a}$$

With increasing frequency, however, the density of states increases very rapidly until it finally reaches a singular infinity at the boundary of the first Brillouin zone ($|k| \rightarrow \pi/a$; $\omega \rightarrow \omega_{\max}$) (Fig. 6.10C). The frequency spectrum must certainly also be normalized over the N degrees of freedom of the finitely long chain. We obtain with (6.148)

$$\int_0^{\omega_{\max}} \varrho(\omega)\,\mathrm{d}\omega = N\,.$$

If the real spectrum (6.148) is replaced by the spectrum (6.149a) of the quasi-elastic waves (the so-called Debye or Tarasov approximation; see Sect. 6.5.2), this normalization can only be satisfied by expanding the elastic spectrum beyond the maximum frequency $\omega_{\max}$ up to a maximum frequency $\omega_{\max}^D$. With (6.149a), one obtains

$$\int_0^{\omega_{\max}^D} \varrho_{el}(\omega)\,\mathrm{d}\omega = \frac{2N}{\pi\omega_{\max}}\,\omega_{\max}^D \equiv N\,,$$

i.e.,

$$\omega_{\max}^D = \frac{\pi}{2}\,\omega_{\max} \tag{6.149b}$$

(Fig. 6.10 C).

The equations of motion (6.128) are linear differential equations. Arbitrary linear combinations of the solutions (6.130) are, therefore, also solutions of the equation (6.128). The actual state of motion of the chain is additively composed of solutions of the form (6.130). The Fourier transforms

$$q_k = \left(\frac{m}{N}\right)^{1/2} \sum_n s_n \mathrm{e}^{-ikan} \tag{6.150a}$$

are of particular importance among these linear combinations. The factor $(m/N)^{1/2}$ is a normalization factor. Consideration of N suggests that we confine ourselves to the consideration of a segment of the infinitely long chain selected by the periodic boundary condition. In (6.150a), we must sum over the N free lattice units of this segment. If we further assume real displacements s_n, the equation:

$$q_k^* = q_{-k} \tag{6.150b}$$

must necessarily be valid where q_k^* denotes the conjugate-complex quantity of q_k. Moreover, in order to form the conjugate-complex quantity of q_k for $k = \pi/a$, one must switch from the right to the left boundary of the first Brillouin zone, a state which is not contained in the set of states $\{q_k\}$. This asymmetry between $\{q_k\}$ and $\{q_k^*\}$ can easily be avoided if one changes over to an odd-numbered N. Inversion of the transformation (6.150) to:

$$\sum_k q_k \mathrm{e}^{ikan} = \left(\frac{m}{N}\right)^{1/2} \sum_k \sum_{n'} s_{n'} \mathrm{e}^{ika(n-n')}$$

leads to

$$s_n = \frac{1}{(mN)^{1/2}} \sum_k q_k \mathrm{e}^{ikan}. \tag{6.151}$$

For $n = n'$, it is obvious that

$$\sum_k \mathrm{e}^{ika(n-n')} = N.$$

For $n \neq n'$, we obtain with (6.145) the double geometric progression

$$\sum_k \mathrm{e}^{ika(n-n')} = \sum_{\ell=0}^{N/2} [\mathrm{e}^{2\pi i(n-n')/N}]^{\ell} + \sum_{\ell=0}^{\frac{N}{2}-1} [\mathrm{e}^{-2\pi i(n-n')/N}]^{\ell} - 1$$

$$= -\frac{2i\mathrm{e}^{2\pi i(n-n')/N}}{1-\mathrm{e}^{2\pi i(n-n')/N}} \sin[\pi(n-n')] = 0.$$

The expression vanishes because $n - n' \neq 0$ is always an integer, $(n - n')/N$, on the other hand, a fractional number. Hence, for arbitrary differences $n - n'$, one obtains

$$\sum_k \mathrm{e}^{ika(n-n')} = N\delta_{nn'} \tag{6.152}$$

and from this follows

$$\sum_k q_k \mathrm{e}^{ikan} = \left(\frac{m}{N}\right)^{1/2} \sum_{n'} s_{n'} N\delta_{nn'} = (mN)^{1/2} s_n ,$$

which, furthermore, leads to (6.151).

The q_k turn out to be the normal coordinates of the system which separate the coupled differential equations (6.128); and the Hamiltonian function (6.129) splits into N mutually independent additive terms. According to (6.151), we have

$$\dot{s}_n = \frac{1}{(mN)^{1/2}} \sum_k \dot{q}_k \mathrm{e}^{ikan} .$$

For the kinetic energy in (6.129), we thus obtain

$$\mathcal{H}_{\mathrm{kin}} = \frac{m}{2} \sum_n \dot{s}_n^2 = \frac{1}{2N} \sum_n \sum_k \sum_{k'} \dot{q}_k \dot{q}_{k'} \mathrm{e}^{i(k+k')an} .$$

As in Eq. (6.152), one can show that

$$\sum_n \mathrm{e}^{i(k+k')an} = N\delta_{k,-k'} \tag{6.153}$$

is valid. With (6.150b), one thus obtains

$$\mathcal{H}_{\mathrm{kin}} = \frac{1}{2} \sum_k \sum_{k'} \dot{q}_k \dot{q}_{k'} \delta_{k,-k'} = \frac{1}{2} \sum_k \dot{q}_k \dot{q}_k^* .$$

By transposition of the terms, one can, to begin with, also write

$$\mathcal{H}_{\mathrm{pot}} = \frac{\beta}{2} \sum_n (s_n - s_{n-1})^2 = -\frac{\beta}{2} \sum_n (s_{n+1} s_n - 2s_n^2 + s_{n-1} s_n)$$

for the potential energy in (6.129). Insertion of (6.151) leads to

$$\mathcal{H}_{\mathrm{pot}} = -\frac{\beta}{2mN} \sum_n \sum_k \sum_{k'} q_k q_{k'} \mathrm{e}^{i(k+k')an} (\mathrm{e}^{-ika} - 2 + \mathrm{e}^{ika}) .$$

i.e., with (6.153), (6.150b) and (6.135)

$$\mathcal{H}_{\mathrm{pot}} = \frac{1}{2} \sum_k \omega_k^2 q_k q_k^* .$$

The Hamiltonian function (6.129)

$$\mathcal{H} = \mathcal{H}_{\text{kin}} + \mathcal{H}_{\text{pot}}$$

separates into a sum

$$\mathcal{H} = \sum_k \mathcal{H}_k \tag{6.154a}$$

of N mutually independent Hamiltonian functions

$$\mathcal{H}_k = \frac{1}{2}\left(\dot{q}_k \dot{q}_k^* + \omega_k^2 q_k q_k^*\right). \tag{6.154b}$$

If one introduces the momentum conjugated to the coordinate q_k

$$p_k \equiv \dot{q}_k^* = \dot{q}_{-k} = p_{-k}^*,$$

then the following is also valid:

$$\mathcal{H}_k = \frac{1}{2}\left(p_k p_k^* + \omega_k^2 q_k q_k^*\right). \tag{6.154c}$$

Within the scope of classical mechanics, (6.154) can be represented as

$$\mathcal{H} = \frac{1}{2}\sum_k \left(p_k^2 + \omega_k^2 q_k^2\right). \tag{6.155}$$

The N equations of motion

$$\ddot{q}_k + \omega_k^2 q_k = 0 \tag{6.156}$$

can be obtained by differentiation of the energy conservation law $\mathcal{H}$ = const. with respect to time using (6.154) as an equivalent to the mutually coupled equations of motion (6.128). These are the equations of motion of N mutually independent simple harmonic oscillators of which two, respectively, vibrate with the frequency ω_k. The coupled motion of the lattice units with their neighbours can thus be reduced to the motion of N non-interacting oscillators *via* (6.156) and (6.151).

When determining the classical partition function (6.21), one should take into account that the oscillators are distinguishable statistical individuals corresponding to the states of the chain distinguished by k. The permutation factor $1/N!$ in (6.21) should, therefore, be omitted. With (6.155) and $f = N$, (6.21) then leads to the classical partition function of the linear chain

$$Q = \frac{1}{h^N} r^N \prod_k u_k$$

with

$$r \equiv \int_{-\infty}^{+\infty} e^{-p_k^2/2RT}\, dp_k = (2\pi RT)^{1/2}$$

and

$$u_k \equiv \int_{-\infty}^{+\infty} e^{-\omega_k^2 q_k^2/2RT}\, dq_k = (2\pi RT)^{1/2}/\omega_k$$

[see Eq. (6.76a)]. In order to avoid confusions with the wave vector k, we have used R for the Boltzmann constant. R can be understood as the Boltzmann constant or as the gas constant if one refers the extensive quantities to one mole of the chain links. With the abbreviation $\hbar \equiv h/2\pi$, insertion of the integral values into the partition function yields

$$Q = \prod_k \frac{RT}{\hbar\omega_k} . \tag{6.157}$$

According to (6.16), (6.22), (6.25), and (6.28), it follows for the free energy, entropy, internal energy, and heat capacity of the linear chain that:

$$F = RT \sum_k \ln(\hbar\omega_k/RT) , \tag{6.158}$$

$$S = RN - R \sum_k \ln(\hbar\omega_k/RT) , \tag{6.159}$$

$$U = RTN , \tag{6.160}$$

$$C_V = RN . \tag{6.161}$$

Equation (6.161) corresponds to the Dulong–Petit law $C_V = 3RN$ for a three-dimensional solid [Eq. (6.227)]. This law applies to many solids to a quite good approximation in a middle temperature range. However, it fails in the temperature range where the harmonic approximation should apply best: According to experiment, the heat capacity of all solids decreases considerably at low temperatures with decreasing temperature until it finally vanishes in the limit $T \rightarrow 0$. According to Einstein (1907), one has to refer to the quantum theory to explain this fact.

The Schrödinger equation, related to the Hamiltonian function (6.154c), yields [after a longer and by no means trivial calculation; *e.g.*, see Pauling, Wilson (1935)]

$$\varepsilon_{k,n} = \hbar\omega_k\left(n + \frac{1}{2}\right), \quad \text{with} \quad n = 0, 1, 2 \ldots \tag{6.162}$$

as energy eigenvalues of the individual harmonic oscillators. First of all, one can observe that even in the ground state ($n = 0$), the quantum-mechanical oscillators are not at rest (so-called zero-point vibrations). Contrary to the assumption of classical mechanics (with the normalization $\Phi_0 = 0$), the lattice units have in the ground state the energy

$$E_0 = \sum_k \varepsilon_{k,0} = \frac{\hbar}{2} \sum_k \omega_k \,, \tag{6.163}$$

The cause can be found in the Heisenberg uncertainty principle which prevents the kinetic energy and the potential energy of the system from being fixed simultaneously at the value zero. Within the framework of the harmonic approximation, in which the frequencies ω_k do not depend on the amplitude of the lattice units, Eq. (6.163) represents, however, only a constant contribution to the energy which vanishes as a result of a corresponding shift of the energy scale.

By inserting (6.162) into (6.17), one obtains as quantum-mechanical partition function Q_k of the individual harmonic oscillators the geometric progression

$$Q_k = \sum_{n=0}^{\infty} \mathrm{e}^{-\hbar\omega_k\left(n+\frac{1}{2}\right)/RT} = \mathrm{e}^{-\hbar\omega_k/2RT} \sum_{n=0}^{\infty} \left(\mathrm{e}^{-\hbar\omega_k/RT}\right)^n = \frac{\mathrm{e}^{-\hbar\omega_k/2RT}}{1-\mathrm{e}^{-\hbar\omega_k/RT}} \tag{6.164}$$

and as quantum-mechanical partition function of the linear chain

$$Q = \prod_k Q_k = \prod_k \frac{\mathrm{e}^{-\hbar\omega_k/2RT}}{1-\mathrm{e}^{-\hbar\omega_k/RT}} \,, \tag{6.165}$$

From this, one obtains, for example, according to (6.16), (6.25), and (6.28), the free energy of the chain

$$F = \frac{\hbar}{2} \sum_k \omega_k + RT \sum_k \ln\left(1 - \mathrm{e}^{-\hbar\omega_k/RT}\right) , \tag{6.166}$$

the internal energy

$$U = \sum_k \hbar\omega_k \left(\frac{1}{\mathrm{e}^{\hbar\omega_k/RT} - 1} + \frac{1}{2} \right) \tag{6.167}$$

and the heat capacity

$$C_V = R \sum_k \left(\frac{\hbar\omega_k}{RT} \right)^2 \frac{\mathrm{e}^{\hbar\omega_k/RT}}{\left(\mathrm{e}^{\hbar\omega_k/RT} - 1\right)^2} \,, \tag{6.168a}$$

The heat capacity now vanishes for $T \to 0$ as the experiment requires. For higher temperatures, *i.e.*, for

$$\hbar\omega_k \ll RT \,, \quad \mathrm{e}^{\hbar\omega_k/RT} \approx 1 + \frac{\hbar\omega_k}{RT}$$

the quantum-mechanical expressions (6.165–6.168) transform into the corresponding classical expressions (6.157–6.161). If N is very large and if the wave numbers k or rather the frequencies ω_k, are practically dense, the sums can, according to (6.146), further be replaced by integrals, *i.e.*,

$$\sum_k \ldots \quad \text{by} \quad \int_0^{\omega_{\max}} \ldots \varrho(\omega)\,\mathrm{d}\omega\,.$$

One then finds, for example, that the heat capacity

$$C_V = R \int_0^{\omega_{\max}} \left(\frac{\hbar\omega_k}{RT}\right)^2 \frac{\varrho(\omega)\,\mathrm{e}^{\hbar\omega/RT}}{(\mathrm{e}^{\hbar\omega/RT}-1)^2}\,\mathrm{d}\omega\,. \tag{6.168b}$$

Besides, Einstein (1907) assumed that the motions of the lattice units can be characterized by a single frequency $\omega_k \equiv \omega_E$. With this assumption, (6.168) leads to

$$C_V = RN\,E\left(\frac{\hbar\omega_E}{RT}\right)\,. \tag{6.169a}$$

The function

$$E(x) = \frac{x^2\,\mathrm{e}^x}{(\mathrm{e}^x-1)^2} = \frac{x^2\,\mathrm{e}^{-x}}{(1-\mathrm{e}^{-x})^2} \tag{6.169b}$$

is designated as the Einstein function. The contribution of individual oscillators to the heat capacity of the chain as a function of the temperature is represented in Fig. 6.13 for different frequencies ω_k.

According to experiment (especially inelastic X-ray and neutron scattering by solids), a certain particle aspect is associated with the field of lattice vibrations. In the quantum field theory of harmonic oscillators, one proceeds from the combination of coordinates

$$b_k = \frac{1}{(2\hbar\,\omega_k)^{1/2}}\,(\omega_k q_k + ip_{-k}) \tag{6.170a}$$

$$b_k^* = \frac{1}{(2\hbar\,\omega_k)^{1/2}}\,(\omega_k q_{-k} - ip_k) \qquad \text{6.(170b)}$$

[*e.g.*, see also with regard to the following, Ziman (1969) or Haken (1973)]. Interpreted as operators, b_k, b_k^* prove to be annihilation or creation operators of quasi-particles. As the field of lattice vibrations, at least in the range of not too high frequencies, is identical with the field of sound waves, these particles are called phonons. The operators (6.170) comply with the commutation relations

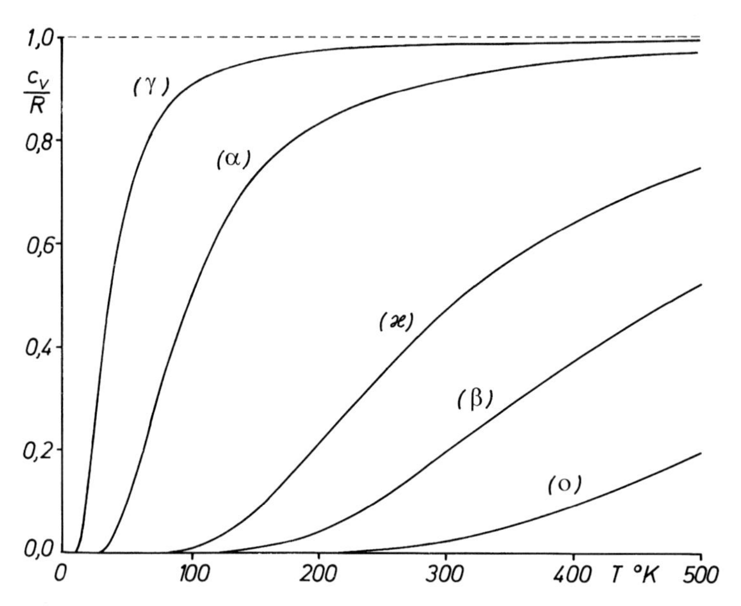

Fig. 6.13. Contribution $C_V/RN = c_V/R$ of individual harmonic oscillators to the heat capacity as a function of the temperature T according to (6.169). The frequencies $\nu \equiv \omega/2\pi$ of the oscillators and the associated temperatures $\Theta \equiv h\nu/R$ (Sect. 6.5.2) are

(γ): $\nu = 2.29 \cdot 10^{12}\ \mathrm{s}^{-1}$; $\Theta = 110$ K,
(α): $\nu = 6.25 \cdot 10^{12}\ \mathrm{s}^{-1}$; $\Theta = 300$ K,
(κ): $\nu = 1.96 \cdot 10^{13}\ \mathrm{s}^{-1}$; $\Theta = 940$ K,
(β): $\nu = 3.00 \cdot 10^{13}\ \mathrm{s}^{-1}$; $\Theta = 1440$ K,
(o): $\nu = 5.00 \cdot 10^{13}\ \mathrm{s}^{-1}$; $\Theta = 2400$ K.

The frequencies (γ) ... (β) more or less correspond to the mean values of the singularities in Table 2 of the Stockmayer–Hecht model (Sect. 6.5.3, in particular Figs. 6.27 and 6.28) with $\beta = 208$ N/m, $\kappa/\beta = 0.1$, $\alpha/\beta = 0.04$, $\gamma/\beta = 0.002$, and $m = 23.4 \cdot 10^{-27}$ kg. The dashed line corresponds to the Dulong–Petit law (6.161)

$$\left.\begin{aligned} b_k\, b_{k'}^* - b_{k'}^*\, b_k &= \delta_{kk'}\,, \\ b_k\, b_{k'} - b_{k'}\, b_k &= 0\,, \\ b_k^*\, b_{k'}^* - b_{k'}^*\, b_k^* &= 0\,, \end{aligned}\right\} \tag{6.171}$$

which are characteristic for all bosons. (Bosons are particles which can occupy the energy eigenvalues of the Hamiltonian operator in arbitrary numbers.)

Inversion of Eq. (6.170) results in

$$q_k = \left(\frac{\hbar}{2\omega_k}\right)^{1/2} (b_k + b_{-k}^*)\,,$$

$$p_k = i\left(\frac{\hbar\omega_k}{2}\right)^{1/2} (b_k^* - b_{-k})\,.$$

Insertion of these expressions into (6.154c) yields, together with (6.171), the Hamiltonian operator of a harmonic oscillator in the quantum field theory

$$\mathcal{H}_k = \hbar\omega_k \left(b_k^* b_k + \frac{1}{2} \right), \tag{172}$$

where $b_k^* b_k$ is an operator whose eigenvalues indicate the number of particles in the state k. Hence, (6.172) directly shows that (6.162) gives the eigenvalues of the Hamiltonian operator. The Schrödinger quantum numbers $n = 0, 1, 2 \ldots$ can now be interpreted as the phonon numbers in the state k. The energy $\hbar\omega_k$ and the momentum $\hbar k$ can, similar to the photons in the electromagnetic field, be assigned to the individual phonons (whereby $\hbar k$ should, however, not be confused with the real momentum $m\dot{z}_n$ of the lattice units!). In the dispersionless range $|k| \ll \pi/a$ in which the proportionality (6.141) exists and the lattice waves correspond to sound waves, the energy of the phonons, like the energy of the photons, is proportional to the momentum. An additional analogy between photons and phonons is that both are bosons, *i.e.*, quasi-particles which follow the Bose–Einstein statistics. As a result of this [*e.g.*, see Huang (1963), Pathria (1972)], the mean number $\langle n_k \rangle$ of the phonons in the state k at the temperature T is given by the Bose-Einstein distribution function

$$\langle n_k \rangle = \frac{1}{e^{\hbar\omega_k/RT} - 1} . \tag{6.173}$$

The contribution of the phonons in state k to the internal energy of the system is

$$\langle \varepsilon_k \rangle = \hbar\omega_k \left[\langle n_k \rangle + \frac{1}{2} \right] \tag{6.174}$$

in agreement with (6.167). The contribution to the heat capacity is

$$c_{V,k} = \frac{\partial}{\partial T} \langle \varepsilon_k \rangle = \hbar\omega_k \frac{\partial}{\partial T} \langle n_k \rangle = R \left(\frac{\hbar\omega_k}{RT} \right)^2 \frac{e^{\hbar\omega_k/RT}}{(e^{\hbar\omega_k/RT} - 1)^2} \tag{6.175}$$

in agreement with (6.168).

Within the harmonic approximation, the field of lattice vibrations is equivalent to an "ideal gas" of phonons without any interaction. In general, this approximation is certainly useful at not too high temperatures (and not too high stress). Nevertheless, it does not provide an explanation for "second order" effects, such as the differences between C_V and C_p or κ_T and κ_S, or the coefficients of thermal expansion α_i (see Sect. 5.2). Moreover, an explanation of heat conduction and the approach to thermal equilibrium is not conceivable without phonon interactions. (The mean free path of phonons is infinitely large in the harmonic approximation.) An explanation of these effects rather requires consideration of the higher terms $\Phi_i\,(i > 2)$ in the expansion (6.124) (anharmonic approximations; see Sect. 6.5.4). In the quantum-field theory, the higher order terms Φ_3, $\Phi_4 \ldots$ correspond to creation and annihilation processes in which 3,

4... phonons are involved. The energy conservation law, but not necessarily the momentum conservation law, holds for these processes. The latter is due to the fact that the sum of two or several momenta $\hbar k$ can lead out of the first Brillouin zone, but the resultant momentum must then, in order to maintain the unambiguity of the description, be brought back to the first Brillouin zone [see the comments on Eq. (6.133)]. In general, one must point out here that the validity of the momentum conservation law of mechanics is linked to the homogeneity of the space. For the real displacements z_n of the lattice units which occur along the homogeneous z-axis, the momentum conservation law naturally remains valid. The total momentum of the chain vanishes completely for $k \neq 0$, as with $m \sum_n s_n$, the centre of mass of the chain is maintained. The phonons, on the other hand, are only, as are the lattice waves, defined in the inhomogeneous space of the mass points, so that the validity of a momentum conservation law for the $\hbar k$ cannot be expected in the least. For the 3-phonon process contained in Φ_3, in which two phonons [1] and [2] are destroyed and a third phonon [3] is created, for example, the energy conservation law is valid:

$$\hbar\omega_1 + \hbar\omega_2 = \hbar\omega_3 \,. \tag{6.176}$$

On the other hand, the "momentum law" is in the form

$$\hbar k_1 + \hbar k_2 = \hbar k_3 + \frac{2\pi}{a}\hbar\ell \tag{6.177}$$

[see Eq. (6.133)]. This is actually a generalization of the Bragg reflection law [Brillouin (1953)]. If $\ell = 0$, *i.e.*, if the momentum of the phonons is conserved, one speaks about normal processes. If $\ell \neq 0$, one speaks about Umklapp-processes [*e.g.*, see Ziman (1967)].

6.4.2 The Linear Chain Composed of Two Different, Alternating Masses

The linear chain of Fig. 6.9 crystallographically forms a so-called primitive Bravais lattice. The simplest example of a complicated lattice (a so-called Bravais lattice with a basis) is given by the linear chain in which mass points with two different masses m_A and m_B are alternately arranged with an equal spacing a (Fig. 6.14). The unit cell of this lattice now comprises two mass points, with the masses m_A and m_B, respectively. Their length amounts to $2a$. If we choose a mass point with the mass m_A as the origin of the coordinate axis z, the coordinates of the mass points m_A are given with

$$n = 0, \pm 1, \pm 2, \pm 3\,, \ldots$$

by

$$z_{2n} = z^o_{2n} + s_{2n}$$

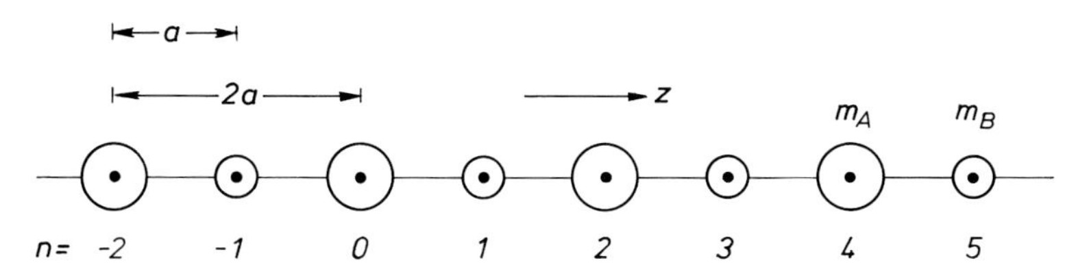

Fig. 6.14. Equilibrium positions of the lattice units of a linear lattice with alternately arranged masses m_A and m_B

and the coordinates of the mass points m_B by

$$z_{2n+1} = z^o_{2n+1} + s_{2n+1} \, .$$

In the harmonic approximation [*cf.* Eq. (6.128)], the equations of motion of the mass points are, when considering the interaction of the nearest neighbours:

$$m_A \ddot{s}_{2n} = \beta(s_{2n-1} - 2s_{2n} + s_{2n+1}) \, , \tag{6.178a}$$

$$m_B \ddot{s}_{2n+1} = \beta(s_{2n} - 2s_{2n+1} + s_{2n+2}) \, . \tag{6.178b}$$

Corresponding to (6.130), the starting solutions for these equations are

$$s_{2n} = A\,\mathrm{e}^{i(2kan - \omega t)} \, , \tag{6.179a}$$

$$s_{2n+1} = B\,\mathrm{e}^{i[ka(2n+1) - \omega t]} \, . \tag{6.179b}$$

In order to understand given vibrational situations, one can, corresponding to (6.130*), proceed from the specific solutions (standing waves)

$$s_{2n} = A \cos 2\,kan \cos \omega t \, ,$$

$$s_{2n+1} = B \cos\,[ka\,(2n+1)] \cos \omega t \, .$$

The assumption is that the lattice units with different masses are also associated with different amplitude factors A and B. Because of the expansion of the unit cell of the lattice to $2a$, the first Brillouin zone must now be restricted to

$$-\frac{\pi}{2a} < k \le \frac{\pi}{2a} \, . \tag{6.180}$$

The smallest, physically significant wavelength is $\lambda = 4a$ [*cf.* Eq. (6.131)]. For the determination of the ratio of amplitudes and the dispersion relation, insertion of (6.179) into (6.178) leads to the system of equations

$$-\,m_A \omega^2 A = \beta(B\,\mathrm{e}^{-ika} - 2A + B\,\mathrm{e}^{ika}) \, ,$$

$$-\,m_B \omega^2 B = \beta(A\,\mathrm{e}^{-ika} - 2B + A\,\mathrm{e}^{ika})$$

or

$$(m_A\omega^2 - 2\beta)\,A + 2\beta\cos ka\,B = 0\,, \tag{6.181a}$$

$$2\beta\cos ka\,A + (m_B\omega^2 - 2\beta)\,B = 0\,. \tag{6.181b}$$

This system of equations only has a solution (A, B) differing from zero if the determinant of the coefficients vanishes, *i.e.*, if

$$(m_A\omega^2 - 2\beta)\,(m_B\omega^2 - 2\beta) - 4\beta^2\cos^2 ka = 0\,,$$

is valid. With

$$2(1 - \cos^2 ka) = 2\sin^2 ka = 1 - \cos 2ka\,,$$

$$\omega^2 = \frac{\beta}{m_A m_B}\,[m_A + m_B \pm (m_A^2 + m_B^2 + 2m_A m_B\cos 2ka)^{1/2}] \tag{6.182}$$

results as the relation between frequency and wave number. The relation is symmetric with respect to the masses m_A and m_B. Therefore, it does not matter which of the masses is considered as the smaller or the greater mass. We assume in the following

$$\frac{m_A}{m_B} \equiv \mu > 1\,.$$

With this, we can also replace (6.182) by

$$\omega^2 = \frac{\beta}{m_A}\,[(\mu + 1) \pm ((\mu + 1)^2 - 4\mu\sin^2 ka)^{1/2}]\,. \tag{6.183}$$

According to (6.181), the ratio of amplitudes is given by

$$\frac{A}{B} = \frac{\dfrac{2\beta}{m_A}\cos ka}{\omega^2 - \dfrac{2\beta}{m_A}} = -\frac{\omega^2 - \dfrac{2\beta}{m_B}}{\dfrac{2\beta}{m_B}\cos ka}\,. \tag{6.184}$$

Equal frequencies are, of course, attributed to lattice waves travelling along the chain to the right or left with the same wave number. Every wave number, however, depending on the sign of the square root in (6.182) or (6.183), has two different frequencies. The dispersion relation is no longer unique. For $k = 0$ (*i.e.*, $\sin^2 ka = 0$),

$$\omega^2 = \frac{\beta}{m_A}\,[(\mu + 1) \pm (\mu + 1)]\,,$$

with the two solutions

$$\omega = \omega_{A,\min} \equiv 0$$

$$\omega = \omega_{O,\max} \equiv \left[\frac{2\beta}{m_A}(1+\mu)\right]^{1/2} = \left[2\beta\left(\frac{1}{m_A}+\frac{1}{m_B}\right)\right]^{1/2}$$

results from (6.183). For $k = \pi/2a$ (*i.e.*, $\sin^2 ka = 1$) at the boundary of the first Brillouin zone, we obtain according to (6.183)

$$\omega^2 = \frac{\beta}{m_A}\left[(\mu+1) \pm (\mu-1)\right],$$

with the two solutions

$$\omega = \omega_{A,\max} \equiv \left(\frac{2\beta}{m_A}\right)^{1/2},$$

$$\omega = \omega_{O,\min} \equiv \left(\frac{2\beta}{m_A}\mu\right)^{1/2} = \left(\frac{2\beta}{m_B}\right)^{1/2}.$$

Over the total range of the first Brillouin zone, the dispersion relation disintegrates into two frequency branches

$$0 \le \omega \le \omega_{A,\max} \tag{6.185}$$

and

$$\omega_{O,\min} \le \omega \le \omega_{O,\max}, \tag{6.186}$$

which are separated by a finite frequency gap

$$\omega_{O,\min} - \omega_{A,\max} = \omega_{A,\max}\left(\sqrt{\mu} - 1\right) \tag{6.187}$$

(Fig. 6.15).

According to (6.184), it is obvious that $A/B > 0$ is always valid for the vibrations of the frequency branch (6.185). Neighbouring lattice units are displaced in the same direction. In the range $|k| \ll \pi/2a$, this branch contains quasi-elastic waves, which are comparable to longitudinally polarized sound waves. The frequency branch (6.185) with

$$\omega = \omega_{A,\max}\left[\frac{1}{2}(\mu+1) - \left(\frac{1}{4}(\mu+1)^2 - \mu\sin^2 ka\right)^{1/2}\right]^{1/2} \tag{6.188a}$$

is, therefore, designated as the acoustical branch, the vibrational states of this branch are referred to as acoustic modes. With $\omega = 0$, the acoustical branch also contains the rigid translational motion of the lattice as a whole ($A/B = 1$). At the boundary of the first Brillouin zone ($k = \pi/2a$; $\omega = \omega_{A,\max}$) we have $B = 0$, so that only the sublattice of the heavier masses m_A oscillates, whereas the sublattice of the lighter masses m_B is at rest.

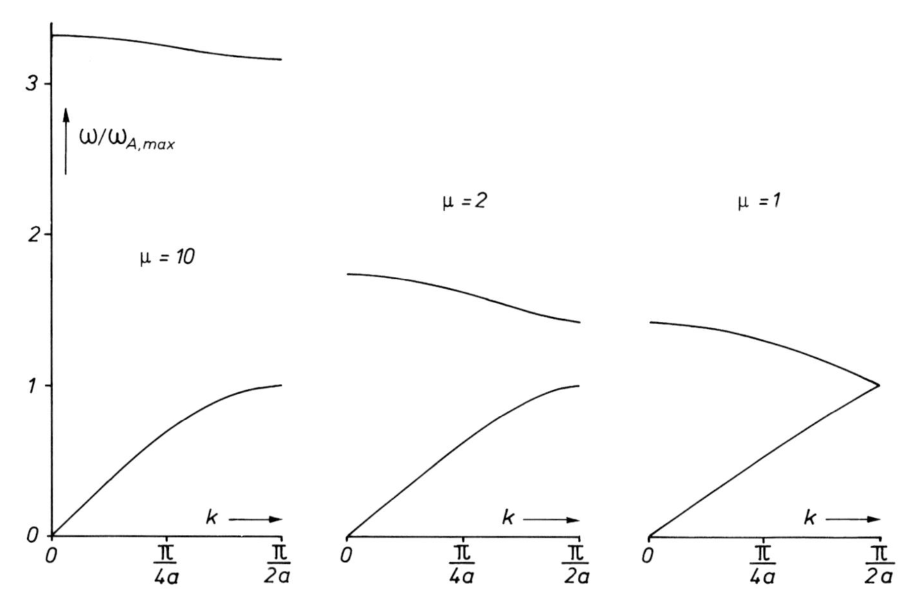

Fig. 6.15. Dispersion relation $\omega(k)$ of the linear chain of Fig. 6.14 according to (6.183) for different ratios of masses $\mu = m_A/m_B$. The case $\mu = 1$ is identical with the dispersion relation (6.136) of the linear chain of equal masses. The dispersion relation (6.136) appears folded at the boundary of the first Brillouin zone $k = \pi/2a$ because of the doubling of the length of the unit cell from a to $2a$

For the frequency branch (6.186) with

$$\omega = \omega_{A,\max}\left[\frac{1}{2}(\mu+1) + \left(\frac{1}{4}(\mu+1)^2 - \mu\sin^2 ka\right)^{1/2}\right]^{1/2} \tag{6.188b}$$

(6.184) leads to $A/B < 0$. Hence, neighbouring lattice units always oscillate with an oppositely directed amplitude. With a suitable electrical charge distribution, this branch contains vibrations which can be electromagnetically stimulated. The branch is thus designated as the optical branch, the vibrational states of this branch (independent of the actual electromagnetic conditions) are referred to as optical modes. According to (6.184) (with $k = \pi/2a$ and $\omega = \omega_{O,\min}$), $A = 0$ holds at the boundary of the first Brillouin zone. That is, only the sublattice of the lighter masses m_B oscillates, whereas the sublattice of the heavier masses m_A is at rest. For $k = 0$, $\omega = \omega_{O,\max}$, (6.184) leads to

$$m_A A + m_B B = 0\,.$$

We are dealing with vibrations within the unit cells in which the units of every cell are displaced with respect to each other in such a way that the centre of mass of the cell is conserved. Considering the lattice as a whole, the lattice of the heavier masses m_A rigidly oscillates against the equally rigid lattice of the lighter masses m_B, so that the wavelength $\lambda = \infty$ is attributed to the motions of the lattice units with the distance $2a$ (the length of the unit cell) [*cf.* Eq. (6.131)].

With increasing ratio of masses μ, the gap (6.187) between the acoustical and the optical branch becomes ever larger. The band width of the optical branch

$$\omega_{O,\max} - \omega_{O,\min} = \left(\sqrt{\mu+1} - \sqrt{\mu}\right) \omega_{A,\max} ,$$

on the other hand, becomes ever smaller. The condition $\mu = 1$, naturally, leads back to the situation in the linear chain with equal masses (Sect. 6.4.1). However, if one maintains the distance $2a$ as the length of the unit cell, the dispersion relation (Fig. 6.10 A) appears folded at the value $k = \pi/2a$ (Fig. 6.15, right).

According to (6.183) or (6.188), the group velocity (6.138) of the lattice waves propagating along the chain with alternating masses is given by

$$v = \mp \frac{a\omega_{A,\max}^2}{\omega} \left(\frac{\mu \sin^2 ka \,(\mu - \mu \sin^2 ka)}{(\mu+1)^2 - 4\mu \sin^2 ka} \right)^{1/2} , \tag{6.190}$$

whereby the minus sign is assigned to the optical branch and the plus sign to the acoustical branch, and the sign of the square root additionally distinguishes between waves travelling to the right or left. With the abbreviation

$$\Omega \equiv \omega / \omega_{A,\max}$$

and

$$\mu \sin^2 ka = (\mu+1)\,\Omega^2 - \Omega^4 ,$$

according to (6.188), one can also write

$$v(\Omega) = \mp \frac{a\omega_{A,\max}}{\mu + 1 - 2\Omega^2} \left[(\mu + 1 - \Omega^2)\,(\mu - \Omega^2\,(\mu + 1 - \Omega^2)) \right]^{1/2} \tag{6.191 a}$$

in place of (6.190). Here, we have with (6.185) for the

$$\text{acoustical branch: } 0 \leq \Omega \leq 1 \tag{6.191 b}$$

and with (6.186) for the

$$\text{optical branch: } \sqrt{\mu} \leq \Omega \leq \sqrt{\mu+1} . \tag{6.191 c}$$

For $\Omega = 0$ and the quasi-elastic waves contained in the acoustical branch, which are characterized by $\Omega \ll 1$, the following is valid:

$$v = a\omega_{A,\max} \left(\frac{\mu}{\mu+1} \right)^{1/2} = a \left(\frac{2\beta}{m_A + m_B} \right)^{1/2} \equiv c_\ell = \text{const.} \tag{6.192}$$

Once again,

$$\omega \sim |k| \quad \text{or} \quad \lambda\nu = c_\ell$$

holds in the range of the quasi-elastic waves. As the waves with a very large wavelength do not "perceive" the specific structure of the lattice, only the mean mass $(m_A + m_B)/2$ appears in the sound velocity c_ℓ [compare (6.192) with (6.142)]. The group velocity vanishes at $\Omega^2 = 1, \mu, \mu + 1$, *i.e.*, at the boundaries of the first Brillouin zone, and in the case of the optical branch also for $k = 0$ ($\Omega^2 = \mu + 1$, $\omega = \omega_{O,\max}$). The magnitude of the group velocity of the acoustic modes is generally considerably larger than that of the optical modes (see Fig. 6.16). Moreover, with increasing ratio of masses μ, the group velocity of the optical modes becomes ever smaller until it finally vanishes practically completely for $\mu \gg 1$ over the whole range of the wave numbers. Hence, mainly the acoustic modes and in this case especially the quasi-elastic modes contribute to the transport of energy along the chain (*e.g.*, to heat conduction).

If we now confine ourselves again by means of the periodic boundary conditions to a chain with N free links (N even-numbered), every branch possesses $N/2$ k-modes. The length of the first Brillouin zone (6.180) is π/a, the density of the modes of a branch on the k-axis thus amounts to $aN/2\pi$. Therefore, corresponding to the Eq. (6.147), the density of states on the frequency scale is given by

$$\varrho(\omega) = \frac{aN}{\pi}\left(\frac{1}{|v_A|} + \frac{1}{|v_O|}\right) = \varrho_A(\omega) + \varrho_O(\omega)\,, \tag{6.193}$$

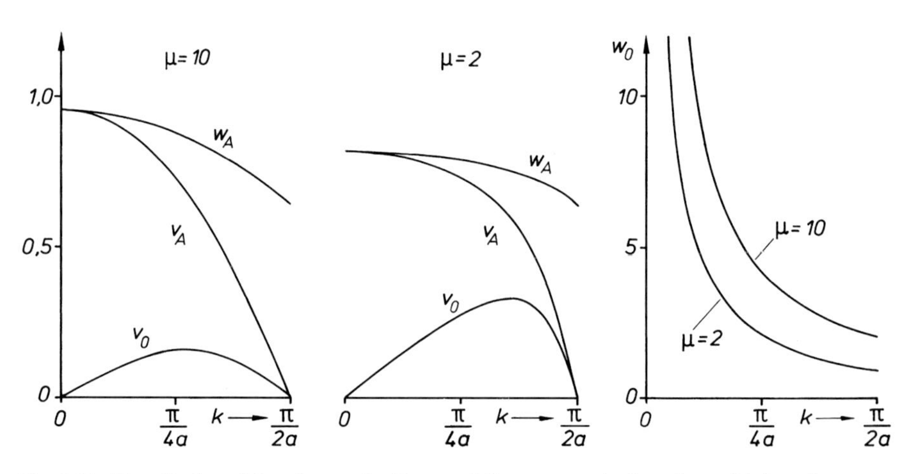

Fig. 6.16. Magnitudes of the phase velocity w and the group velocity v in multiples of $a\omega_{A,\max}$ according to (6.190), (6.183), and (6.132) for different ratios of masses μ. Index A: acoustical branch. Index O: optical branch. In the case of the optical modes with $k = 0$ and $\omega = \omega_{O,\max} \neq 0$, the phase instantaneously spreads over the chain with $w_O = \infty$ (left figure). These are oscillations in which the sublattices with the masses m_A or m_B are rigidly displaced against one another

where v_A designates the group velocity of the acoustic modes [Eq. (6.191 a) with the positive sign and the range of values (6.191 b)], v_O the group velocity of the optical modes [Eq. (6.191 a) with the negative sign and the range of values (6.191 c)]. The frequency spectrum (6.193) is represented in Fig. 6.17 for $\mu = 2$ and $\mu = 10$. For quasi-elastic waves with $\Omega \ll 1$, one obtains from (6.193) and (6.192) the constant frequency distribution

$$\varrho_{el}(\omega) = \frac{N}{\pi \omega_{A,\max}} \left(\frac{1+\mu}{\mu} \right)^{1/2} . \tag{6.194 a}$$

One can see that replacing the total spectrum (6.193) by the constant spectrum of elastic waves would be a very bad approximation for the case of the chain with two different, alternately arranged masses, especially when, with a greater ratio of masses μ, the gap between the acoustical branch and the optical branch

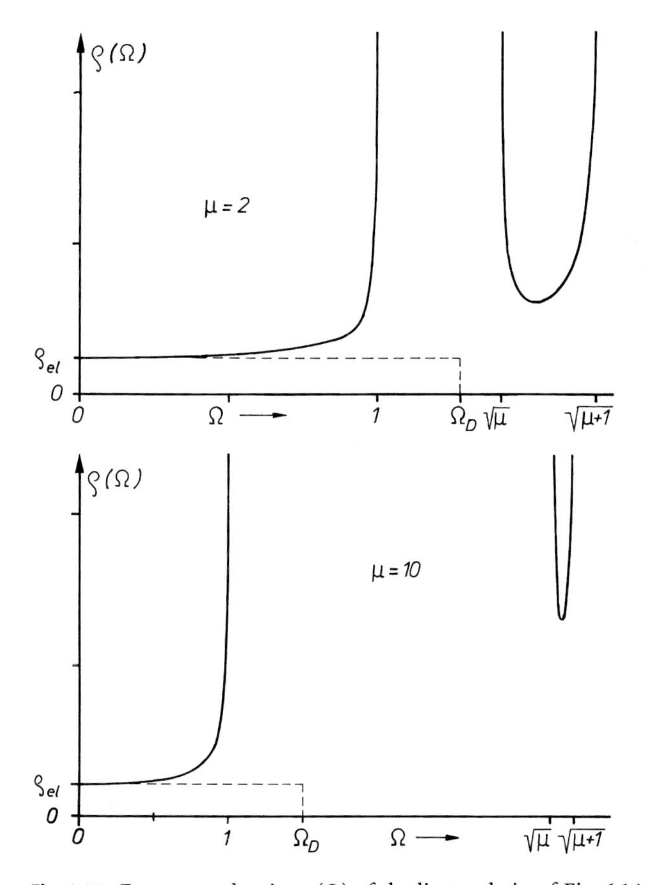

Fig. 6.17. Frequency density $\varrho(\Omega)$ of the linear chain of Fig. 6.14 for different ratios of masses μ ($\Omega = \omega/\omega_{A,\max}$). Solid curves: according to (6.193) and (6.191). Dashed line: continuum approximation according to (6.194)

is large and the width of the optical spectrum is very narrow. However, one can only approximate the spectrum of the acoustical branch by using the spectrum (6.194a). With the normalization

$$\int_0^{\omega_{A,\max}^D} \varrho_{el}(\omega)\,d\omega = N/2$$

the cut-off frequency $\omega_{A,\max}^D$ of the acoustic band is determined by

$$\omega_{A,\max}^D = \frac{\pi}{2}\,\omega_{A,\max}\left(\frac{\mu}{1+\mu}\right)^{1/2} \tag{6.194b}$$

[*cf.* Eq. (6.149b)]. With a greater ratio of masses μ, the optical spectrum can be replaced very well by a single mean frequency

$$\omega_E = \frac{1}{2}\,(\omega_{O,\max} + \omega_{O,\min}) = \frac{1}{2}\,(\sqrt{\mu+1} + \sqrt{\mu})\;\omega_{A,\max}\,.$$

The equation

$$\varrho_O(\omega) = N\,\delta(\omega - \omega_E)/2$$

then holds for the density of the optical frequency band (δ: Dirac's delta function). Here, the contribution C_V^O of the optical modes to the heat capacity of the chain can very well be described by the Einstein term

$$C_V^O = \frac{RN}{2}\,E\left(\frac{\hbar\omega_E}{RT}\right)$$

[*cf.* Eq. (6.169)]. In general, the frequencies of the optical modes are of the order of magnitude of the infrared light frequencies. Their contribution to the equilibrium thermodynamics of the linear lattice can, therefore, usually be neglected at lower temperatures [see Fig. 6.13 (o)].

Qualitatively equal results can be achieved if the equilibrium positions of equal mass points in the linear chain are alternately arranged at two different distances, or rather two different spring constants β_1 and β_2 alternately act between the mass points [Maradudin, Weiss (1958)]. If the unit cell of the linear lattice contains more than two mass points, every additional mass point in the unit cell produces an additional optical branch in the frequency spectrum of the lattice. The frequency spectrum of the strictly one-dimensional … AABAAB … – or … ABCABC … – chain is thus composed of one acoustical and two optical branches. The frequency spectrum of the chain whose unit cell contains Z mass points possesses one acoustical and $Z-1$ optical branches (see Sect. 6.5.1). The linear chain composed of an equal number of mass points m_A and m_B, in which the different masses are statistically distributed over the chain, represents a special case [Dean (1960), (1961); see also Martin (1960),

(1961)]. This chain constitutes the simplest example of a disordered lattice. A comparison of its spectrum with that of the periodically ordered ...ABAB...-chain allows a first insight into the effects of disorder and defect structures on the distribution of the vibrational frequencies. As compared to the spectrum of the chain with alternating masses, the gap between the acoustic and the optical modes is filled up in the spectrum of the statistical AB-chain. A multitude of new maxima develops in the high-frequency range, *i.e.*, in the range of optical modes. These maxima are partly due to so-called local modes, *i.e.*, to vibrations which do not propagate over the whole lattice, but are linked to specific local structures of the lattice units.

6.4.3
The Bending Modes of the Linear Chain

The strictly one-dimensional chain particularly serves as a transparent model for the considerably more complicated case of the three-dimensional lattice. However, if one considers the one-dimensional chain as a model of a polymer molecule, arbitrary displacements $\vec{s}_n$ of the lattice units from their equilibrium position in three-dimensional space corresponding to the real possible movements of the chain links must be allowed. In a rectangular system of coordinates (x, y, z) whose z-axis points in the chain direction, the displacements $\vec{s}_n$ can be split into three components: into a displacement s_n^z in the chain direction (the only one we have regarded so far) and two displacements s_n^x and s_n^y parallel to the x- and y-axes. Accordingly, one can expect that a lattice wave which propagates along the linear polymer chain in the z-direction is composed of three different vibrational types: of a longitudinally polarized wave in which the lattice units are displaced in the direction of propagation, *i.e.*, in the z-direction and of two transversely polarized waves in which the lattice units are displaced perpendicularly to the direction of propagation in the x- and y-directions. The rotational symmetry of the one-dimensional chain with respect to the z-axis demands, however, that the two transversely polarized waves only differ in the direction of their amplitudes.

At first, we will confine ourselves to a chain of equal mass points (Fig. 6.9). If we assume the origin of the system of coordinates again to be the equilibrium position of one of the mass points, the equilibrium position of the nth mass point is given by the three-dimensional position vector

$$\vec{r}_n^{\,o} = (0, 0, z_n^o)$$

with

$$z_n^o = an\,, \qquad n = 0, \pm 1, \pm 2, \ldots\,.$$

The actual positions of the mass points displaced from their equilibrium position are described by the position vectors

$$\vec{r}_n \equiv (x_n, y_n, z_n) = (s_n^x, s_n^y, an + s_n^z)\,.$$

The distance of two arbitrary mass points n and $n + m$ is

$$|\vec{r}_{n+m} - \vec{r}_n| = [(x_{n+m} - x_n)^2 + (y_{n+m} - y_n)^2 + (z_{n+m} - z_n)^2]^{1/2} \tag{6.195}$$

$$= am\left[1 + \frac{2\,(s^z_{n+m} - s^z_n)}{am} + \left(\frac{s^z_{n+m} - s^z_n}{am}\right)^2 + \left(\frac{s^x_{n+m} - s^x_n}{am}\right)^2 + \left(\frac{s^y_{n+m} - s^y_n}{am}\right)^2\right]^{1/2} .$$

If we suppose that the displacements from the equilibrium positions are small as compared with the distance between the mass points ($|\vec{s}_n| \ll a$), (6.195) can be expanded in a Taylor series and the series broken off after the linear terms. This leads to

$$|\vec{r}_{n+m} - \vec{r}_n| = am + (s^z_{n+m} - s^z_n) + \frac{(s^z_{m+n} - s^z_n)^2}{2\,am} + \frac{(s^x_{m+n} - s^x_n)^2}{2\,am} + \frac{(s^y_{n+m} - s^y_n)^2}{2\,am}$$

$$\approx am + (s^z_{n+m} - s^z_n) = z_{n+m} - z_n \,. \tag{6.196}$$

This means, however: If we assume (as before) that central forces act between the mass points, no transversely polarized vibrations are possible within the harmonic approximation, as the restoring forces only have a component in the z-direction in this approximation. In continuum mechanics, this corresponds to the result that a (one-dimensional) string, which is not exposed to external forces, is not capable of transversely polarized vibrations (bending modes). A string can only be stimulated to produce bending modes if it is subject to external stress [*e.g.*, see Schaefer (1922)]. In the linear chain, in which only central forces act, transversely polarized vibrations are only possible in the harmonic approximation under the influence of intermolecular interaction forces (*cf.* Sect. 6.5.3). It thus becomes clear that there is primarily a physical difference between the longitudinally polarized waves (stretching modes) and the transversely polarized waves (bending modes) of an isolated polymer chain.

The links of a real polymer chain are usually coupled by covalent bonding forces. Covalent bonds, however, not only resist a change in the bond length but also a change in the bond angle (valence angle). Covalent bonds can thus not be described by central forces. Nevertheless, within the scope of the harmonic approximation, the restoring force $\vec{F}_n$, which acts on the nth mass point in the field of interaction of the nearest neighbours, can be divided into two parts $\vec{F}_n(r)$ and $\vec{F}_n(\varphi)$. $\vec{F}_n(r)$ is a central force which is caused by a change in the bond lengths. According to (6.196), their x- and y-components vanish. Their z-component is still given by the Eq. (6.127), *i.e.*, more precisely by

$$F^z_n(r) = -\beta[(s^z_n - s^z_{n-1}) + (s^z_n - s^z_{n+1})] \,. \tag{6.197}$$

$\vec{F}_n(\varphi)$ is due to a perturbation of the valence angles. As the valence angles of the one-dimensional extended chain are 180° in a state of equilibrium, this force cannot have a component in the z-direction. Corresponding to Hooke's law, the components in the x- and y-directions can be expected to be proportional to the deviations of the valence angles from the equilibrium state in the (x, z)- or

(y, z)-plane. The constant of proportionality is a measure of the bending resistance of the chain. One must take into account that the displacement of the nth mass point in the x- or y-directions not only causes a change in the nth valence angle, but also a change in the neighbouring valence angles at $n-1$ and $n+1$.

Consideration of only s_n^x or s_n^y is sufficient because of the rotational symmetry of the chain around the z-axis. In the following, we will confine ourselves to the displacements in the x-direction and neglect the subscript x in order to simplify matters. The displacement s_n of the nth mass point in the x-direction in an otherwise rigid chain causes a deviation of the valence angles from the "equilibrium state" 180° by $\varphi_n = 2\alpha$ at the nth mass point and by $\varphi_{n-1} = \alpha$ or $\varphi_{n+1} = \alpha$ at the $(n-1)$th or $(n+1)$th mass point (compare to Fig. 6.18 with $\gamma = \delta = 0$, *i.e.*, $\alpha = \beta$). The displacement produces a restoring force which is proportional to $2\varphi_n = 4\alpha$. If the lattice units at $n-1$ and $n+1$ are already displaced by s_{n-1} and s_{n+1}, the deviations of the valence angles are

$$\varphi_{n-1} = \alpha + \gamma\,, \quad \varphi_n = \alpha + \beta\,, \quad \varphi_{n+1} = \beta + \delta\,,$$

(Fig. 6.18). The restoring force $F_n(\varphi)$ then is proportional to

$$F_n(\varphi) \sim 2(\alpha - \gamma) + 2(\beta - \delta) = 2(2\varphi_n - \varphi_{n-1} - \varphi_{n+1})\,.$$

With κ' as a measure of the flexural stiffness, it follows analogously to (6.197) that

$$F_n(\varphi) = -2\kappa'[(\varphi_n - \varphi_{n-1}) + (\varphi_n - \varphi_{n+1})]\,. \tag{6.198}$$

Furthermore,

$$\alpha = \arcsin\frac{s_n - s_{n-1}}{b}\,; \quad \beta = \arcsin\frac{s_n - s_{n+1}}{c}$$

hold for the angles α and β in Fig. 6.18. As we assume displacements $|\vec{s}_n| \ll a$ in the harmonic approximation, we have $a \approx b \approx c$ and $\sin\alpha \approx \alpha$, $\sin\beta \approx \beta$, *i.e.*,

$$\alpha = \frac{s_n - s_{n-1}}{a}\,; \quad \beta = \frac{s_n - s_{n+1}}{a}$$

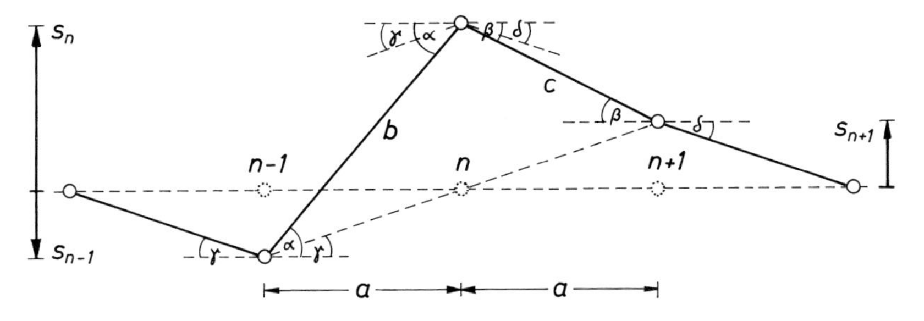

Fig. 6.18. Characteristic parameters of the linear chain of equal masses (Fig. 6.9) with respect to displacements s_{n-1}, s_n, s_{n+1} of the lattice units perpendicular to the chain

and

$$\varphi_n = \alpha + \beta = \frac{1}{a}(2s_n - s_{n-1} - s_{n+1}) .$$

With the spring constant $\kappa \equiv 2\kappa'/a$, insertion of this expression into (6.198) results in the restoring force

$$F_n = -\kappa(s_{n-2} - 4s_{n-1} + 6s_n - 4s_{n+1} + s_{n+2}) . \tag{6.199}$$

The force acting on the nth mass point necessarily also depends on the displacements of the second neighbours, as the valence angles in the direct neighbours depend on these displacements. In the case of transversely polarized bending modes of the one-dimensional chain with covalent bonds (Fig. 6.9), the equations of motion

$$m\ddot{s}_n = \kappa(-s_{n-2} + 4s_{n-1} - 6s_n + 4s_{n+1} - s_{n+2}) \tag{6.200}$$

appear in place of the equations of motion (6.128) valid for the longitudinally polarized stretching modes.

As the presumed transversely polarized waves are also only capable of propagating in the chain direction, one can proceed again from the equation (6.130) for the solution of the differential equation (6.200), whereby s_n now describes displacements in the x-direction. Insertion of (6.130) into (6.200) leads to

$$-m\omega^2 s_n = \kappa s_n(-\mathrm{e}^{-i2ka} + 4\,\mathrm{e}^{-ika} - 6 + 4\,\mathrm{e}^{ika} - \mathrm{e}^{i2ka}) ,$$

i.e., to

$$m\omega^2 = 4\kappa(\cos^2 ka - 2\cos ka + 1) = 16\kappa \sin^4 \frac{ka}{2} .$$

Hence, the dispersion relation

$$\omega(k) = \left(\frac{16\kappa}{m}\right)^{1/2} \sin^2 \frac{ka}{2} \tag{6.201}$$

holds for the transversely polarized bending modes. Moreover, the wave number should again be restricted to the first Brillouin zone (6.134). The frequency spectrum ranges from $\omega = 0$ to the maximum frequency

$$\omega_{\max} = \left(\frac{16\kappa}{m}\right)^{1/2} , \tag{6.202}$$

which is attained with $k = \pm\pi/a$ at the boundary of the first Brillouin zone. With $k \geq 0$,

$$|v| = a\omega_{\max} \sin \frac{ka}{2} \cos \frac{ka}{2} = \frac{a}{2}\, \omega_{\max} \sin ka \tag{6.203}$$

follows from (6.201) as the magnitude of the group velocity (6.138).

If the chain links are capable of vibrating in an arbitrary direction in the space $\vec{s}_n = (s_n^x, s_n^y, s_n^z)$, three mechanical degrees of freedom are assigned to each chain link. The finitely long polymer chain of N free links thus possesses $3N$ degrees of freedom. The finitely long chain segment selected from the infinitely long chain by means of the periodic boundary condition (6.144) is capable of altogether $3N$ mutually independent vibrational modes. Of these, N modes are attributed to the differently polarized branches s_n^x, s_n^y and s_n^z, respectively. The density of states of the transverse vibrations (6.200) linked with a finitely long polymer chain must be normalized to N degrees of freedom. Thus, the relation

$$\varrho(\omega) = \frac{\mathrm{N}}{\pi} \frac{1}{[\omega(\omega_{\max} - \omega)]^{1/2}} \tag{6.204}$$

follows from (6.147) with (6.201–6.203) as the density of states of a branch of the transversely polarized bending modes. The complete density of states of $3N$ vibrational modes of a finitely long linear polymer chain is given by

$$\varrho(\omega) = \varrho_s(\omega) + 2\varrho_b(\omega)\,, \tag{6.205}$$

whereby $\varrho_s(\omega)$ is the density of states (6.148) of the stretching modes and $\varrho_b(\omega)$ the density of states (6.204) of the bending modes. The factor 2 in front of $\varrho_b(\omega)$ is due to the degeneration of the bending modes s_n^x and s_n^y caused by the rotational symmetry of the chain. The dispersion relation (6.201), the group velocity (6.203), and the phase velocity (6.132) with (6.201) as a function of the wave number as well as the density of states (6.204) of transversely polarized waves are depicted in Fig. 6.19.

A comparison of the Figs. 6.10 and 6.19 shows a considerable physical difference between the stretching and bending modes of an isolated linear polymer chain. This difference is particularly pronounced in the area of quasi-elastic waves with small wave numbers (large wavelengths) $|k| \ll \pi/a$ and relatively low frequencies $\omega \ll \omega_{\max}$.

$$\omega(k) = \frac{1}{4}\, a^2 \omega_{\max} k^2 \sim k^2 \tag{6.206}$$

follows from (6.201, 6.202) for the quasi-elastic bending modes. In continuum mechanics, this corresponds to the dispersion relation

$$\omega = \left(\frac{EI}{Q\varrho_m}\right)^{1/2} k^2$$

which exists between the frequency and the wave number of the bending vibrations of a homogeneous thin rod with the cross section Q, the moment of

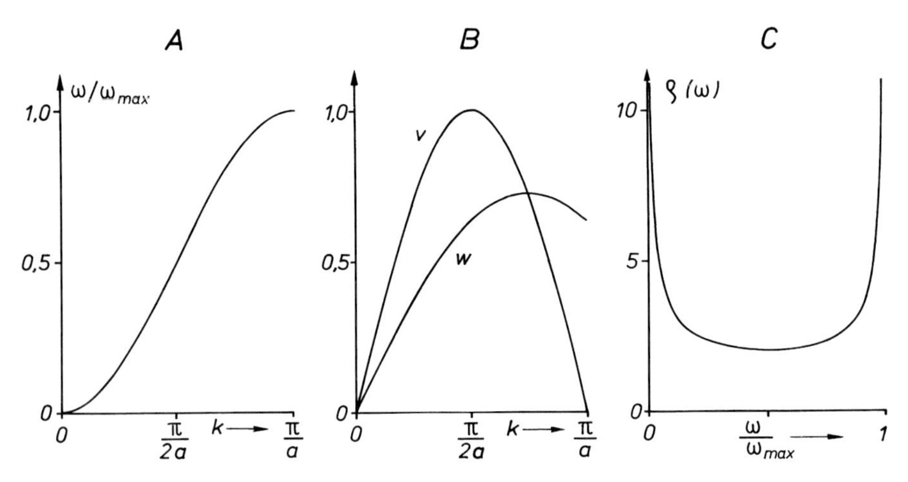

Fig. 6.19. Bending modes of the linear chain of Fig. 6.9 in the harmonic approximation. A: dispersion relation $\omega(k)$ according to (6.201) and (6.202). B: phase velocity w and group velocity v as a function of the wave number k according to (6.132)/(6.201) or (6.203). The ordinate values are multiples of $a\omega_{\max}/2$. C: frequency distribution $\varrho(\omega)$ according to (6.204). The ordinate values are multiples of $N/\pi\omega_{\max}$

flexure I, Young's modulus E and the mass density ϱ_m [*e.g.*, see Schaefer (1922)]. According to (6.203) or (6.206), the group velocity of quasi-elastic bending modes is given by

$$v = \frac{1}{2} a^2 \omega_{\max} k \sim k \tag{6.207}$$

and the magnitude of the phase velocity (6.132) by

$$|w| = \frac{1}{4} a^2 \omega_{\max} |k| = \frac{1}{2} |v| . \tag{6.208}$$

$$\varrho_{el}(\omega) = \frac{N}{\pi} \frac{1}{(\omega \omega_{\max})^{1/2}} \sim \omega^{-1/2} \tag{6.209}$$

results from (6.204) with $\omega/\omega_{\max} \ll 1$ for the density of states. The decisive difference between the quasi-elastic stretching modes and the quasi-elastic bending modes is that the frequency increases proportionally to the wave number in the case of the stretching modes [*cf.* Eq. (6.141)], but increases proportionally to the square of the wave number in the case of the bending modes. Thus, in the phonon picture, only the longitudinally polarized phonons are analogous to photons with respect to the energy-momentum relation. The transversely polarized phonons, in which, according to (6.206), the energy $\hbar\omega$ is proportional to the square of the momentum $\hbar k$, behave like the acoustical magnons [quasi-particles of the spin-waves; *e.g.*, compare Kittel (1966)] or like non-interacting inert (mass-charged) particles [Baur (1972)]. In the quasi-elastic range, the group velocity (6.139) of the stretching modes assumes its maximum value (6.142) and is constant and equal to the phase velocity. Signals from quasi-elastic stretching modes are transferred with maximum velocity without dispersion. The group velocity (6.203), on the other hand, vanishes with $k = 0$ and then, according to (6.207), gradually increases proportionally to the wave number. Signals from quasi-elastic bending modes are transferred extremely slowly and are always subject to dispersion. In the quasi-elastic range, the phase velocity of the bending modes is only half of the group velocity. The fact that ω does not converge towards zero with $|k|$ but with k^2 finally leads to the occurrence of an additional point of accumulation at $\omega = 0$ in the frequency spectrum (6.204). While the density of states of the stretching modes (6.148) is constant in the quasi-elastic range, the density of states of the bending modes (6.204) decreases with $\omega^{-1/2}$ from the singular point $\omega = 0$. These differences certainly have a decisive influence on the contributions of the stretching and bending modes to the thermodynamic response functions of a polymer. However, this will first be addressed in Sect. 6.5, as one can see that there is an additional important difference between the stretching and bending modes propagating along an extended polymer chain: whereas the properties of the stretching modes are hardly affected by the intermolecular interaction forces, which are relatively weak compared to the covalent bonds, the properties of the acoustic bending modes at the limit $\omega \to 0$ are extremely dependent on the intermolecular interaction field.

Hence, the $3N$ vibrational modes of a linear chain composed of identical covalently bound lattice units which are arranged at an equal distance (Fig. 6.9) are divided into three frequency branches with N different modes, respectively: one branch of longitudinally polarized stretching modes and two mutually degenerate branches of transversely polarized bending modes. In a linear chain composed of two different, alternately arranged masses (Fig. 6.14) or rather, a bit more generally, in every linear chain with two lattice units within the unit cell, each of these frequency branches splits into an acoustical and an optical branch (see Fig. 6.15). The $3N$ vibrational modes of the AB-chain with alternately arranged, covalently bound lattice units are distributed over six frequency branches with $N/2$ different modes, respectively: one branch of longitudinally polarized acoustic stretching modes, two mutually degenerate branches of transversely polarized acoustic bending modes, one branch of longitudinally polarized optical stretching modes and two mutually degenerate branches of transversely polarized optical bending modes. The frequency spectrum of the linear chain whose unit cell contains Z mass points possesses, in addition to the three acoustical branches, $Z-1$ longitudinally polarized optical branches and $2(Z-1)$ doubly degenerate transversely polarized optical branches. As the optical branches in the case of covalent (relatively strong) bonding of the lattice units mostly have high frequencies, the three acoustical branches are particularly important for the thermodynamic properties, whereby in the range of quasi-elastic waves

$$\omega \sim |k| \quad \text{and} \quad \varrho(\omega) = \text{const.} \tag{6.210a}$$

hold for the longitudinally polarized stretching modes and

$$\omega \sim k^2 \quad \text{and} \quad \varrho(\omega) \sim \omega^{-1/2} \tag{6.210b}$$

for the transversely polarized bending modes.

6.4.4
The Planar Zigzag Chain

Real polymer molecules rarely occur as strictly linear chains. In the ideal extended state, for example, they assume the form of a zigzag chain or that of a helix. In the extended polyethylene chain (Sect. 6.2) with a valence angle $\gamma \approx 110°$, the equilibrium positions of the lattice units (CH_2-groups) along the chain direction (z-axis) are arranged in a zigzag pattern on a narrow, two-dimensional planar band (Fig. 6.20). Three differences become immediately obvious when comparing the zigzag chain to the strictly one-dimensional linear chain:

1) The displacements s_n^x and s_n^y in the x- and y-directions perpendicular to the z-axis perturb the bonds of the lattice units in a completely different manner. The degeneracy of the bending modes is broken.
2) The displacements s_n^z in the z-direction are not only capable of changing the bond lengths but also the valence angles. The displacements s_n^y are not only capable of changing the valence angles, but also the bond lengths, as the bond lengths now possess a component in the y-direction. One should expect a possible coupling of the stretching and bending modes.
3) The band in the (y, z)-plane, on which the equilibrium positions of the mass points are arranged, exhibits a moment of inertia around the z-axis. In addition to the stretching and bending modes, torsional modes should also be possible.

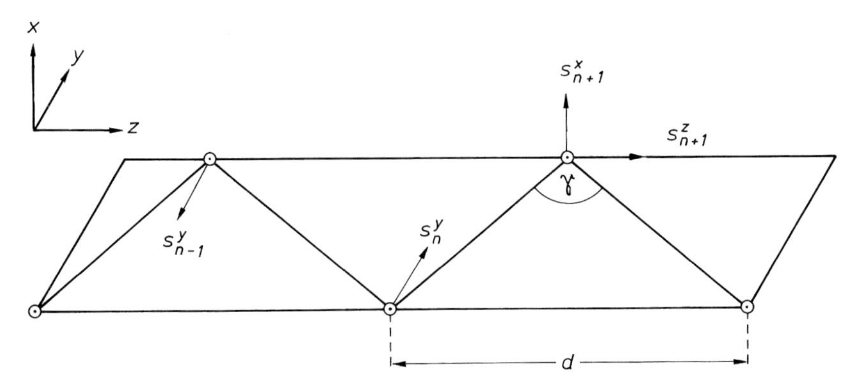

Fig. 6.20. Equilibrium positions of the lattice units of a zigzag chain on a two-dimensional planar band in the (y, z)-plane. Length of the unit cell: d; valence angle in the equilibrium position: γ; displacements in the three directions of space $x, y, z : s$

Furthermore, the geometry of the zigzag chain demands that the exact analogue of the displacement s_n^y of a lattice unit in the y-direction must be designated with $s_{n-1}^y = s_{n+1}^y = -s_n^y$ for its neighbours. If δ describes the difference in phase of two neighbouring lattice units, we have

$$\frac{s_{n+1}^z}{s_n^z} = \mathrm{e}^{i\delta} \quad \text{and} \quad \frac{s_{n+1}^x}{s_n^x} = \mathrm{e}^{i\delta}, \tag{6.211 a}$$

but

$$\frac{s_{n+1}^y}{s_n^y} = -\mathrm{e}^{i\delta} = \mathrm{e}^{i(\delta-\pi)}. \tag{6.211 b}$$

In the literature, the phase difference is often used in place of the wave number $|k|$. Apart from that, one can see that the unit cell of the zigzag lattice contains two lattice units which do not differ materially, but differ in their geometrical position. The frequency spectrum of the chain composed of equal mass points with valence angles $\gamma \neq 180°$ also contains optical branches which are separated from the acoustical branches by a gap.

The dispersion relations of the stretching and bending modes, whose displacements remain restricted to the (y, z)-plane (these are the so-called in-plane modes), can be summarized with a single expression using the phase difference δ. According to Kirkwood (1939), we have

$$\omega^2 = \omega_1^2 \pm (\omega_1^4 - \omega_2^4)^{1/2}, \tag{6.212 a}$$

$$\omega_1^2 \equiv \frac{\beta}{m}(1 + \cos\gamma\cos\delta) + 2\frac{\kappa}{m}(1 + \cos\delta)(1 - \cos\gamma\cos\delta), \tag{6.212 b}$$

$$\omega_2^4 \equiv 8\left(\frac{\beta}{m}\right)\left(\frac{\kappa}{m}\right)(1 + \cos\delta)^2(1 - \cos\delta), \tag{6.212 c}$$

where γ is the valence angle of the chain in the equilibrium position of the lattice units. With $\gamma = \pi$, the plus sign in (6.212a) and $\delta = ka$, (6.212) transforms into the dispersion relation (6.136) of the stretching modes of the linear chain, and with $\gamma = \pi$, the minus sign in (6.212a) and $\delta = ka + \pi$ [see Eq. (6.211b)], (6.212) transforms into the dispersion relation (6.201) of the bending modes of the linear chain. Generally, (*i.e.*, for $\gamma \neq \pi$), the plus sign in (6.212a) corresponds only to an optical branch (in which neighbouring lattice units are displaced in the opposite direction) and the minus sign in (6.212a) to an acoustical branch (in which neighbouring lattice units are displaced in the same direction), whereby both branches are separated by a gap. The simultaneous occurrence of the spring constant β for the perturbation of the distance of the lattice units and the spring constant κ for the perturbation of the valence angles indicates that with $\gamma \neq \pi$, a clear distinction between the stretching and bending modes is no longer possible. With $\delta = 0$, $\omega = 0$, the acoustical branch contains the rigid translation of the lattice in the z-direction, and with $\delta = \pi$, $\omega = 0$ the rigid translation of the lattice in the y-direction. In between, the orientation of the displacements $\vec{s}_n$ gradually changes with an increasing phase difference from the z-direction to the y-direction. The in-plane modes generally have a z-component s_n^z as well as a y-component s_n^y: $\vec{s}_n = (0, s_n^y, s_n^z)$. The waves with the frequencies (6.212) propagating in the z-direction are generally neither transversely nor longitudinally polarized.

Within the harmonic approximation, the vibrations in the x-direction perpendicular to the (y, z)-plane (the so-called out-of-plane modes) can be regarded as being independent of the vibrations in the (y, z)-plane. In order to describe them, however, an additional spring constant κ_x must be introduced, which is a measure of the flexural stiffness of the two-dimensional band. According to Pitzer (1940), the dispersion relation for these vibrations, which are independent of γ, is

$$\omega^2 = \frac{\kappa_x}{m}(4 - 2\cos\delta - 4\cos 2\delta + 2\cos 3\delta) . \qquad (6.213)$$

The values $\delta = 0$, $\omega = 0$ correspond to a rigid translation of the lattice in the x-direction. With $0 < \delta < \pi/2$, neighbouring lattice units are displaced in the same direction leading to the development of transversely polarized bending modes of the band. With $\pi/2 < \delta < \pi$, neighbouring lattice units are displaced in the opposite direction. These are the torsional modes of the band. With $\delta = \pi$, $\omega = 0$, the torsional modes transform into a rigid rotation of the bandshaped lattice around the z-axis.

When one directly compares the dispersion relations (6.212), (6.213) of the zigzag chain with the dispersion relations (6.136) and (6.201) of the strictly linear chain, all the relations must be folded at $\delta = \pi/2$ or $|k| = \pi/2a$, as the unit cell of the zigzag chain contains two mass points and the length of the unit cell thus amounts to $d = 2a$ (see Fig. 6.15). The connection between the phase difference δ and the wave number $|k|$ in the range

$\delta = 0 \ldots \pi/2$ is given by $|k| = \delta/a$

and in the range

$\delta = \pi/2 \ldots \pi$ by $|k| = (\pi - \delta)/a$.

The folded dispersion relations are represented by $\gamma = 180°$ and $\gamma = 110°$ in Fig. 6.21 [the transition from the one to the other case becomes even clearer if one selects, for example, $\gamma = 160°$ as an intermediate value]. In detail, the comparison shows the following:

In the zigzag chain, the stretching modes $s(\beta)$ of the linear chain split up into an acoustical branch $sa(\beta, \kappa)$ and an optical branch $so(\beta, \kappa)$, whereby both branches also become dependent on the flexural stiffness κ. In the range $|k| \ll \pi/d$ of the quasi-elastic waves, the difference between $s(\beta)$ and $sa(\beta, \kappa)$ is small. In the case of long wavelengths, large regions of neighbouring lattices units are displaced in the z-direction, so that the valence angles in the zigzag chain are hardly perturbed. The smaller the wavelength with increasing wave

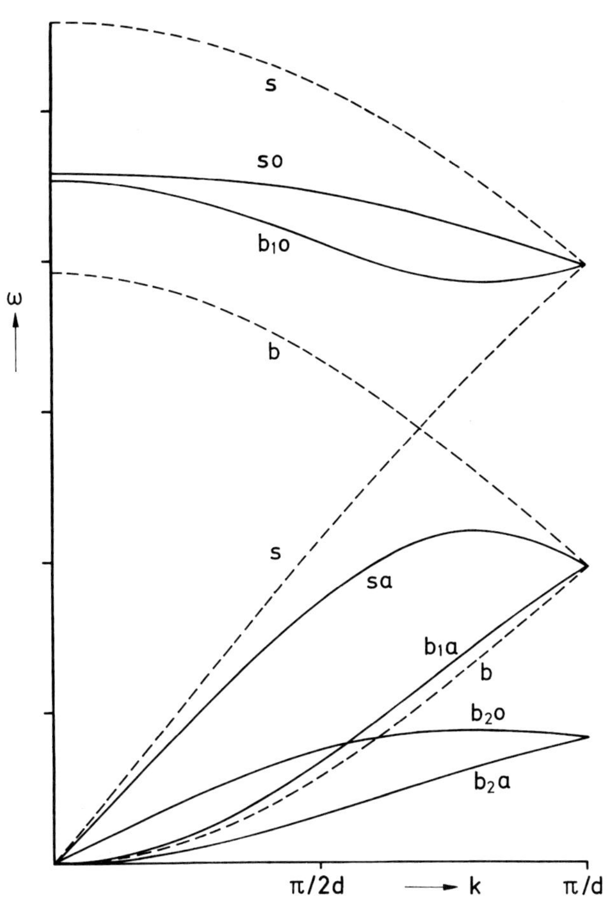

Fig. 6.21. Solid curves: dispersion relations $\omega(k)$ of the oscillations in the planar zigzag chain with a valence angle $\gamma = 110°$ according to (6.212) and (6.213). *so*, b_1o: optical in-plane modes; *sa*, b_1a: acoustical in-plane modes. b_2o: torsional modes; b_2a: acoustical out-of plane modes. Dashed curves: dispersion relations of the linear chain ($\gamma = 180°$) according to (6.136) and (6.201). *s*: stretching modes; *b*: bending modes. The dispersion relations of the linear chain were folded at $k = \pi/2a = \pi/d$ (see Fig. 6.15). [From Baur (1972)]

number, the smaller the regions in which neighbouring lattice units are displaced in the same direction. The valence angles are more and more perturbed. The displacements acquire a component in the y-direction and with increasing wave number, $sa(\beta, \kappa)$ increasingly approaches the branch $b(\kappa)$ of the bending modes of the linear chain. In the zigzag chain, the bending modes $b(\kappa)$ of the linear chain also split into an acoustical branch $b_ya(\beta, \kappa) \equiv b_1a$ and an optical branch $b_yo(\beta, \kappa) \equiv b_1o$, whereby the difference between $b(\kappa)$ and b_1a remains small in the entire range of the first Brillouin zone. It is remarkable that even in the zigzag chain in the range of quasi-elastic waves, in which the stretching and bending modes can still quite clearly be distinguished, the stretching modes still satisfy the relations (6.210a), and the bending modes the relations (6.210b).

The vibrations (6.213) of the two-dimensional band naturally do not have an analogue in the linear chain. The folding of the relation (6.213) at $\delta = \pi/2$ or $|k| = \pi/2a = \pi/d$ leads to the development of two branches $b_xa(\kappa_x) \equiv b_2a$ and $b_xo(\kappa_x) \equiv b_2o$. The branch of the bending modes in the x-direction b_2a, in which neighbouring lattice units are displaced in the same direction, must be classified as the third acoustical branch of the zigzag chain. It also fulfils the relations (6.210b) in the quasi-elastic range. The branch b_2o, in which neighbouring lattice units are displaced in the opposite x-direction, includes the torsional modes of the zigzag chain and must be classified as an optical branch. As such, this branch represents an anomaly in so far as one obtains $\omega = 0$ in the limit $|k| \to 0$ (rigid rotation of the chain around the z-axis), as for the acoustical branches. This anomaly, however, is cancelled in the field of intermolecular interaction. In the three-dimensional solid, b_2o occurs as an optical branch with relatively low frequencies for which $\omega \neq 0$ in the whole first Brillouin zone.

Consideration of the internal degrees of freedom of the lattice units, *i.e.*, in the case of polyethylene consideration of the degrees of freedom of motion within the CH_2-groups, leads to further splitting and the occurrence of new branches. However, this only affects the range of optical modes, whereas the acoustic modes are practically not affected [Tasumi *et al.* (1962)]. The effects of the intermolecular interaction field on the dispersion relations of the lattice waves propagating along the CH_2-zigzag chain were examined by Tasumi, Shimanouchi (1965) and Tasumi, Krimm (1967). In chain molecules which form helices, the strong covalent binding can also be subject to a certain "softening" in the range of optical modes: With an increasing number of units per turn, optical modes of the helices can drop to a relatively low frequency range [*e.g.*, see Telezhenko, Sukharevskii (1982)].

6.5 The Ideal Polymer Crystal

Consideration of the intermolecular interaction effective in a three-dimensional solid is important for a variety of reasons when examining the lattice vibrations of a polymer chain. Moreover, the lattice waves in a three-dimensional solid are not only capable of propagating along the chain molecules but also in every arbitrary direction of the space occupied by mass points.

6.5.1 Born–Von Kármán (Harmonic) Approximation

At first, we will generally consider the three-dimensional solid to be an aggregation of N free structureless mass points without considering a specific structure of the aggregation. If m_n denotes the mass of the nth mass point, $\vec{s}_n = (s_n^x, s_n^y, s_n^z)$ its displacement from the equilibrium position, and $\vec{F}_n = (F_n^x, F_n^y, F_n^z)$ the total force acting on the nth mass point, the equations of motion of the mass points, corresponding to the equations (6.121), are

$$m_n \ddot{\vec{s}}_n = \vec{F}_n \quad (n = 1 \dots N) \,. \tag{6.214a}$$

As long as there are no assumptions regarding the arrangement of the mass points, a consecutive numbering of the coordinates $s_1^x \dots s_n^z$ and the components of the forces $F_1^x \dots F_n^z$ with $j = 1 \dots 3N$ is advantageous. If one sets $m_{3n-2} = m_{3n-1} = m_{3n}$, one can also write

$$m_j \ddot{s}_j = F_j \quad (j = 1 \dots 3N) . \tag{6.214b}$$

in place of (6.214a). The system of equations (6.214) is formally identical with the equations (6.121) of the linear chain except for the discrimination of different masses as well as the greater number of considered degrees of freedom. In the same way, one can use the formalism of the equations (6.122–6.126) and thus obtain the equations of motion of the mass points of the three-dimensional solid in the harmonic approximation [Born–von Kármán theory; Born, v. Kármán (1912, 1913), see also Born, Huang (1954)]

$$m_j \ddot{s}_j = -\sum_k \Phi_{jk} s_k , \quad (j = 1 \dots 3N) . \tag{6.215a}$$

whereby

$$\Phi_{jk} \equiv \left(\frac{\partial^2 \Phi}{\partial s_j \, \partial s_k} \right)_0 , \tag{6.215b}$$

was used to abbreviate the coefficients of the quadratic term (6.124d) of the potential energy Φ. Equation (6.215) leads to the Hamiltonian function of the system of mass points

$$\mathcal{H} = \frac{1}{2} \sum_j m_j \dot{s}_j^2 + \frac{1}{2} \sum_j \sum_k \Phi_{jk} s_j s_k . \tag{6.216}$$

As long as there are no assumptions regarding the structure of the solid and the masses m_j are regarded as being different, the solutions of the equations (6.215) have to be formulated somewhat more generally when compared to (6.130) with

$$s_j = A_j \, e^{i(\omega t + \alpha)} . \tag{6.217}$$

The mass points oscillate around their equilibrium position with the amplitude

$$\vec{A}_n = (A_n^x , A_n^y , A_n^z) = (A_{3n-2}, A_{3n-1}, A_{3n})$$

of the frequency $\nu \equiv \omega/2\pi$ and an arbitrary phase α, common to all mass points. In order to determine the amplitudes and the angular frequency, insertion of (6.217) into (6.215) yields the system of equations

$$\sum_k (\Phi_{jk} - m_j \omega^2 \delta_{jk}) \, A_k = 0 \tag{6.218}$$

(δ_{jk}: Kronecker delta). With the mean mass

$$m = \frac{1}{3N} \sum_j m_j$$

and the transformed amplitudes

$$B_j \equiv \left(\frac{m_j}{m}\right)^{1/2} A_j ,$$

(6.218) becomes

$$\sum_k \left(\frac{m}{(m_j m_k)^{1/2}} \Phi_{jk} - m\omega^2 \delta_{jk} \right) B_k = 0 . \qquad (6.219)$$

This is the eigenvalue equation of the matrix $m\Phi_{jk}/(m_j m_k)^{1/2}$ with the eigenvalues $m\omega^2$ and the eigenvectors $(B_1 \dots B_{3N})$ [e.g., see Margenau, Murphy (1956) or Courant, Hilbert (1965)]. It only has a solution different from zero if its determinant vanishes:

$$\left| \frac{m}{(m_j m_k)^{1/2}} \Phi_{jk} - m\omega^2 \delta_{jk} \right| = 0 \qquad (6.220)$$

The so-called characteristic determinant (6.220) forms an algebraic equation of the 3*N*th order, which generally possesses 3*N* different solutions ω_ℓ ($\ell = 1 \dots 3N$). An eigenvector $(B_1^\ell \dots B_{3N}^\ell)$ is attributed to each of these solutions, so that there are $(3N)^2$ equations

$$\sum_k \left(\frac{m}{(m_j m_k)^{1/2}} \Phi_{jk} - m\omega_\ell^2 \delta_{jk} \right) B_k^\ell = 0 . \qquad (6.221)$$

In addition, because of

$$\Phi_{jk} = \Phi_{kj} = \text{real} ,$$

the 3*N* eigenvectors necessarily form a so-called orthonormal system, *i.e.*, we have

$$\sum_k B_k^j B_k^\ell = \sum_k B_j^k B_\ell^k = \delta_{j\ell} \qquad (6.222)$$

[the equations (6.152) and (6.153) in Sect. 6.4.1 represent a special case of the equations (6.222)].

By means of the eigenvectors, the 3*N* coupled equations of motion (6.215) can be transformed into their normal form, *i.e.*, into a system of mutually completely independent equations. If one introduces the so-called normal coordinates q_ℓ *via*

$$s_j = \left(\frac{m}{m_j}\right)^{1/2} \sum_\ell B_j^\ell q_\ell , \tag{6.223}$$

one first of all obtains from (6.215)

$$m \sum_\ell B_j^\ell \ddot{q}_\ell = - \sum_\ell \sum_k \frac{m}{(m_j m_k)^{1/2}} \Phi_{jk} B_k^\ell q_\ell$$

or with (6.221)

$$\sum_\ell B_j^\ell \ddot{q}_\ell = - \sum_\ell \omega_\ell^2 B_j^\ell q_\ell .$$

If one multiplies this equation by B_j^k and sums over j, one also gets

$$\sum_\ell \left(\sum_j B_j^\ell B_j^k\right) \ddot{q}_\ell = - \sum_\ell \omega_\ell^2 \left(\sum_j B_j^\ell B_j^k\right) q_\ell ,$$

that is with (6.222)

$$\ddot{q}_k + \omega_k^2 q_k = 0 \quad (k = 1 \dots 3N) . \tag{6.224}$$

These are the equations of motion of $3N$ mutually independent harmonic oscillators, as we have already found them with the equations (6.156) under considerably more specific conditions for the linear chain. In contrast to (6.156), however, k in (6.224) does not have the meaning of a wave number but only of a running subscript. (The concept of a wave number only becomes relevant if a certain structure is assumed in the aggregation of mass points; see further below.) The inversion of the equation (6.223) shows that the oscillations described by (6.224) are not vibrations of individual mass points but collective motions of all the mass points in the solid, the so-called normal modes of the solid as a whole.

With the transformation (6.223), the Hamiltonian function (6.216) decomposes into $3N$ mutually independent additive terms. Insertion of (6.223) into (6.216) yields with (6.221), (6.222) and the momentum $p_k = m\dot{q}_k$ conjugated to the coordinate q_k

$$\mathcal{H} = \sum_k \mathcal{H}_k .$$

$$\mathcal{H}_k = \frac{1}{2m} p_k^2 + \frac{1}{2} m \omega_k^2 q_k^2 . \tag{6.225}$$

The determination of the classical or quantum-mechanical partition function of the solid is carried out as in the equations (6.158) or (6.165), with the only difference that k is a running subscript ranging from 1 to $3N$. Because of the increased number of degrees of freedom, we now obtain in place of (6.160) within the framework of the classical calculation

$$U = 3RTN \tag{6.226}$$

for the internal energy of the solid and in place of (6.161) the Dulong–Petit law

$$C_V = 3RN \tag{6.227}$$

for the heat capacity. Within the framework of quantum mechanics, the thermodynamic functions of the F-representation of the three-dimensional harmonic solid are given by the equations (6.165–6.168). The equation (6.168b), here repeated as equation (6.228),

$$C_V = R \int_0^{\omega_{\max}} \left(\frac{\hbar\omega}{RT}\right)^2 \frac{\varrho(\omega)\, e^{\hbar\omega/RT}}{(e^{\hbar\omega/RT} - 1)^2}\, d\omega\,, \tag{6.228}$$

also holds specifically for the heat capacity of the solid, whereby the frequency distribution function $\varrho(\omega)$ must now be normalized to $3N$ degrees of freedom. As we have not made any assumptions so far regarding the structure of the solid, the equations (6.165–6.168) or (6.226–6.228) hold for every arbitrary solid in the harmonic approximation, *i.e.*, also, at least in principle, for glasses or for any mixed forms such as partially crystalline polymers. One must take into account, however, that the harmonic approximation, in which only the quadratic term Φ_2 of the potential energy is considered, postulates a parabolic potential around a given equilibrium position. The region of validity of this presumption definitely depends on the structure of the ensemble of mass points. In glasses, the anharmonic terms Φ_i $(i > 2)$ in (6.124) are already important at very low temperatures, so that the harmonic approximation can only provide a first, rough idea.

Hence, the main problem with respect to the thermodynamics of solids consists, within the framework of the harmonic approximation, in the determination of the density of states $\varrho(\omega)$. This is, of course, a very difficult problem in the case of an amorphous arrangement of mass points. But even in the case of a periodic arrangement of mass points only approximations are usually possible.

Let us again consider an infinitely extended point lattice. With a periodic arrangement of the mass points, this lattice is composed of an infinite number of mutually identical unit cells, each of which can be described by three noncoplanar primitive translation vectors (basis vectors)

$$\vec{a}_j = (a_j^x,\ a_j^y, a_j^z), \quad j = 1, 2, 3$$

[see any textbook on crystallography or Jagodzinski (1955)]. In the rectangular coordinate system (x, y, z), the positions of the unit cells are then characterized by the position vectors

$$\vec{r}_{\vec{n}}^{\,o} = n_1\vec{a}_1 + n_2\vec{a}_2 + n_3\vec{a}_3\,. \tag{6.229a}$$

The n_j are arbitrary positive or negative whole integers

$$n_j = 0, \pm 1, \pm 2, \pm 3, \ldots \quad (j = 1, 2, 3)\,,$$

which can be condensed as number triplets (vectors) $\vec{n} = (n_1, n_2, n_3)$. If

$$\mathcal{A} = \begin{pmatrix} a_1^x & a_2^x & a_3^x \\ a_1^y & a_2^y & a_3^y \\ a_1^z & a_2^z & a_3^z \end{pmatrix}$$

designates the matrix of the basis vectors $\vec{a}_j$, one can also abbreviate (6.229a) with

$$\vec{r}_{\vec{n}}^{\,o} = \mathcal{A}\vec{n}\,. \tag{6.229b}$$

If the origin of the coordinate system (x, y, z) occurs at the equilibrium position of a lattice point, and if we have a simple Bravais lattice with a single lattice unit per unit cell, (6.229) simultaneously describes the equilibrium position of the lattice points. In a cubic primitive lattice with the lattice constant a, we obtain, for example,

$$\mathcal{A} = \begin{pmatrix} a & 0 & 0 \\ 0 & a & 0 \\ 0 & 0 & a \end{pmatrix} \tag{6.230}$$

If the unit cell contains Z lattice units, the equilibrium position of every unit within the cell must be characterized by an individual position vector $\vec{r}_\mu^{\,o}$ ($\mu = 1 \ldots Z$). The equilibrium position of an arbitrary lattice unit in the crystal is then given by

$$\vec{r}_{\vec{n},\mu}^{\,o} = \mathcal{A}\vec{n} + \vec{r}_\mu^{\,o}\,. \tag{6.231}$$

If the lattice units are subject to displacements

$$\vec{s}_{\vec{n},\mu} = (s_{\vec{n},\mu}^x, s_{\vec{n},\mu}^y, s_{\vec{n},\mu}^z)$$

from their equilibrium positions, their positions are determined by the position vectors

$$\vec{r}_{\vec{n},\mu} = \vec{r}_{\vec{n},\mu}^{\,o} + \vec{s}_{\vec{n},\mu}\,. \tag{6.232}$$

Due to the periodicity of the lattice, the solutions of the equations of motion (6.215) are

$$\vec{s}_{\vec{n},\mu} = \vec{A}_\mu \, e^{i(\vec{k}\vec{r}_{\vec{n},\mu}^{\,o} - \omega t)} = \vec{A}_\mu'(\vec{k})\, e^{i(\vec{k}\mathcal{A}\vec{n} - \omega t)} \tag{6.233}$$

[*e.g.*, see, also with respect to the following, Maradudin et al. (1963) or Leibfried (1955)]. They correspond to the solutions (6.130) or (6.179) of the linear chain, but now describe spatially polarized plane waves which are capable of propagating in every arbitrary direction in the crystal occupied by lattice units. (An example is shown in Fig. 6.22.) The scalar product of the two vectors $\vec{k}$ and $\vec{r}^{\,o}_{\vec{n},\mu}$ is expressed by:

$$\vec{k}\,\vec{r}^{\,o}_{\vec{n},\mu} = k_x x^o_{\vec{n},\mu} + k_y y^o_{\vec{n},\mu} + k_z z^o_{\vec{n},\mu} .$$

The direction of propagation of the waves is given by the wave vector $\vec{k} = (k_x, k_y, k_z)$. The magnitude of the wave vector

$$k \equiv |\vec{k}| = (k_x^2 + k_y^2 + k_z^2)^{1/2} ,$$

the wave number, is still connected with the wavelength λ of the waves *via* the relation (6.131). The periodicity of the lattice reflected in (6.229) has the effect that the wave vectors

$$\vec{k}' = \vec{k} + 2\pi \tilde{\mathcal{A}}^{-1}\vec{\ell} \tag{6.234a}$$

with

$$\vec{\ell} = (\ell_1, \ell_2, \ell_3)$$

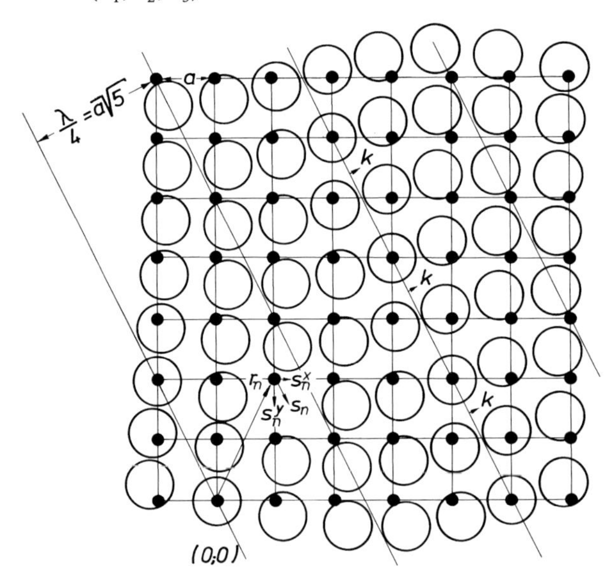

Fig. 6.22. Snapshot of the (001)-plane of a cubic primitive lattice traversed by a transversely polarized planar lattice wave (6.233). Black dots: equilibrium positions of the centres of mass of the lattice units; circles: instantaneous positions of the lattice units. The planar wave fronts are perpendicular to the plane of the drawing and intersect this plane at the diagonal thin lines [*i.e.*, run parallel to the (210)-planes of the lattice]. Direction of propagation of the waves (wave vector): k; lattice constant: a; wave length: λ; arbitrarily chosen reference point: (0;0); distance of the equilibrium position of the centre of mass of the nth lattice unit from this reference point: r_n; displacement of the centre of mass of the nth lattice unit: s_n; respective components in the (001)-plane ($s_n^z = 0$): s_n^x, s_n^y. [From Wunderlich, Baur (1970)]

and

$$\ell_j = 0, \pm 1, \pm 2, \pm 3, \ldots \quad (j = 1, 2, 3) \tag{6.234b}$$

describe the same physical situation as the wave vector $\vec{k}$ [*cf.* Eq. (6.133)]. $\tilde{\mathcal{A}}^{-1}$ is the reciprocal of the transposed matrix $\tilde{\mathcal{A}}$ of $\mathcal{A}$; that is the matrix of the basis vectors in the reciprocal lattice. In the case of the cubic primitive lattice, (6.230) leads to

$$\tilde{\mathcal{A}}^{-1} = \begin{pmatrix} \frac{1}{a} & 0 & 0 \\ 0 & \frac{1}{a} & 0 \\ 0 & 0 & \frac{1}{a} \end{pmatrix}.$$

Uniqueness in the description can again be achieved by restricting the wave vector $\vec{k}$ to the first Brillouin zone, determined by the unit cell of the reciprocal lattice. Corresponding to (6.134), the following holds in the cubic primitive lattice:

$$-\frac{\pi}{a} < k_x \le \frac{\pi}{a},$$

$$-\frac{\pi}{a} < k_y \le \frac{\pi}{a},$$

$$-\frac{\pi}{a} < k_z \le \frac{\pi}{a}.$$

Insertion of the solutions (6.233) into the equations of motion when determining the amplitude vectors $\vec{A}_\mu$ (and the polarization of the waves) as well as when determining the dispersion relations $\omega = \omega(\vec{k})$ leads to a system of equations of the form (6.218) or (6.219). Due to the periodicity of the lattice, however, its characteristic determinant (6.220) reduces to $3Z$ rows and $3Z$ columns. Hence, in the case of periodic lattices, (6.220) represents an algebraic equation of the $3Z$th order which generally yields $3Z$ different solutions $\omega_\sigma(\vec{k}) \ge 0$ ($\sigma = 1 \ldots 3Z$; Z: number of lattice units in the unit cell). The dispersion relations split up into $3Z$ generally different branches $\omega_\sigma(\vec{k})$, and

$$\lim_{k \to 0} \omega_\sigma(\vec{k}) = 0$$

is valid for three of these branches. These are the acoustical branches, whose waves in the range of very small wave numbers (long wavelengths) correspond more or less to the plane waves

$$\vec{s}(\vec{r},t) = \vec{A}\, e^{i(\vec{k}\vec{r} - \omega t)}$$

which propagate in a three-dimensional elastic continuum. The limiting case $k = 0$, $\omega = 0$ describes the rigid translation of the lattice as a whole in the direction determined by $\vec{A}$. The remaining $3(Z-1)$ branches – in the sense explained in more detail in Sect. 6.4.2 – have to be classified as optical branches, and

$$\lim_{k \to 0} \omega_\sigma(\vec{k}) \neq 0$$

is valid for these branches. In the limiting case $k = 0$ (*i.e.*, $\lambda = \infty$), the primitive lattices, of which the complicated lattice in the case $Z > 1$ is composed, oscillate rigidly with respect to each other (*cf.* Sect. 6.4.2).

In order to formulate the so-called periodic boundary condition (*cf.* Sect. 6.4.1), a parallelepiped with the edge lengths $L\vec{a}_1$, $L\vec{a}_2$, $L\vec{a}_3$ must be marked in the infinitely large crystal (L: a positive integer). This is an enlarged copy of the unit cell. $L^3 \equiv N_c$ is the number of unit cells, $N = ZN_c$ the number of lattice units contained in this parallelepiped. The periodic boundary condition now requires that the displacements of the lattice units in the unit cells on one of the surfaces of the parallelepiped correspond to the displacements of the lattice units in the unit cells on the opposite surface. The periodic boundary condition, thus, requires

$$\vec{s}_{n_j,\mu} = \vec{s}_{n_j + L,\mu} \quad (j = 1, 2, 3) \tag{6.235}$$

Insertion of (6.233) leads to

$$e^{i\vec{k}L\vec{a}_j} = 1 \,.$$

These equations are only fulfilled if

$$\vec{k}L\vec{a}_j = 2\pi\, \ell_j$$

is valid and ℓ_j is an arbitrary positive or negative integer. Corresponding to Eq. (6.145a), one obtains for the wave vectors with $\vec{\ell} = (\ell_1, \ell_2, \ell_3)$ the restricting condition

$$\vec{k} = \frac{2\pi}{L} \mathcal{A}^{-1} \vec{\ell} \,. \tag{6.236a}$$

The wave vector lies in the first Brillouin zone if the numbers ℓ_j, corresponding to (6.145b), are restricted to

$$\ell_j = 0, \pm 1, \pm 2, \dots \pm \left(\frac{L}{2} - 1\right), \frac{L}{2} \tag{6.236b}$$

(L : an even number). Altogether $L^3 = N_c$ different wave vectors are found in the first Brillouin zone, in the selected parallelepiped under the periodic boundary condition. As stated above, $3Z$ frequencies $\omega \geq 0$ are attributed to each of these vectors. Altogether, the parallelepiped is capable of $3ZN_c = 3N$ different vibrational states. As already mentioned in Sect. 6.4.1, this parallelepiped, under the periodic boundary condition with sufficiently large L, is fully equivalent to a finite, real crystal of the same structure consisting of N free lattice units and being subject to the boundary condition of constant volume [Born, Huang (1954), Appendix IV]. In order to simulate the external form of the real crystal, one can, in addition, naturally proceed from a parallelepiped with the edge lengths $L_1\vec{a}_1, L_2\vec{a}_2, L_3\vec{a}_3 (L_1 \neq L_2 \neq L_3$: positive integers).

In the case of a propagation in three-dimensional space, the group velocity (6.138) of the waves characterized by $\vec{k}$ is given by the vector

$$\vec{v} = \left(\frac{\partial\omega}{\partial k_x}, \frac{\partial\omega}{\partial k_y}, \frac{\partial\omega}{\partial k_z}\right) \equiv \text{grad}_k\,\omega\,. \tag{6.237}$$

To determine the density of states $\varrho(\omega)$ of the vibrations, one can proceed as in the case of Eqs. (6.146 and 6.147). As every fixed ω is linked to waves capable of propagating in all possible directions of the space, the factor 2 in (6.146) (which took into account that two directions of propagation were possible for every fixed frequency ω) needs to be replaced by an integration over the surfaces $\omega(\vec{k}) = \text{const.}$ in the (practically dense) three-dimensional $\vec{k}$-space. One thus obtains for the density of the individual frequency branches $\omega_\sigma(\vec{k})$ ($\sigma = 1 \ldots 3Z$)

$$\varrho_\sigma(\omega) = \left(\frac{L}{2\pi}\right)^3 |\mathcal{A}| \int \frac{\mathrm{d}A_\omega}{|\vec{v}_\sigma|}, \tag{6.238}$$

where $\mathrm{d}A_\omega$ is a surface element on the surface $\omega_\sigma = \text{const.}$ in the $\vec{k}$-space over which one must integrate. The determinant of $\mathcal{A}$, $|\mathcal{A}|$, is equal to the volume V_c of the unit cell, *i.e.*, $L^3|\mathcal{A}| = N_cV_c = V$, the volume of the crystal. According to (6.236), $L^3|\mathcal{A}|/(2\pi)^3$ indicates the homogeneous density of the $\vec{k}$-vectors in the $\vec{k}$-space. The total frequency spectrum of the crystal is additively composed of the spectra (6.238) of the individual branches:

$$\varrho(\omega) = \sum_\sigma \varrho_\sigma(\omega)\,. \tag{6.239}$$

Evaluation of the integrals (6.238) requires that the dispersion relations $\omega = \omega_\sigma(\vec{k})$ in every direction $\vec{k}/k$ of the space be known, a problem which can only be handled approximatively. The logically simplest method for determining the density of states $\varrho(\omega)$ is the so-called root sampling method (for which, however, if one wishes to obtain useful results, extensive calculations are definitely necessary) [see Maradudin et al. (1963)]: One takes a quite large number of allowable $\vec{k}$-values and determines the respective frequencies $\omega_\sigma(\vec{k})$ for these $\vec{k}$-values from the characteristic determinant (6.220). The dispersion relations and densities $\varrho_\sigma(\omega)$ are then interpolated using the points in $\vec{k}$-space thus obtained. In this respect, the crystal symmetry often allows a considerable restriction of the ranges from which the $\vec{k}$-values must be selected. A disadvantage of this method is that critical points on the surfaces ω = const. [which are necessarily present; van Hove (1953)] and the resulting singularities of $\varrho(\omega)$, which are characteristic for the spectra, may possibly be lost.

6.5.2
Continuum Approximations

Thermodynamic quantities such as the Helmholtz free energy (6.166), the internal energy (6.167), or the heat capacity (6.168), (6.228) depend *via* a sum or integral on the density of the lattice vibrations of the solid. This means that the subtleties of the density of states are not reflected in these quantities and that relatively rough approximations for $\varrho(\omega)$ are sufficient for a description of the development of the thermodynamic quantities as a function of temperature. On the other hand, these approximations do not yield a large amount of physical information on the real molecular situation. An exception in this regard is the range of very low temperatures in which, for energetic reasons, only lattice waves with a low frequency and wave number or a long wavelength are thermally excited. As the displacements of neighbouring lattice units practically only differ by infinitesimal amounts in cases of a low frequency and long wavelength, the specific structure of the lattice cannot play a decisive role in this range. Simple interrelationships displaying a certain general validity, irrespective of the solid structure, for the development of thermodynamic equilibrium quantities at very low temperatures can be expected within the framework of the harmonic approximation.

The lattice vibrations with a low frequency and small wave number excited at low temperatures originate from the acoustical branches and are more or less identical with the running elastic waves (sound waves) which propagate in a continuum. A simple proportionality to a power of the frequency:

$$\varrho(\omega) \sim \omega^{n-1}, \quad n > 0 \tag{6.240}$$

is generally valid for the frequency distribution of these waves [*e.g.*, see Eqs. (6.149), (6.194), (6.209), or (6.210) in Sect. 6.4 with $n = 1$ and $n = 1/2$]. If one inserts (6.240), for example, into (6.228), one obtains an expression for the heat capacity which is proportional to the so-called generalized Debye function

$$D_n\left(\frac{\Theta_m}{T}\right) \equiv n\left(\frac{T}{\Theta_m}\right)^n \int_0^{\Theta_m/T} \frac{x^{n+1}\,\mathrm{e}^x}{(\mathrm{e}^x-1)^2}\,\mathrm{d}x$$

$$= n(n+1)\left(\frac{T}{\Theta_m}\right)^n \int_0^{\Theta_m/T} \frac{x^n}{\mathrm{e}^x-1}\,\mathrm{d}x - n\left(\frac{\Theta_m}{T}\right)\frac{1}{\mathrm{e}^{\Theta_m/T}-1}, \tag{6.241 a}$$

where

$$\Theta_m \equiv \hbar\omega_m/R \tag{6.241 b}$$

is a characteristic temperature fixed by the upper limit ω_m of the frequency interval in which (6.240) should be valid. According to (6.241), the simple proportionality of the heat capacity to

$$D_n\left(\frac{\Theta_m}{T}\right) = n\,(n+1)\,\Gamma(n+1)\,\varsigma(n+1)\left(\frac{T}{\Theta_m}\right)^n \tag{6.242}$$

follows for temperatures $T \ll \Theta_m$. Here, Γ denotes the so-called gamma function and ς the so-called Riemann's zeta function [*e.g.*, see Gröbner, Hofreiter (1961)]. Hence, with an exponential law of the form (6.240), the heat capacity is proportional to T^n at very low temperatures. In addition, D_n was normalized with the factor $n\,\omega_m^{-n}$ in such a way that

$$D_n\left(\frac{\Theta_m}{T}\right) = 1 \tag{6.243}$$

results for $T \gg \Theta_m$.

Debye (1912) proceeded from the argument that every solid, irrespective of its structure, should resemble a homogeneous elastic three-dimensional continuum with regard to the frequency spectrum of its collective vibrational modes at low temperatures. Three different branches must be distinguished among the possible vibrational modes in an isotropic continuum: a branch of longitudinally polarized drift waves (ℓ) and two branches of transversely polarized shear waves (t). These waves propagate in all directions of the space with the same wavelength-independent sound velocity

$$c_\ell = \left[\frac{E\,(1-\nu)}{\varrho_m(1+\nu)\,(1-2\nu)}\right]^{1/2} \quad \text{or} \quad c_t = \left[\frac{E}{2\varrho_m(1+\nu)}\right]^{1/2} \tag{6.243}$$

[E : Young's modulus; ν: Poisson's ratio; ϱ_m: mass density; compare Schaefer (1922)]. They fulfil the dispersion relations

$$\omega = c_\ell\,|k| \quad \text{or} \quad \omega = c_t\,|k|\,. \tag{6.244}$$

Their group velocity (6.237) is thus given by

$$\vec{v}_\ell = c_\ell\,\vec{k}/k\,, \quad i.e.,\quad |\vec{v}_\ell| = c_\ell = \text{const.} \tag{6.245 a}$$

or

$$\vec{v}_t = c_t\,\vec{k}/k\,, \quad i.e.,\quad |\vec{v}_t| = c_t = \text{const.} \tag{6.245 b}$$

The phase and group velocities are equal, the waves do not undergo dispersion. According to (6.244), the integration in (6.238) extends over the surface of a sphere with the radius $|k|$, whereby the constant group velocity can be placed in front of the integral. This leads to

$$\int \mathrm{d}A_\omega = 4\pi k^2 = 4\pi \frac{\omega^2}{c_\ell^2} \quad \text{or} \quad = 4\pi \frac{\omega^2}{c_t^2} .$$

For the frequency distribution function (6.238, 6.239) of the sound waves in a three-dimensional homogeneous isotropic continuum, therefore, one obtains

$$\varrho(\omega) = \frac{V}{2\pi^2}\left(\frac{1}{c_\ell^3} + \frac{2}{c_t^3}\right)\omega^2 . \tag{6.246}$$

The factor 2 in the last term is due to the degeneration of the two transversely polarized branches in an isotropic medium. Even though the polarization and velocities of propagation c_ℓ and c_t are dependent on the direction of propagation of the waves in an anisotropic continuum, the proportionality of the spectrum $\varrho(\omega)$ with respect to ω^2 is maintained. Hence, the exponent n in (6.240) is generally $n = 3$ in the case of a three-dimensional elastic continuum. For $T \ll \Theta_m$ and with $\Gamma(4) = 6, \varsigma(4) = \pi^4/90$, one obtains *via* (6.242) Debye's famous T^3-law

$$C_V \sim D_3\left(\frac{\Theta_m}{T}\right) = \frac{4\pi^4}{5}\left(\frac{T}{\Theta_m}\right)^3 , \tag{6.247}$$

which should be valid for all three-dimensional solids irrespective of their structure at low temperatures. As a matter of fact, this law has largely been confirmed [*e.g.*, see to Blackman (1955)]. As long as the heat capacity is solely determined by the elastic vibrations of the solid, deviations can only occur if the vibrational spectrum is superimposed by low-frequency local or optical modes which, for example, make an exponential contribution to the heat capacity *via* (6.169), or if from $T = 0$ upwards anharmonic effects are important. In glasses, a deviation from Debye's T^3-law can be found below 1 K. This so-called glass anomaly can be explained by the fact that the molecular units of the glasses do not have a defined equilibrium position [Hunklinger, Sussner, Dransfeld (1976); Phillips (1981)]. There are, of course, always deviations if, in addition to the elastic vibrations, other mechanisms noticeably contribute to the heat capacity (*e.g.*, the conduction electrons in metals or the spin waves in ferromagnetic substances).

For crystals, the proportionality of $\varrho(\omega)$ with respect to ω^2 and thus the proportionality of the heat capacity with respect to $D_3(\Theta_m/T)$ or T^3 is only physically justified in the temperature range in which only the dispersionless quasi-elastic vibrations of the lattice units are excited, *i.e.*, for low or very low temperatures. However, Debye (1912) had already observed that in some substances a good approximation can be achieved by expanding the validity of the proportionality over the whole range of the frequency spectrum. One reason for this success is that, as already mentioned, the subtleties of the real frequency spectrum are not reflected in the heat capacity. The major reason, however, is that the complicated spectra of crystals with an only slightly pronounced anisotropy can often be averaged quite well by means of parabolically increasing density of states (see Fig. 6.26 in Sect. 6.5.3). If one accordingly normalizes the density of states (6.246) over all $3N$ degrees of freedom, which a finitely extended crystal composed of N free discrete lattice units possesses, one obtains with

$$\int_0^{\omega_3^D} \varrho(\omega)\,\mathrm{d}\omega = 3N$$

as the upper limit of the Debye spectrum

$$\omega_3^D = \left(3N\frac{6\pi^2}{V}\frac{c_t^3\,c_\ell^3}{c_t^3 + 2c_\ell^3}\right)^{1/3} \tag{6.248}$$

and as the characteristic Debye temperature

$$\Theta_3 \equiv \hbar\omega_3^D/R\,. \tag{6.249}$$

For the heat capacity (6.228), one obtains *via* (6.241) the equation

$$C_V = 3NR\,D_3\left(\frac{\Theta_3}{T}\right). \tag{6.250a}$$

According to (6.247), Debye's law holds then at low temperatures in the form:

$$T \ll \Theta_3 : C_V(T) = NR\frac{12\pi^4}{5}\left(\frac{T}{\Theta_3}\right)^3, \tag{6.250b}$$

and, according to (6.243), the Dulong–Petit law holds at high temperatures:

$$T \gg \Theta_3 : C_V(T) = 3NR\,. \tag{6.250c}$$

Deviations from (6.250) which may possibly occur at mean temperatures can be corrected by understanding the characteristic temperature Θ_3 itself to be a function of the temperature. The variation of $\Theta_3 = \Theta_3(T)$ can give orientating information not only on the real distribution $\varrho(\omega)$ of the states of the lattice vibrations, but also on the interaction forces between the lattice units as well as on the characteristic elastic quantities of the lattice [Blackman (1955)].

Polymer crystals are characterized by strong covalent bonds between the lattice units along the chain molecules, weak van-der-Waals bonds perpendicular to the chain molecules, or dipole bonds which, compared with the covalent bonds, are still weak. Hence, polymers develop highly anisotropic crystals. For this reason alone, the Debye approximation in the form (6.250) cannot apply to polymers. The large energetic difference between the binding forces along and at right angles to the chain molecules has the effect that particularly those lattice waves which propagate along the chain molecules contribute to the high-frequency range of the distribution function $\varrho(\omega)$. The high-frequency lattice waves propagating along the chains stress the strong covalent bonds. The intermolecular interaction forces, on the other hand, only represent weak disturbances for these waves (*cf.* Sect. 6.5.3). Tarasov (1950, 1953) concluded from this that the density of states of the lattice vibrations of a polymer crystal can only be replaced by the density of states (6.246) of the three-dimensional continuum in, compared with (6.248), a highly restricted interval $0 \le \omega \le \omega_3$. In an adjacent interval $\omega_3 \le \omega \le \omega_1$, on the other hand, the real density of states must be approximated by the distribution (6.149) of the one-dimensional continuum.

Hence, one proposes in the so-called Tarasov approximation that

$$0 \le \omega \le \omega_3 \colon \varrho(\omega) = a_3 \omega^2 \tag{6.251 a}$$

holds in a first frequency interval and

$$\omega_3 \le \omega \le \omega_1 \colon \varrho(\omega) = a_1 = \text{const.} \tag{6.251 b}$$

in an adjacent frequency interval. If f_3 independent vibrational modes are possible in the first interval and f_1 independent vibrational modes in the second interval, the density of states has to be normalized by the integrals

$$\int_0^{\omega_3} \varrho(\omega)\,\mathrm{d}\omega = f_3\,, \quad \int_{\omega_3}^{\omega_1} \varrho(\omega)\,\mathrm{d}\omega = f_1\,,$$

whereby

$$f_3 + f_1 = 3N \tag{6.251 c}$$

must be valid. This normalization leads to

$$a_3 = 3f_3/\omega_3^3\,; \quad a_1 = f_1/(\omega_1 - \omega_3)\,. \tag{6.251 d}$$

If one inserts (6.251) into (6.228), one obtains in place of the Debye approximation (6.250) the following continuum approximation for the heat capacity of a polymer crystal:

$$C_V = R\left[f_3 D_3\left(\frac{\Theta_3}{T}\right) + \frac{f_1}{\Theta_1 - \Theta_3}\left[\Theta_1 D_1\left(\frac{\Theta_1}{T}\right) - \Theta_3 D_1\left(\frac{\Theta_3}{T}\right)\right]\right]. \tag{6.252 a}$$

According to (6.241), D_3 or D_1 are the three- or one-dimensional Debye functions, respectively, and Θ_3 or Θ_1, the characteristic temperatures corresponding to the upper limits of the frequency intervals (6.251):

$$\Theta_3 = \hbar\omega_3/RT \quad \text{or} \quad \Theta_1 = \hbar\omega_1/RT\,. \tag{6.252 b}$$

If one further assumes with Tarasov for the distribution of the vibrational modes

$$f_3 = 3N\frac{\Theta_3}{\Theta_1}\,, \quad f_1 = 3N\left(1 - \frac{\Theta_3}{\Theta_1}\right),$$

(6.252) results in the Tarasov equation

$$C_V = 3NR\left[D_1\left(\frac{\Theta_1}{T}\right) + \frac{\Theta_3}{\Theta_1}\left[D_3\left(\frac{\Theta_3}{T}\right) - D_1\left(\frac{\Theta_3}{T}\right)\right]\right]. \tag{6.253a}$$

This leads with (6.242) and (6.243) in the range $T \ll \Theta_3 < \Theta_1$ again to Debye's T^3-law in the form

$$C_V = NR\frac{12\pi^4}{5\Theta_1\Theta_3^2}T^3 \tag{6.253b}$$

and in the range $\Theta_3 < \Theta_1 \ll T$ to the Dulong–Petit law (6.250c). If, furthermore, $\Theta_3 \ll \Theta_1$, *i.e.*, $\omega_3 \ll \omega_1$ holds (which can be assumed to be true in many cases due to the strong anisotropy of the interaction forces), an intermediate range $\Theta_3 \ll T \ll \Theta_1$ should also exist, in which, according to (6.241), (6.242) with $n = 1$, we have a polymer-specific T-law

$$C_V = 3NR\frac{\pi^2}{2\Theta_1}T\,. \tag{6.253c}$$

In a way, the Tarasov approximation takes into account the anisotropy of polymer crystals, but not the fact that a constant density of states (6.251b) is only given for the quasi-elastic stretching modes of the polymer chains, whereas the quasi-elastic bending modes, which produce at least two thirds of the vibrational modes of an isolated polymer chain, satisfy a relation of the form (6.209) [Lifshitz (1952); Stockmayer, Hecht (1953); Genensky, Newell (1957); Baur (1970), (1972)]. The equations (6.252) or (6.253) can, therefore, only be valid for the contribution C_V^s of the stretching modes to the heat capacity. If one normalizes the density of the stretching modes over N degrees of freedom, one obtains corresponding to (6.252)

$$C_V^s = R\left[f_{3s}D_3\left(\frac{\Theta_{3s}}{T}\right) + \frac{f_{1s}}{\Theta_{1s} - \Theta_{3s}}\left[\Theta_{1s}D_1\left(\frac{\Theta_{1s}}{T}\right) - \Theta_{3s}D_1\left(\frac{\Theta_{3s}}{T}\right)\right]\right]. \tag{6.254a}$$

with

$$f_{3s} + f_{1s} = N\,; \quad \Theta_{3s} = \hbar\omega_{3s}/R\,; \quad \Theta_{1s} = \hbar\omega_{1s}/R\,. \tag{6.254b}$$

According to (6.209), the exponent in (6.240) amounts to $n = 1/2$ for the bending modes of isolated chain molecules. If one normalizes the density of the bending modes over $2N$ degrees of freedom, one gets according to Tarasov's arguments,

$$C_V^b = R\left[f_{3b} D_3\left(\frac{\Theta_{3b}}{T}\right) + \frac{f_{1b}}{\Theta_{1b}^{1/2} - \Theta_{3b}^{1/2}}\left[\Theta_{1b}^{1/2} D_{1/2}\left(\frac{\Theta_{1b}}{T}\right) - \Theta_{3b}^{1/2} D_{1/2}\left(\frac{\Theta_{3b}}{T}\right)\right]\right]. \tag{6.254c}$$

as a contribution to the heat capacity with

$$f_{3b} + f_{1b} = 2N\,; \quad \Theta_{3b} = \hbar\omega_{3b}/R\,; \quad \Theta_{1b} = \hbar\omega_{1b}/R\,. \tag{6.254d}$$

In this approximation, the heat capacity of the polymer crystal is given by

$$C_V = C_V^s + C_V^b\,. \tag{6.254e}$$

In an intermediate range $\Theta_{3s}, \Theta_{3b} \ll T \ll \Theta_{1s}, \Theta_{1b}$, the polymer-specific relation

$$C_V(T) = c_s T + c_b T^{1/2} \tag{6.254f}$$

should now be valid instead of (6.253c), as is seen, for example, in the case of linear polyethylene between *ca.* 100 K and 200 K [Baur (1970)]. The representation of the contribution of the bending modes in the range $0 \le \omega \le \omega_{3b}$ by means of a Debye function D_3, however, remains somewhat problematic in the approximation (6.254). If the proportionality $\omega \sim k^2$, *i.e.*, $|\vec{v}| \sim k$ or $|\vec{v}| \sim \omega^{1/2}$ holds for the bending modes [see (6.206) or (6.207)], one obtains, in the case of the three-dimensional wave propagation *via* (6.238), a proportionality $\varrho_b(\omega) \sim \omega^{1/2}$, *i.e.*, in (6.240) an exponent $n = 3/2$. According to (6.242), there should be a range at low temperatures in which the bending modes yield a contribution

$$C_V^b(T) \sim T^{3/2} \tag{6.255}$$

to the heat capacity. This would correspond to Bloch's $T^{3/2}$-law valid for spin waves [Kittel (1966)]. In the case of the lattice waves of a polymer crystal, however, the validity of Bloch's law certainly does not apply to very low temperatures as the proportionality $\varrho_b(\omega) \sim \omega^{1/2}$ in the limit $\omega \to 0$ results in an instability of the lattice (Sect. 6.5.3). At very low temperatures, the true Debye's T^3-law in the form (6.247) also holds in all cases for polymer crystals. With increasing temperature, however, the contribution of the bending modes to the heat capacity can change from a proportionality with respect to D_3 into a proportionality with respect to $D_{3/2}$. Hence, the restriction of the validity of Debye's T^3-law to a very narrow temperature range in polymers is not only due to the strong anisotropy of the crystals but also due to the particularities of the bending modes of the polymer chain molecules. The difficulties which occur when compiling a simple continuum approximation are caused by the strong dependence of the relation between the frequency ω and the wave number k on the direction of propagation of the waves in the bending modes (Sect. 6.5.3). In a pure form, the proportionality $\omega \sim k^2$ (with $k > 0$) only holds, provided that there is a sufficiently strong anisotropy of the interaction forces, for the quasi-elastic bending modes propagating in the direction of the chain molecules. For the quasi-elastic bending modes propagating perpendicularly to the chain molecules, on the other hand, the proportionality $\omega \sim k$ supposed by Debye and Tarasov is maintained. In the other directions of propagation, a gradual transition takes place between these two proportionalities [see Eq. (6.269), below]. With respect to the bending modes, a consistent continuum approximation would thus have to take into account the different directions of propagation. Moreover, one should then also point out that because of the anisotropy, different maximum frequencies are generally attributed to the vibrations (*i. e.*, also to the stretching modes) in the different directions of propagation. The cut-off frequencies and the characteristic Θ-temperatures would thus become dependent on the direction of propagation. Such a complicated continuum approximation does not seem very useful if only because of the multitude of parameters that are needed. Simple lattice models are, therefore, preferable for an approximative determination of the density of states $\varrho(\omega)$.

6.5.3 The Stockmayer–Hecht Model

Stockmayer and Hecht (1953) studied an infinitely large crystal composed of parallelly aligned linear chains whose links consist of equal mass points m and form a simple tetragonal Bravais lattice in the equilibrium position (Fig. 6.23). The equilibrium position of an arbitrary mass point is selected as the origin of the rectangular coordinate system (x, y, z), in which the z-axis points in the direction of the chain axis. The distance of the nearest neighbours in the z-direc-

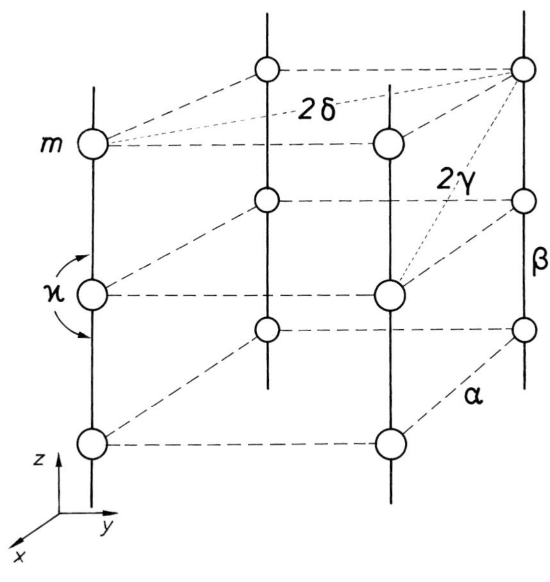

Fig. 6.23. Stockmayer–Hecht model. The mass points m of the linear chains (unbroken lines) form a primitive tetragonal Bravais lattice. Parameters of interaction: $\alpha \ldots \kappa$ (see text)

tion is c, the distance of the nearest neighbours in the x- and y-directions $a = b$. In the harmonic approximation, the interaction between the lattice units is described by five spring constants:

β for the changes in distance of the covalently bound direct neighbours in the chains,
κ for the flexural stiffness of the covalent bonds,
α for the changes in distance of direct neighbours in adjacent chains,
$2\gamma, 2\delta$ for the changes in distance of diagonally neighbouring chain links in adjacent chains.

Compared with the more complicated crystal structures which exist in real polymers, the Stockmayer–Hecht model is without doubt a highly simplified model. With a single lattice unit in the unit cell, for example, the gaps which generally exist between the acoustical and optical branches (*cf.* Sect. 6.4.2) are disregarded. With valence angles of 180° in the chains, the real torsional modes possible (Sect. 6.4.4) are ignored. With the suppositions $\beta \gg \alpha, \gamma, \delta$ and $\kappa \neq 0$, however, the model contains the two most important characteristics of all polymer crystals: the strong anisotropy of the interaction forces and the flexural stiffness of the chain molecules. Furthermore, a cubic primitive lattice whose lattice units are only exposed to central forces can also easily be described with $c \to a, \beta \to \alpha, \kappa \to 0$, and $\gamma \to \delta$. The model thus allows with increasing β and κ a simple demonstration of the development of polymer-specific properties from the quasi-isotropic cubic lattice. The model further allows a direct comparison with the simple model of the isolated linear chain (Sects. 6.4.1 and 6.4.3) and thus a first simple insight into the effect of intermolecular interaction forces. The properties of the model in the range of quasi-elastic waves should, after all, have a certain general validity, so that the model becomes suitable for controlling the predictive power and limits of validity of the various continuum approximations (Sect. 6.5.2).

The characteristic determinant (6.220) reduces to 3 rows and 3 columns since the unit cell of the tetragonal Bravais lattice only contains a single mass point. We obtain in particular

$$\begin{aligned}\Phi_{11} &= 2\alpha(1 - \cos k_x a) + 4\delta(1 - \cos k_x a \cos k_y a) \\ &\quad + 4\gamma(1 - \cos k_x a \cos k_z c) + 4\kappa(1 - \cos k_z c)^2 ,\end{aligned}$$

$$\begin{aligned}\Phi_{22} &= 2\alpha(1 - \cos k_y a) + 4\delta(1 - \cos k_y a \cos k_x a) \\ &\quad + 4\gamma(1 - \cos k_y a \cos k_z c) + 4\kappa(1 - \cos k_z c)^2 ,\end{aligned}$$

$$\begin{aligned}\Phi_{33} &= 2\beta(1 - \cos k_z c) + 4\gamma(1 - \cos k_z c \cos k_x a) \\ &\quad + 4\gamma(1 - \cos k_z c \cos k_y a) ,\end{aligned}$$

$$\begin{aligned}\Phi_{12} &= \Phi_{21} = 4\delta \sin k_x a \sin k_y a , \\ \Phi_{13} &= \Phi_{31} = 4\gamma \sin k_x a \sin k_z c , \\ \Phi_{23} &= \Phi_{32} = 4\gamma \sin k_y a \sin k_z c .\end{aligned}$$

The components k_x, k_y, k_z of the wave vector $\vec{k}$ should be restricted to the first Brillouin zone

$$
\begin{aligned}
-\frac{\pi}{a} &< k_x \leq \frac{\pi}{a}, \\
-\frac{\pi}{a} &< k_y \leq \frac{\pi}{a}, \\
-\frac{\pi}{c} &< k_z \leq \frac{\pi}{c}.
\end{aligned}
\tag{6.256}
$$

If one selects a subvolume of the infinitely large lattice, which contains $N_x N_y$ parallel chains with N_z chain links, as the periodicity volume, the values of the components are further restricted to

$$
k_x = \frac{2\pi}{aN_x}\ell_x\,; \quad k_y = \frac{2\pi}{aN_y}\ell_y\,; \quad k_z = \frac{2\pi}{cN_z}\ell_z\,,
\tag{6.257 a}
$$

in which the ℓ_i $(i = x, y, z)$ are integers which lie in the intervals

$$
-\frac{N_i}{2} < \ell_i \leq \frac{N_i}{2},
\tag{6.257 b}
$$

[see Eq. (6.145) in Sect. 6.4.1].

If we specifically consider a polymer crystal in which van-der-Waals forces act as intermolecular interaction forces,

$$
\beta > \kappa \gg \alpha \gg \gamma, \delta
$$

must be presupposed. The left part of this inequality expresses that the intramolecular covalent bonds are considerably stronger than the intermolecular van-der-Waals bonds. The right part expresses that the van-der-Waals forces do not extend very far. Under these conditions, the non-diagonal terms $\Phi_{ij}(i \neq j)$ of the characteristic determinant can, as a good approximation, be neglected. Moreover, $\gamma \approx \delta$ can approximately be assumed. As solutions of the characteristic determinant, we then obtain the three frequency branches

$$
\begin{aligned}
\omega_{bx}^2 = \frac{1}{m}\,[&2\alpha(1 - \cos k_x a) + 4\gamma(2 - \cos k_x a \cos k_y a - \cos k_x a \cos k_z c) \\
&+ 4\kappa(1 - \cos k_z c)^2]\,,
\end{aligned}
\tag{6.258 a}
$$

$$
\begin{aligned}
\omega_{by}^2 = \frac{1}{m}\,[&2\alpha(1 - \cos k_y a) + 4\gamma(2 - \cos k_y a \cos k_x a - \cos k_y a \cos k_z c) \\
&+ 4\kappa(1 - \cos k_z c)^2]\,,
\end{aligned}
\tag{6.258 b}
$$

$$
\omega_s^2 = \frac{1}{m}\,[2\beta(1 - \cos k_z c) + 4\gamma(2 - \cos k_z c \cos k_x a - \cos k_z c \cos k_y a)]
\tag{6.258 c}
$$

The omission of the non-diagonal terms of the characteristic determinant has the consequence that the amplitudes of the lattice vibrations are exactly oriented in the direction of the coordinate axes. During the vibrations of the branches ω_{bx} and ω_{by}, the mass points are displaced in the direction of the x- or y-axes, respectively, *i.e.*, perpendicularly to the chain direction (bending modes of the chains). During the vibrations of the branch ω_s, the mass points are displaced in the direction of the z-axis, *i.e.*, in chain direction (stretching modes of the chains). This, however, does not allow a statement about the polarization of the lattice waves.[6] The waves of the branches ω_{bx} and ω_{by} are transversely and the waves of the branch ω_s longitudinally polarized if the waves propagate in the chain direction ($k_x = k_y = 0$). The waves of the branch ω_{bx} are longitudinally and the waves of the branches ω_{by} and ω_s transversely polarized if the waves propagate in the x-direction ($k_y = k_z = 0$) *etc.* In general, the lattice waves (6.258) are neither longitudinally nor transversely polarized. It should be mentioned that the branches ω_{bx} and ω_{by} are also interchanged following the interchange of k_x and k_y, whereas the branch ω_s is not affected. This reflects the specific symmetry of the tetragonal Bravais lattice, which would be broken, *e.g.*, with a $\neq b$. The interchangeability of k_x and k_y means that the two branches ω_{bx} and ω_{by} lead to an identical frequency spectrum $\varrho_b(\omega)$. It does not mean, however, that the branches ω_{bx} and ω_{by} in a given direction $\vec{k}/k$ are mutually degenerate. A degeneration only occurs if $k_x = k_y$ is valid, *i.e.*, if the lattice waves propagate in the z-direction ($k_x = k_y = 0$) or in a diagonal direction (with $k_x = k_y \neq 0$ and arbitrary k_z).

Let us more closely consider the lattice waves which propagate along the chain molecules ($k_x = k_y = 0$). If the intermolecular interaction with $\alpha \approx \gamma \approx 0$ can be neglected, the relation (6.258 c) transforms into the dispersion relation (6.136) of the stretching modes and the relations (6.258 a, b) into the dispersion relation (6.201) of the bending modes of the isolated linear chain. When considering the intermolecular interaction, (6.258 c) leads to

$$\omega_s = \left[\frac{4(\beta + 4\gamma)}{m}\right]^{1/2} \left|\sin\frac{k_z c}{2}\right| . \tag{6.259 a}$$

for the stretching modes. Because of $\beta \gg \gamma$, this equation is also practically identical with (6.136). For the longitudinally polarized stretching modes propagating along the chain, the intermolecular interaction forces only represent a

6 In the polymer literature, the lattice waves in which the displacement of the chain links occurs perpendicularly to the chain axes are often designated as transversely polarized and the lattice waves in which the displacement of the chain links occurs in the direction of the chain axes as longitudinally polarized. Nevertheless, this does not correspond to the practice in physics according to which the polarization of plane waves is defined by the ratio of the directions of the amplitudes to the direction of propagation.

weak disturbance over the whole range of the first Brillouin zone. In addition, the maximum frequency of these vibrations

$$\omega_{\max} = \left[\frac{4(\beta + 4\gamma)}{m}\right]^{1/2} \tag{6.259b}$$

is at the same time the maximum frequency of the whole spectrum of lattice vibrations. For the frequency (6.259b), however, a point of accumulation no longer occurs in the three-dimensional lattice, as was the case in the one-dimensional model (Fig. 6.10C). The density $\varrho_s(\omega)$ of the stretching modes, which has to be determined *via* (6.238), rather approaches zero with $\omega \to \omega_{\max}$ (see further below). That the interaction parameter α does not play a role in (6.258c) and (6.259) is again due to the specific symmetry of the tetragonal Bravais lattice. But even when this symmetry breaks (*e.g.*, when the chain molecules are no longer perpendicular to the (x, y)-plane), the statement that the dispersion relation (6.136) of the isolated chain is practically maintained for stretching modes propagating along the chains remains valid because of $\beta \gg \alpha$.

From (6.258a, b) with $k_x = k_y = 0$, one obtains

$$\begin{aligned}\omega_{bx} = \omega_{by} &= \left[\frac{1}{m}\left[4\gamma(1 - \cos k_z c) + 4\kappa(1 - \cos k_z c)^2\right]\right]^{1/2} \\ &= \left[\frac{8}{m}\left(\gamma + 2\kappa \sin^2 \frac{k_z c}{2}\right)\right]^{1/2} \left|\sin \frac{k_z c}{2}\right|\end{aligned} \tag{6.260}$$

for the bending modes propagating along the chain molecules. The statement that the intermolecular interaction forces merely represent a weak disturbance for the lattice waves only applies here in the range of sufficiently large wave numbers. If

$$\gamma \ll 2\kappa \sin^2 \frac{k_z c}{2},$$

(6.260) becomes practically identical with the dispersion relation (6.201) of the isolated chain. In the range of quasi-elastic waves with $|k_z| \ll \pi/c$, (6.260) under the condition $\gamma \ll \kappa$ leads to

$$\omega_{bx} = \omega_{by} = \left[\frac{1}{m}(2\gamma + \kappa c^2 k_z^2)\right]^{1/2} c|k_z|. \tag{6.261}$$

If $|k_z|$ is still sufficiently large so that $2\gamma \ll \kappa c^2 k_z^2$ is valid, we get the relation (6.206, 6.202)

$$\omega_{bx} = \omega_{by} = \left(\frac{\kappa}{m}\right)^{1/2} c^2 k_z^2 \sim k^2, \tag{6.262}$$

which is valid for the quasi-elastic bending waves of the isolated chain. For very small wave numbers for which $2\gamma \gg \kappa c^2 k_z^2$, however, we obtain

$$\omega_{bx} = \omega_{by} = \left(\frac{2\gamma}{m}\right)^{1/2} c\,|k_z| \sim k\,. \tag{6.263}$$

The characteristic of quasi-elastic bending waves, the proportionality of the frequency ω to the square k^2 of the wave number, which we particularly emphasized in Sect. 6.4.3 and 6.4.4, is thus cancelled in the field of intermolecular interaction in the limit $k \to 0$. This is a physically necessary result, as the lattice would become unstable with the proportionality (6.262) in the limit $k \to 0$ (see below). Nevertheless, the "anomaly" of the bending modes, as compared to the stretching modes, is not lost. This becomes most evident if one compares the group velocity of the waves (Fig. 6.24):

$$\vec{v} = \left(0, 0, \frac{\partial \omega}{\partial k_z}\right).$$

From (6.259 a) one obtains for the quasi-elastic stretching modes

$$|\vec{v}_s| = |\vec{w}_s| = c\left(\frac{\beta + 4\gamma}{m}\right)^{1/2}, \tag{6.264}$$

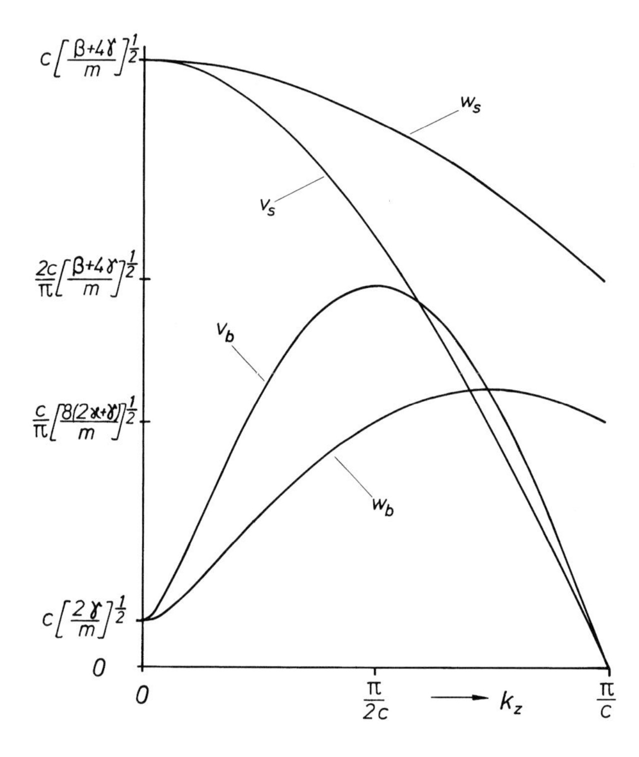

Fig. 6.24. Phase velocity w and group velocity v of the stretching modes (s) and bending modes (b) of the waves which propagate in the chain direction ($k_x = k_y = 0$) in the Stockmayer–Hecht model. According to (6.258), (6.237), and (6.132) with $\beta/\kappa = 10$ and $\gamma/\kappa = 0.03$

from (6.263) for the quasi-elastic bending modes with very small wave numbers

$$|\vec{v}_b| = |\vec{w}_b| = c\left(\frac{2\gamma}{m}\right)^{1/2}. \tag{6.265}$$

In both cases, the group velocity becomes constant and equal to the phase velocity, as is required by Debye's T^3-law (Sect. 6.5.2). The group velocity of the bending waves, however, only depends on the intermolecular interaction parameters (in the case of the tetragonal Bravais lattice on γ, in a more general case on γ and α) and because of $\gamma, \alpha \ll \beta$ thus stays considerably smaller than the group velocity of the stretching waves, which has a β-determined maximum value in the range of small wave numbers. This means, for example, that essentially only the stretching modes contribute to the heat conduction in the chain direction at very low temperatures, whereas the bending modes make the main contribution to the heat capacity, as the smaller group velocity is connected with a higher density of states. The accumulation point of the density of states $\rho_b(\omega)$ of the bending modes, which occurs in the limit $\omega \to 0$ in the isolated chain (Fig. 6.19C), however, is suppressed in the three-dimensional case by (6.265) or shifted towards a finite frequency $\omega(\alpha,\gamma)$ (see Fig. 6.28). Apart from that, the constancy of the group velocity of the bending modes given by (6.265) is restricted to a considerably smaller range of wave numbers as in the case (6.264) of the stretching modes. With increasing wave number (decreasing wavelength), the quadratic term multiplied with the large factor κ rather quickly becomes predominant in (6.261). This is the main reason for the fact that in polymers the validity of Debye's T^3-law is restricted to a very small temperature range.

Equations (6.258) results in

$$\omega_s = \omega_{by} = \left(\frac{8\gamma}{m}\right)^{1/2} \left|\sin\frac{k_x a}{2}\right|, \tag{6.266a}$$

$$\omega_{bx} = \left[\frac{4}{m}(\alpha + 4\gamma)\right]^{1/2} \left|\sin\frac{k_x a}{2}\right| \tag{6.266b}$$

for the lattice waves propagating along the x-axis, *i.e.*, perpendicular to the chain molecules ($k_y = k_z = 0$). The transversely polarized branch ω_{by} of the bending modes is degenerate with the equally transversely polarized branch ω_s of the stretching modes. As a result of the anisotropy of the interaction forces, the maximum frequencies of the waves (6.266) (with $|k_x| = \pi/a$) lie at considerably lower frequencies than the maximum frequencies of the waves (6.259, 6.260) propagating in the z-direction. This is the second reason for the restriction on the validity of Debye's T^3-law to relatively low temperatures in polymers. In the range $|k_x| \ll \pi/a$ of the quasi-elastic waves, $\omega \sim k$ holds for all three branches (6.266). Hence, the characteristic of the quasi-elastic bending waves, the proportionality $\omega \sim k^2$, is completely lost in the bending waves propagating in the x- and y-directions. The group velocity of the quasi-elastic waves (6.266), like the group velocity (6.265), only depends on the intermolecular interaction forces. One can generally say that in the case of very small wave numbers the bending waves behave like the lattice waves in a quasi-isotropic lattice. (In the special case of the tetragonal Bravais lattice, this fact, provided that $\alpha \gg \gamma$, is somewhat blurred as the waves propagating in the z-direction become independent of α).

From (6.258) we generally obtain with $k_x a, k_y a, k_z c \ll \pi$ for the quasi-elastic stretching modes

$$m\omega_s^2 = 2\gamma[(k_x a)^2 + (k_y a)^2] + (\beta + 4\gamma)(k_z c)^2 \tag{6.267 a}$$

and (with the assumption $\gamma, \alpha \ll \kappa$) for the quasi-elastic bending modes

$$m\omega_{bx}^2 = (a + 4\gamma)(k_x a)^2 + 2\gamma[(k_y a)^2 + (k_z c)^2] + \kappa(k_z c)^4 \,, \tag{6.267 b}$$

$$m\omega_{by}^2 = (a + 4\gamma)(k_y a)^2 + 2\gamma[(k_x a)^2 + (k_z c)^2] + \kappa(k_z c)^4 \,. \tag{6.267 c}$$

After introduction of the polar angles in the $\vec{k}$-space (Fig. 6.25),

$$0 \leq \vartheta \leq \pi\,; \quad 0 \leq \varphi \leq 2\pi$$

we get with

$$k_x = k \sin\vartheta \cos\varphi\,, \quad k_y = k \sin\vartheta \sin\varphi\,, \quad k_z = k\cos\vartheta\,,$$

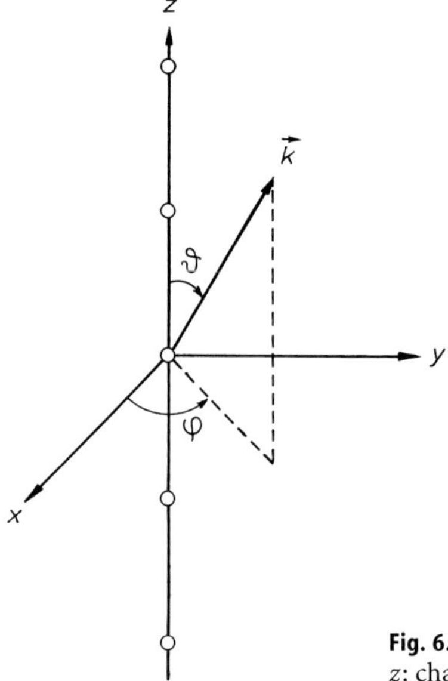

Fig. 6.25. Definition of the polar coordinates $|\vec{k}| = k, \varphi, \vartheta$. z: chain direction

$$m\omega_s^2 = [(\beta + 4\gamma)\, c^2 \cos^2 \vartheta + 2\gamma\, a^2 \sin^2 \vartheta]\, k^2 \,, \tag{6.268a}$$

$$m\omega_{bx}^2 = [(\alpha + 4\gamma)\, a^2 \sin^2 \vartheta \cos^2 \varphi + 2\gamma(a^2 \sin^2 \vartheta \sin^2 \varphi + c^2 \cos^2 \vartheta)]\, k^2 + \kappa c^4 \cos^4 \vartheta k^4 \,, \tag{6.268b}$$

$$m\omega_{by}^2 = [(\alpha + 4\gamma)\, a^2 \sin^2 \vartheta \sin^2 \varphi + 2\gamma(a^2 \sin^2 \vartheta \cos^2 \varphi + c^2 \cos^2 \vartheta)]\, k^2 + \kappa c^4 \cos^4 \vartheta k^4 \,. \tag{6.268c}$$

The dispersion relations of the quasi-elastic waves can thus be represented in the form

$$\omega_s = \left(\frac{C}{m}\right)^{1/2} k \,, \tag{6.269a}$$

$$\omega_{bx} = \left[\frac{1}{m}\,(A_x + Bk^2)\right]^{1/2} k \,, \tag{6.269b}$$

$$\omega_{by} = \left[\frac{1}{m}\,(A_y + Bk^2)\right]^{1/2} k \,, \tag{6.269c}$$

whereby

$$B \equiv \kappa c^4 \cos^4 \vartheta \,,$$

$$C \equiv (\beta + 4\gamma)\, c^2 \cos^2 \vartheta + 2\gamma a^2 \sin^2 \vartheta$$

only depend on the angle ϑ, but

$$A_x \equiv (\alpha + 4\gamma)\, a^2 \sin^2 \vartheta \cos^2 \varphi + 2\gamma\,(a^2 \sin^2 \vartheta \sin^2 \varphi + c^2 \cos^2 \vartheta) \,,$$

$$A_y \equiv (\alpha + 4\gamma)\, a^2 \sin^2 \vartheta \sin^2 \varphi + 2\gamma\,(a^2 \sin^2 \vartheta \cos^2 \varphi + c^2 \cos^2 \vartheta)$$

also depend on φ. All the stretching and bending modes propagating in the (x, y)-plane ($\vartheta = \pi/2$; $B = 0$) are, therefore, not subject to dispersion. They fulfil the proportionality $\omega \sim k$ supposed by Debye. This only holds for all the other bending modes ($\vartheta \neq \pi/2$) as long as the wave numbers are small enough, so that $A/B \gg k^2$. With somewhat higher k-values, Bk^2 should, because of $\gamma, \alpha \ll \kappa$, no longer be neglected with respect to A_x, A_y. The dispersion relations of the bending modes acquire a convex curvature with respect to the k-axis, and signals from these vibrations are subject to dispersion. $\omega \sim k^2$ holds in a pure form when, with increasing k, $A/B \ll k^2$ finally results. The convex curvature occurs the faster the closer ϑ is to 0 or π (chain direction). This dependence of the dispersion relations of the bending modes on the direction of propagation prevents a simple physically justifiable continuum approximation which goes beyond the true Debye range.

The complete density of states of the lattice vibrations of the Stockmayer–Hecht model was determined by Genensky and Newell (1957). The singularities of this spectrum, which occur on the surface of the positive quadrant of the first Brillouin zone, are given in Table 6.2 (maxima, minima and saddle points on the areas ω = const. in the $\vec{k}$-space). For symmetry reasons, the same situations exist in the other quadrants. A comparison with the density of states of the lattice vibrations of the cubic primitive lattice, determined by Newell (1953) on the assumption that $\alpha \geq 16\gamma$, is instructive.

The frequency distribution function $\varrho(\omega)$ of the cubic primitive lattice is represented with $\beta = \alpha$, $\kappa = 0$, $c = a$ in Fig. 6.26. One can see that the true Debye range in which $\varrho(\omega) \sim \omega^2$ is valid, already ends below the lowest singularity $\omega(2\gamma)$ of the spectrum. The frequency

$$\omega(2\gamma) = \left(\frac{8\gamma}{m}\right)^{1/2}$$

is the maximum frequency of the waves propagating in the direction of the principal axes of the crystal, *i.e.*, in the x-, y- and z-directions. If the intermolecular interaction forces are weak van-der-Waals bonds, the frequency $\omega(2\gamma)$ and the characteristic temperature (6.241 b) of the true Debye range lie at extremely low values. The validity of Debye's T^3-law is restricted to very low temperatures if only because of the weak binding forces. Figure 6.26 shows further that the whole spectrum of the lattice waves of the cubic crystal can be averaged quite well by means of a parabola $\varrho(\omega) \sim \omega^2$ (dashed curve in Fig. 6.26; the areas marked with + and – are about equal in size). The Debye formulas in the form (6.250), therefore, also provide a good approximation. However, the fact that the real density $\varrho(\omega)$ is first estimated as being a little too low and then a

Table 6.2. Frequency ω and wave vector (k_x, k_y, k_z) of the singularities of the density of states in the Stockmayer–Hecht model according to Genensky and Newell (1957). Above: stretching modes; below: bending modes. The labels refer to Figs. 6.26–6.28

ω	(k_x, k_y, k_z)	Label
$(8\gamma/m)^{1/2}$	$(\pi/a, 0, 0)$, $(0, \pi/a, 0)$	$\omega(2\gamma)$
$(16\gamma/m)^{1/2}$	$(\pi/a, \pi/a, 0)$	$\omega(4\gamma)$
$(4\beta/m)^{1/2}$	$(\pi/a, \pi/a, \pi/c)$	$\omega(\beta)$
$[4(\beta + 2\gamma)/m]^{1/2}$	$(\pi/a, 0, \pi/c)$, $(0, \pi/a, \pi/c)$	$\omega(\beta, 2\gamma)$
$[4(\beta + 4\gamma)/m]^{1/2}$	$(0, 0, \pi/c)$	$\omega(\beta, 4\gamma)$
$(8\gamma/m)^{1/2}$	$(\pi/a, 0, 0)$, $(0, \pi/a, 0)$	$\omega(2\gamma)$
$[4(\alpha + 2\gamma)/m]^{1/2}$	$(\pi/a, \pi/a, 0)$	$\omega(\alpha, 2\gamma)$
$[4(\alpha + 4\gamma)/m]^{1/2}$	$(\pi/a, 0, 0)$, $(0, \pi/a, 0)$	$\omega(\alpha, 4\gamma)$
$[4(4\kappa + 2\gamma)/m]^{1/2}$	$(0, 0, \pi/c)$	$\omega(4\kappa, 2\gamma)$
$[4(4\kappa + 4\gamma)/m]^{1/2}$	$(\pi/a, 0, \pi/c)$, $(0, \pi/a, \pi/c)$	$\omega(4\kappa, 4\gamma)$
$[4(4\kappa + \alpha)/m]^{1/2}$	$(\pi/a, \pi/a, \pi/c)$	$\omega(4\kappa, \alpha)$
$[4(4\kappa + \alpha + 2\gamma)/m]^{1/2}$	$(\pi/a, 0, \pi/c)$, $(0, \pi/a, \pi/c)$	$\omega(4\kappa, \alpha, 2\gamma)$

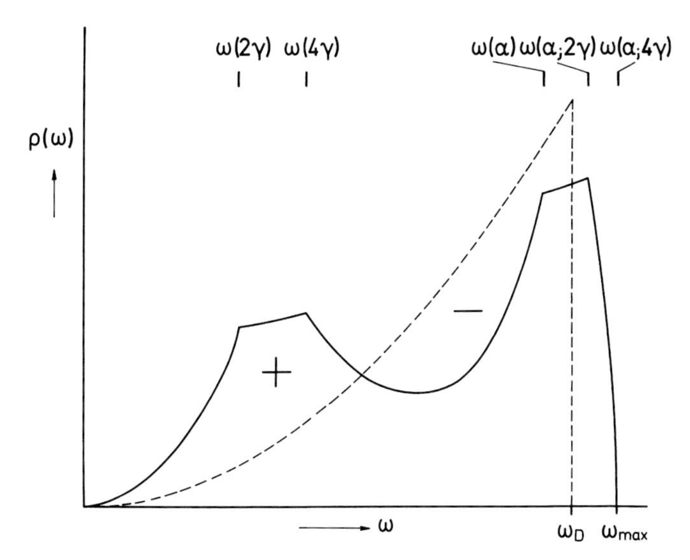

Fig. 6.26. Schematic representation of the frequency density $\varrho(\omega)$ of the oscillations of a primitive cubic lattice according to Newell (1953). Dashed curve: Debye's continuum approximation according to (6.246). The line spectrum above indicates the position of the singularities of Table 6.2 with $\beta = \alpha$, $\kappa = 0$, $a = b = c$. [From Baur (1972)]

little too high by the parabola becomes apparent from a certain temperature dependence of the Θ-temperature.

Let us now consider the density of states $\varrho_s(\omega)$ of the stretching modes of the polymer crystal (Fig. 6.27). From a purely analytical point of view, this spectrum is identical with the spectrum of the cubic crystal, except for the normalization (N instead of $3N$ degrees of freedom). In particular, the true Debye range, provided that γ is equal, extends over the same frequency interval $0 \leq \omega < \omega(2\gamma)$. Because of $\beta \gg \alpha$, however, the maximum frequency of the spectrum given by the maximum frequency of the longitudinally polarized waves propagating in the z-direction lies at considerably higher frequencies in the polymer lattice than in the cubic lattice. The density of states of the stretching modes of the polymer crystal thus appears as a "strongly stretched" density of states of the lattice vibrations of the cubic crystal. This "stretched-out" spectrum can no longer be averaged by means of a parabola. The Debye approximation in the form (6.250) is, therefore, not suitable. The spectrum of the stretching modes of the three-dimensional polymer crystal above the singularities $\omega(2\gamma)$, $\omega(4\gamma)$, on the other hand, is largely similar to the spectrum of the stretching modes of the one-dimensional linear chain (Fig. 6.10C). Hence, the Tarasov-approximation in the form (6.254 a, b) should represent a good approximation for the stretching modes of a polymer crystal. There is, however, a difference in the singularity type. In the three-dimensional case, the singularities $\omega(\beta)$, $\omega(\beta, 2\gamma)$, $\omega(\beta, 4\gamma)$, which lie very close to each other, do not approach infinity and at the

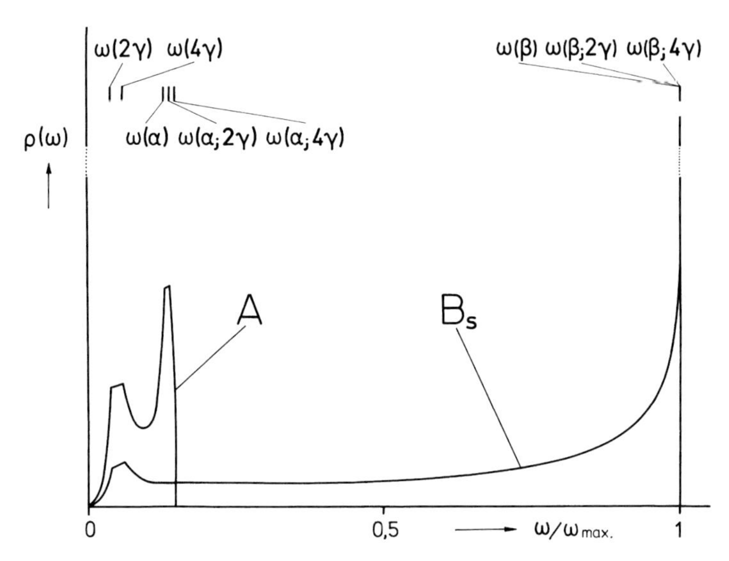

Fig. 6.27. Frequency density $\varrho(\omega)$ of the stretching modes (6.258c) with $\gamma/\beta = 0.002$ according to Genensky, Newell (1957): B_s. In comparison, A, the frequency density (Fig. 6.26) of the cubic crystal with $\alpha = 0.04\beta$. The line spectrum above gives the position of the singularities of Table 6.2. [From Baur (1972)]

maximum frequency $\omega(\beta, 4\gamma)$ even yield a minimum [compare the development of the density of states at the corresponding singularities $\omega(\alpha)$, $\omega(\alpha, 2\gamma)$, $\omega(\alpha, 4\gamma)$ in Fig. 6.26].

More complicated situations are observed in the case of the density of states $\varrho_b(\omega)$ of the bending modes (Fig. 6.28), although one also finds here roughly schematically the spectrum of the isolated linear chain between the α- and κ-singularities (Fig. 6.19C). In particular, the proportionality $\varrho(\omega) \sim \omega^{-1/2}$ is valid between these singularities, as in the isolated chain. The beginning of this proportionality, however, has now been shifted towards a finite frequency $\omega > \omega(\alpha, 4\gamma)$. A completely new spectrum, on the other hand, can be observed below the α-singularities, which is only dominated by intermolecular interaction forces. The true Debye range with $\varrho(\omega) \sim \omega^2$ is restricted to very low frequencies. There is a larger range still below the singularity $\omega(2\gamma)$ in which the proportionality $\varrho(\omega) \sim \omega^{3/2}$ is predominant [Genensky, Newell (1957)]. A less pronounced range, in which the Bloch proportionality $\varrho(\omega) \sim \omega^{1/2}$ is given, exists between $\omega(2\gamma)$ and $\omega(\alpha, 2\gamma)$. Hence, the continuum approximation (6.254c, d) for the bending modes of a polymer crystal provides a useful approximation at higher temperatures where the $D_{1/2}$-terms are a determining factor. In contrast, a transformation of the D_3-term into a $D_{5/2}$-term and finally a $D_{3/2}$-term can be expected with increasing temperature at low temperatures. Hence, the Tarasov approximation in the form (6.252/6.253) no longer has a physical basis. Purely formally, the density of the states $\varrho_b(\omega)$ of the bending modes can also be averaged quite well by a frequency distribution (6.251), particularly when the high-

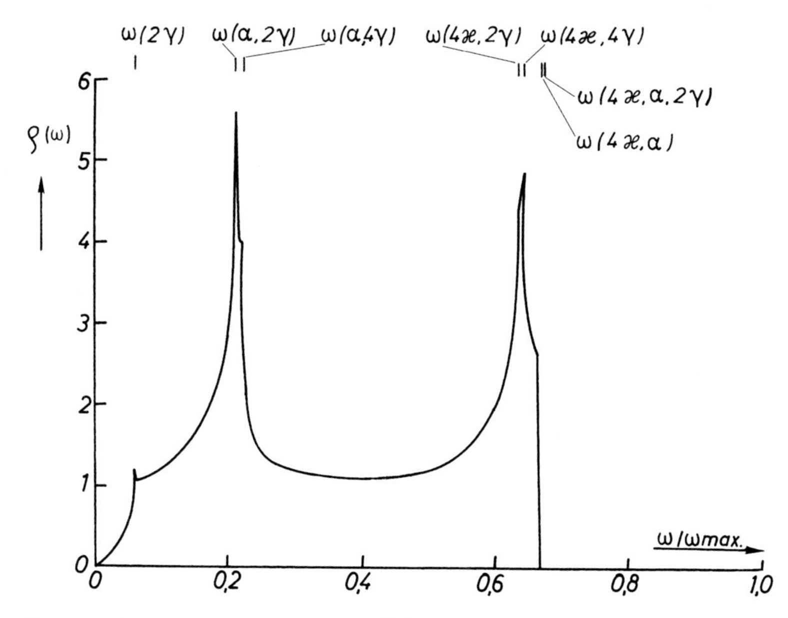

Fig. 6.28. Frequency spectrum $\varrho(\omega)$ of the bending modes (6.258 a/b) with $\gamma/\beta = 0.002$, $\alpha/\beta = 0.04$, and $\kappa/\beta = 0.1$ according to Genensky, Newell (1957). The line spectrum above indicates the position of the singularities of Table 6.2

frequency range of the spectrum (the range of the actual optical modes) is separated and its centres substituted by additional Einstein terms (6.169).

Corresponding to the progression (6.151) or (6.223),

$$s_{\vec{n}}^{j} = \frac{1}{(mN)^{1/2}} \sum_{\vec{k}} \sum_{\sigma} q_{\vec{k},\sigma}\, e_{\sigma}^{j}(\vec{k})\, e^{ik\mathcal{A}\vec{n}} \tag{6.270}$$

holds in a primitive Bravais lattice for the components $s_{\vec{n}}^{j}$ ($j = x, y, z$) of the displacements $\vec{s}_{\vec{n}}$ of the lattice units. The three frequency branches $\omega_\sigma(\vec{k})$ of the Bravais lattice are numbered by $\sigma = 1, 2, 3$. The $\vec{e}_\sigma = (e_\sigma^x, e_\sigma^y, e_\sigma^z)$ are unit vectors with $|\vec{e}_\sigma| = 1$, which characterize the direction of polarization of the waves with the frequency $\omega_\sigma(\vec{k})$. The $q_{\vec{k},\sigma}$ are the normal coordinates of the lattice wave system. They fulfil the mutually independent oscillator equations

$$\ddot{q}_{\vec{k},\sigma} + \omega_\sigma^2(\vec{k})\, q_{\vec{k},\sigma} = 0 \tag{6.271}$$

[compare Eqs. (6.156) and (6.224)]. With these normal coordinates, the Hamiltonian operator (the Hamiltonian function) of the system

$$\mathcal{H} = \sum_{\vec{k}} \sum_{\sigma} \mathcal{H}_{\vec{k},\sigma} \tag{6.272 a}$$

disintegrates into $3N$ additive terms

$$\mathcal{H}_{\vec{k},\sigma} = \frac{1}{2}\,[\dot{q}_{\vec{k},\sigma}\,\dot{q}^{*}_{\vec{k},\sigma} + \omega^2_\sigma(\vec{k})\,q_{\vec{k},\sigma}\,q^{*}_{\vec{k},\sigma}]\,. \tag{6.272b}$$

In the harmonic approximation, the lattice units always oscillate symmetrically around their equilibrium position. That is, the mean values of the displacements vanish:

$$\langle \vec{s}_{\vec{n}} \rangle = \langle s^j_{\vec{n}} \rangle = \langle q_{\vec{k},\sigma} \rangle = 0\,. \tag{6.273}$$

On average, the lattice units are in their predetermined equilibrium positions. The lattice is not capable of a thermal expansion (see Sect. 6.5.4). The mean square displacements are a measure of the thermal fluctuations of the lattice units. According to (6.270), we have

$$\langle (s^j_{\vec{n}})^2 \rangle = \frac{1}{mN} \sum_{\vec{k},\sigma} \sum_{\vec{k}',\sigma'} \langle q_{\vec{k},\sigma}\, q^{*}_{\vec{k}',\sigma'} \rangle\; e^j_\sigma(\vec{k})\, e^j_{\sigma'}(\vec{k}')\, \mathrm{e}^{i(\vec{k}-\vec{k}')\mathcal{A}\vec{n}}\,. \tag{6.274}$$

If $\vec{k}' = \vec{k}$ [or – somewhat more generally – if $\vec{k}'$ fulfils Eq. (6.234)] and $\sigma = \sigma'$, we obtain from (6.272b) with $\dot{q}_{\vec{k},\sigma} = i\omega q_{\vec{k},\sigma}$ and $\dot{q}^{*}_{\vec{k},\sigma} = -i\omega\, q^{*}_{\vec{k},\sigma}$

$$\langle \varepsilon_\sigma(\vec{k}) \rangle = \omega^2_\sigma(\vec{k})\,\langle q_{\vec{k},\sigma}\, q^{*}_{\vec{k},\sigma} \rangle\,.$$

Corresponding to (6.174), the mean energy is given by the expression

$$\langle \varepsilon_\sigma(\vec{k}) \rangle = \hbar\omega_\sigma(\vec{k}) \left[\frac{1}{\mathrm{e}^{\hbar\omega_\sigma/RT} - 1} + \frac{1}{2} \right]. \tag{6.275}$$

If $\vec{k}' \neq \vec{k}$ or $\sigma \neq \sigma'$, we further obtain, as the normal coordinates are mutually independent, with (6.273)

$$\langle q_{\vec{k},\sigma}\, q^{*}_{\vec{k}',\sigma'} \rangle = \langle q_{\vec{k},\sigma} \rangle\, \langle q^{*}_{\vec{k}',\sigma'} \rangle = 0\,.$$

Hence,

$$\langle q_{\vec{k},\sigma}\, q^{*}_{\vec{k}',\sigma'} \rangle = \frac{1}{\omega^2_\sigma(\vec{k})}\, \langle \varepsilon_\sigma(\vec{k}) \rangle\; \delta_{\vec{k}\vec{k}'}\, \delta_{\sigma\sigma'}$$

results for arbitrary $\vec{k}'$, σ'. Insertion of this expression into (6.274) yields

$$\langle (s^j_{\vec{n}})^2 \rangle = \frac{1}{mN} \sum_{\vec{k}} \sum_{\sigma} \frac{1}{\omega^2_\sigma(\vec{k})}\, \langle \varepsilon_\sigma(\vec{k}) \rangle\, [e^j_\sigma(\vec{k})]^2\,. \tag{6.276}$$

For the mean square displacements of the lattice units

$$\langle \vec{s}^{\,2}_{\vec{n}} \rangle = \langle (s^x_{\vec{n}})^2 \rangle + \langle (s^y_{\vec{n}})^2 \rangle + \langle (s^z_{\vec{n}})^2 \rangle\,,$$

one thus obtains

$$\langle \vec{s}^{\,2} \rangle \equiv \langle \vec{s}^{\,2}_{\vec{n}} \rangle = \frac{1}{mN} \sum_{\vec{k}} \sum_{\sigma} \frac{1}{\omega_\sigma^2(\vec{k})} \langle \varepsilon_\sigma(\vec{k}) \rangle \,. \tag{6.277}$$

If one takes (6.277) as a measure of the thermal fluctuations, the terms with $\vec{k} = 0$, $\omega_\sigma(\vec{k}) = 0$ must be cancelled, as they describe a translation of the lattice as a whole – only the relative motions with respect to the centre of mass of the lattice are important in the thermal fluctuations. If the frequency spectrum is practically dense, one can also write in place of (6.277):

$$\langle \vec{s}^{\,2} \rangle = \frac{1}{mN} \int_0^{\omega_{\max}} \frac{1}{\omega^2} \langle \varepsilon(\omega, T) \rangle \, \varrho(\omega) \, \mathrm{d}\omega \,. \tag{6.278}$$

If one assumes the Debye approximation in (6.278), *i.e.*, the proportionality (6.240) with $n = 3$, one gets

$$\langle \vec{s}^{\,2} \rangle \sim \int_0^{\omega_3^D} \langle \varepsilon(\omega, T) \rangle \, \mathrm{d}\omega \,. \tag{6.278}$$

The thermal fluctuations are proportional to the elastic energy contained in the lattice and increase with rising temperature. On the other hand, if $n < 3$, the integral (6.278) becomes divergent. The fluctuations become infinitely large. A very long isolated polymer chain with $n = 1$ or $n = 1/2$ [see Eq. (6.210)] or a crystal composed of very long polymer chains without any intermolecular bonds are, therefore, unstable [in the case of a finite chain with a discrete frequency spectrum, the divergence of (6.278) corresponds to the fact that the fluctuations of one chain end when compared to the other increase with chain length]. A periodic lattice of harmonically vibrating mass points is thus only stable in a three-dimensional interaction field. In addition, the Eqs. (6.268) make it clear that merely taking into account the interaction of direct neighbours is not sufficient in order to stabilize the lattice. Rather, one should at least take into account the interaction of the second neighbours. This is also true when, as in the Stockmayer–Hecht polymer crystal, $\beta \gg \alpha \gg \gamma$ is assumed.

In the case of the Stockmayer–Hecht model, the density $\varrho(\omega)$ of the lattice vibrations splits into two parts

$$\varrho(\omega) = \varrho_s(\omega) + \varrho_b(\omega)$$

which are due to the stretching modes (s) and the bending modes (b). Consequently, according to (6.278), the mean thermal fluctuations also split into two parts

$$\langle \vec{s}^{\,2} \rangle \equiv \langle \vec{s}^{\,2}_s \rangle + \langle \vec{s}^{\,2}_b \rangle \,. \tag{6.279}$$

The long-wave-length lattice vibrations with low frequencies particularly contribute to the fluctuations because of the denominator ω^2 in (6.278). In this frequency range, however, the density of the bending modes is predominant (see Fig. 6.27/6.28). Hence, the main contribution to the thermal fluctuations of the lattice units of a polymer crystal is due to the bending modes. In the case of elastic scattering of X-rays in polymer crystals, for example, especially those lattice vibrations should be effective in which the chain links are displaced perpendicularly to the chain molecules. The cause for this effect, however, is not found in the anisotropy of the polymer crystals but in the "anomaly" of the bending modes (the high density at low frequencies) [see Baur (1972), (1972a)].

6.5.4 Anharmonic Effects

In the harmonic approximation, according to the Eqs. (6.273), all lattice units assume their equilibrium positions in a time average. The lattice is not capable of a thermal expansion. With the coefficients of thermal expansion α_i, the difference between the heat capacities C_σ and C_ε or C_p and C_V [compare Eqs. (5.29) and (5.50)] and the difference between the isothermal and adiabatic elastic coefficients [Eq. (5.30)] also vanish. The harmonic solid reacts purely energy-elastically (Sect. 5.3). The elastic coefficients do not depend on temperature. These properties apply for most crystals, but only at lower or rather low temperatures. The amplitudes of the displacements $\vec{s}_{\vec{n},\mu}$ of the lattice units increase at higher temperatures, so that the higher terms become successively more important in the progression (6.124). The higher terms destroy the vibrational symmetry around the equilibrium positions, so that the mean displacements (6.273) no longer vanish. Thermal expansion is (at least in the case of ideal crystals) the best indicator for the occurrence of anharmonicity in the vibrations.

With the complete Taylor-series (6.124), the Hamiltonian function of the system of lattice units is

$$\mathcal{H} = \mathcal{H}_{\mathrm{kin}} + \Phi_0 + \Phi_1 + \Phi_2 + \Phi_3 + \Phi_4 + \ldots . \tag{6.280}$$

The term $\Phi_1 \neq 0$ should generally, also in the classical case, be included, because at higher temperatures the equilibrium positions of the lattice units are no longer determined by a minimum in the potential energy, but by a minimum in the free energy. Using (6.280), one can easily also represent the partition function and the free energy of the system by means of a suitable progression (see Sect. 6.1). The Hamiltonian function $\mathcal{H}_0$ of the harmonic approximation was

$$\mathcal{H}_0 = \mathcal{H}_{\mathrm{kin}} + \Phi_0 + \Phi_2 . \tag{6.281}$$

When considering the harmonic approximation as the zeroth approximation of a perturbation expansion, the perturbation function is given by

$$\mathcal{H}' = \Phi_1 + \Phi_3 + \Phi_4 + \ldots . \tag{6.282}$$

By means of a transformation corresponding to (6.150, 6.170), the classical displacement coordinates $\vec{s}_{\vec{n},\mu}$ can be transformed into quantum field creation and annihilation operators $b^*_{\vec{k},\sigma}$, $b_{\vec{k},\sigma}$. The Hamiltonian functions (6.280–6.282) then transform into operators in which the terms Φ_i appear as sums of products with i factors $b^*_{\vec{k},\sigma}$ or $b_{\vec{k},\sigma}$. While the operator (6.281) leads to an "ideal gas" of completely non-interacting phonons (Sect. 6.4.1), the perturbation operator (6.282) now contains the mutual interaction of the phonons. For the energy E_n of the perturbed states, the quantum-mechanical perturbation theory [*e.g.*, see Dirac (1958)] yields the progression

$$E_n = E_n^0 + \langle n|\mathcal{H}'|n\rangle + \sum_{n' \neq n} \frac{\langle n|\mathcal{H}'|n'\rangle \langle n'|\mathcal{H}'|n\rangle}{E_n^0 - E_{n'}^0} + \ldots . \tag{6.283}$$

Here, E_n^0 denotes (with a somewhat reduced indexing) the eigenvalues (energy levels) and $|n\rangle$ the eigenfunctions of the unperturbed operator (6.281) of the harmonic approximation. As the system is not exposed to external forces and is supposed to be in an internal equilibrium, only processes in which the initial and final states are identical can occur in (6.283). A product of an odd number of creation and annihilation operators $b_{\vec{k},\sigma}$, however, never leads back to the initial state. Therefore, correction terms of the first order $\langle n|\Phi_i|n\rangle$ with an odd subscript i cannot appear in (6.283). In a first approximation, the Φ_i with an odd subscript only contribute to the correction terms of the second order. One can see that the correction terms of the second order with an odd subscript i are of about the same order of magnitude as the correction terms of the first order with an even subscript $i + 1$ (every 4-phonon process, for example, can be represented by two consecutive 3-phonon processes). In a first approximation, an expression in the form

$$F = F_0 + F_1(\Phi_1^2) + F_{13}(\Phi_1\Phi_3) + F_3(\Phi_3^2) + F_4(\Phi_4) \tag{6.284}$$

results for the free energy of the perturbed system [Ludwig (1967)]. Corresponding to (6.166), F_0 is the free energy of the harmonic approximation, *i. e.*,

$$F_0 = \Phi_0 + \frac{\hbar}{2} \sum_{\vec{k}} \sum_{\sigma} \omega_\sigma(\vec{k}) + RT \sum_{\vec{k}} \sum_{\sigma} \ln\left[1 - \mathrm{e}^{-\hbar\omega_\sigma(\vec{k})/RT}\right] \tag{6.285a}$$

or if the frequency spectrum of the individual branches $\omega_\sigma(\vec{k})$ is practically dense

$$F_0 = \Phi_0 + \int_0^{\omega_{\max}} \left[\frac{\hbar\omega}{2} + RT \ln\left(1 - \mathrm{e}^{-\hbar\omega/RT}\right)\right] \varrho(\omega)\,\mathrm{d}\omega . \tag{6.285b}$$

F_0 is, however, not completely independent of the correction terms in (6.284). The equilibrium positions $\vec{r}^{\,o}_{\vec{n},\mu}$ and equilibrium distances of the lattice units are now determined by the minimum in F. If the correction terms in F induce a change in these distances with temperature, F_0 must then also be referenced to each of the changed distances. The frequencies ω in (6.285) are then no longer constant, but depend on the variable equilibrium distances. The binding forces between the lattice units, and thus also the vibrational frequencies, usually decrease with increasing lattice distances.

The anharmonic effects are already somewhat considered if one neglects the perturbation terms (6.282), *i.e.*, sets $F = F_0$, but regards Φ_0 and $\omega_\sigma(\vec{k})$ in F_0 as functions of the thermal deformation. This approximation is called the quasi-harmonic approximation. Let us now consider an isotropic medium (or a cubic crystal) for which the Debye approximation.

$$\varrho(\omega) = 9N\omega^2/(\omega_3^D)^3$$

is valid [see Eqs. (6.246) and (6.248)]. We obtain according to (6.285b)

$$F = F_0 = \Phi_0 + 3NRT\,\varphi\left(\frac{\Theta_3}{T}\right) \tag{6.286a}$$

with

$$\varphi\left(\frac{\Theta_3}{T}\right) \equiv 3\left(\frac{T}{\Theta_3}\right)^3 \int_0^{\Theta_3/T} \left[\frac{x}{2} + \ln(1 - \mathrm{e}^{-x})\right] x^2\,\mathrm{d}x\,. \tag{6.286b}$$

In the quasi-harmonic approximation, Φ_0 and $\Theta_3 = \hbar\omega_3^D/R$ have to be regarded as functions of the volume:

$$\Phi_0 = \Phi_0(V)\,, \quad \Theta_3 = \Theta_3(V)\,. \tag{6.287}$$

Here, V only refers to the volume given by the thermal deformation, whereas external influences, such as the atmospheric pressure or, in addition, the vapour pressure, are disregarded. One can observe further that for every function $\varphi(\Theta_3/T)$

$$\Theta_3\left(\frac{\partial\varphi}{\partial\Theta_3}\right)_T = -\,T\left(\frac{\partial\varphi}{\partial T}\right)_{\Theta_3} \tag{6.288}$$

is generally always valid. As equations of state (2.3) and (2.5), (6.286) with (6.287) leads to

$$-\,S(T,V,N) = \left(\frac{\partial F}{\partial T}\right)_{V,N} = 3RN\varphi + 3RNT\left(\frac{\partial\varphi}{\partial T}\right)_{V,N}, \tag{6.289}$$

$$-\,p(T,V,N) = \left(\frac{\partial F}{\partial V}\right)_{T,N} = \frac{\partial\Phi_0}{\partial V} + 3RNT\left(\frac{\partial\varphi}{\partial\Theta_3}\right)_{T,N}\frac{\partial\Theta_3}{\partial V}\,. \tag{6.290}$$

With

$$3RNT\varphi = F - \Phi_0 \quad \text{and} \quad F + TS = U$$

[see Eqs. (6.286a) and (2.18)], we obtain from (6.289)

$$3RNT^2\left(\frac{\partial\varphi}{\partial T}\right)_{V,N} = \Phi_0 - U$$

and with this because of (6.288) from (6.290)

$$\left(\frac{\partial F}{\partial V}\right)_{T,N} = \frac{\partial \Phi_0}{\partial V} + (U - \Phi_0)\frac{\partial \ln \Theta_3}{\partial V}\,.$$

As the coefficient of thermal expansion is [in the F-representation according to (2.45) and (2.58)]

$$\alpha = -\,\kappa_T\,\frac{\partial^2 F}{\partial T\,\partial V}\,, \tag{6.291}$$

we obtain further

$$\alpha = -\,\kappa_T\left(\frac{\partial U}{\partial T}\right)_{V,N}\frac{\partial \ln \Theta_3}{\partial V} = -\,\frac{\kappa_T C_V}{V}\,\frac{\partial \ln \Theta_3}{\partial \ln V}$$

[see Eq. (2.52)]. The expression:

$$\gamma_G \equiv -\,\frac{\partial \ln \Theta_3}{\partial \ln V}\,, \tag{6.292a}$$

which may actually still be dependent on V, is designated as the Grüneisen constant. The thermal expansion of an isotropic solid in the quasi-harmonic Debye approximation is thus given by

$$\alpha = \gamma_G\,\frac{\kappa_T C_V}{V}\,. \tag{6.293}$$

In this approximation, the coefficient of thermal expansion is proportional to the heat capacity. In particular, the thermal expansion coefficient should also obey Debye's T^3-law at low temperatures and be α = const. at higher temperatures. For the difference between the isobaric and isochoric heat capacities, one gets with (6.293) from (2.60)

$$C_p - C_V = \gamma_G^2\,\frac{T\kappa_T C_V^2}{V}\,. \tag{6.294}$$

If all the frequencies ω of the lattice vibrations are equally dependent on the volume, one can also write

$$\gamma_G = -\,\frac{\partial \ln \omega}{\partial \ln V} \tag{6.292b}$$

in place of (6.292a) and, in addition, in the case of a cubic lattice with the lattice constant a and the volume $V = N\,a^3$

$$\gamma_G = -\frac{1}{3}\frac{\partial \ln \omega}{\partial \ln a}\,. \tag{6.292c}$$

Proceeding from $T = 0$, (6.293) often describes quasi-isotropic media quite well over a larger temperature range. Here, the Grüneisen constant is in an order of magnitude of 2 [Born, Huang (1954)]. If one also proceeds from (6.293) in the case of more complicated structures, γ_G can, of course, only be interpreted as a parameter averaged over all directions of deformation, directions of propagation, and frequencies, which can then become dependent not only on the volume but also on the temperature [with regard to polymers see *e.g.* Hartwig (1994)]. If the frequencies exhibit a differing dependence on the volume, one obtains from (6.285) somewhat more generally for isotropic media:

$$\frac{\partial F}{\partial V} = \frac{\partial \Phi_0}{\partial V} + \frac{1}{V}\sum_{\vec{k}}\sum_{\sigma}\left[\frac{1}{2}\hbar\omega_\sigma(\vec{k}) + \frac{\hbar\omega_\sigma(\vec{k})}{(e^{-\hbar\omega_\sigma(\vec{k})/RT}-1)}\right]\frac{\partial \ln \omega_\sigma(\vec{k})}{\partial \ln V}$$

$$= \frac{\partial \Phi_0}{\partial V} - \frac{1}{V}\sum_{\vec{k}}\sum_{\sigma}\gamma_G^\sigma(\vec{k})\,\langle \varepsilon_\sigma(\vec{k})\rangle\,.$$

Here,

$$\gamma_G^\sigma(\vec{k}) \equiv -\frac{\partial \ln \omega_\sigma(\vec{k})}{\partial \ln V} \tag{6.295a}$$

is the Grüneisen parameter assigned to each individual mode $\omega_\sigma(\vec{k})$. The energy contribution of the phonons of the mode $\omega_\sigma(\vec{k})$ is $\langle \varepsilon_\sigma(\vec{k})\rangle$, given by (6.173/6.174) or (6.275). One gets *via* (6.291) for the coefficient of thermal expansion:

$$\alpha = \frac{\kappa_T}{V}\sum_{\vec{k}}\sum_{\sigma}\gamma_G^\sigma(\vec{k})\,c_V^\sigma(\vec{k})\,, \tag{6.295b}$$

where $c_V^\sigma(\vec{k})$ describes the contribution of the phonons to the heat capacity according to (6.175). Hence, the vibrational modes individually provide an additive contribution of the form (6.293) to the thermal expansion coefficient. In the case of an arbitrary, for example anisotropic, lattice, one has to proceed from (5.18) and (5.26). The coefficients of linear thermal expansion are then (in the F-representation) given by

$$\alpha_i = -\frac{1}{V}\sum_j \kappa_{ij}\frac{\partial^2 F}{\partial T\,\partial \varepsilon_j}$$

(ε_j: components of thermal deformation). *Via* (6.285) one obtains the quasi-harmonic approximation

$$\alpha_i = \frac{1}{V} \sum_j \sum_{\vec{k}} \sum_\sigma \kappa_{ij}\, \gamma_j^\sigma(\vec{k})\, c_V^\sigma(\vec{k}) \,. \tag{6.296a}$$

The Grüneisen parameters

$$\gamma_j^\sigma(\vec{k}) = -\frac{\partial \ln \omega_\sigma(\vec{k})}{\partial \varepsilon_j} \tag{6.296b}$$

now form, like the components of the deformation tensor ε_j and the coefficients of thermal expansion α_i, a symmetric tensor of the second rank (see Sects. 5.1 and 5.2).

The smaller the masses of the lattice units and the weaker the binding forces between the lattice units, the lower the temperatures at which anharmonic effects generally occur. Hence, one can expect with regard to polymers that particularly those lattice vibrations which stress only the weak intermolecular bonds contribute to the sum (6.296). In the Stockmayer–Hecht model (Sect. 6.5.3), these are essentially the vibrations which fill the frequency spectrum up to the α-singularities (Table 6.2), *i.e.*, once more above all (but not exclusively) the low-frequency bending modes. In addition, the low-frequency torsional modes (Sect. 6.4.4) certainly also contribute to the anharmonicity. It should be noted that the melting points of polymers are relatively high when compared to the melting points of low-molecular-mass substances with the same (weak) intermolecular binding forces. This is due to the fact that the lattice units of polymers remain bound in chains while melting, and, therefore, with approximately the same enthalpy of melting per lattice unit, the entropy of melting of the polymers is lower than the entropy of melting of low-molecular-mass substances. The relatively high melting point of polymers, relative to the weak intermolecular bonding, leads to the result that the Θ-temperatures $\Theta = \hbar\omega_{\max}/R$ are all below the melting point for the maximum frequencies of the vibrations which only strain the intermolecular bonds (see the example in Fig. 6.29). In polymers, anharmonic effects can already be expected at relatively low temperatures (*e.g.*, in polyethylene from about 70 K upwards; Fig. 6.29). One can probably also expect that, in addition to the perturbation terms Φ_1^2, Φ_3^2, Φ_4 of the first approximation (6.284), the higher perturbation terms with Φ_1^4, Φ_3^4, Φ_4^2, Φ_5^2, Φ_6 become apparent at higher temperatures. In this respect, the "anomalies" of the bending modes also play a decisive role with regard to the phonon interaction [Baur (1971), (1972)].

In the approximation (6.284) [see Ludwig (1967) for the following], the heat capacity usually increases proportionally to the temperature beyond the Dulong–Petit value (Fig. 6.30). The lattice constants also usually increase proportionally to the temperature while the elastic constants decrease proportionally to the temperature. In individual cases, this proportionality to the temperature can definitely have an inverted sign. A lattice constant which decreases

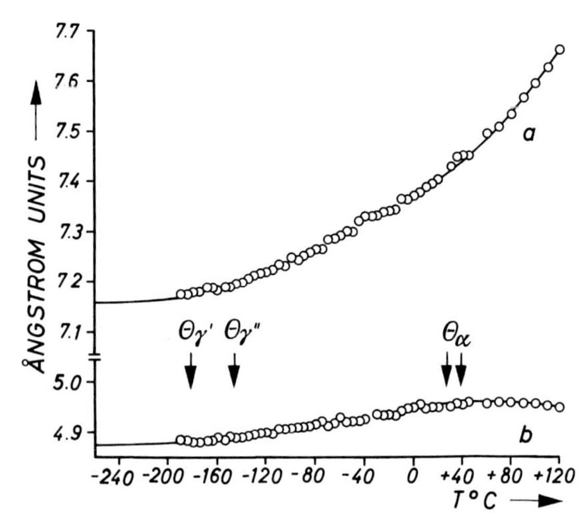

Fig. 6.29. Lattice distances a, b perpendicular to the chain direction of linear, gradually cooled polyethylene as a function of the temperature T. Measured points according to Hendus (unpublished 1962). No change in distances was noticed in the chain direction. One can, using as a basis the variation of a, estimate that the first-order anharmonic terms $\Phi_1^2 \ldots \Phi_4$ become noticeable from about $T = -200\,°C$ and the anharmonic terms $\Phi_1^4 \ldots \Phi_6$ from about $T = 0\,°C$. The arrows indicate the position of the temperatures $\Theta = \hbar\omega/R$ of the singularities $\omega(2\gamma)$, $\omega(4\gamma)$, $\omega(\alpha, 2\gamma)$, $\omega(\alpha, 4\gamma)$ from Table 6.2. $\beta = 208$ N/m, $\alpha = 0.04\,\beta$, $\gamma = 0.002\,\beta$, and $m = 23.4 \cdot 10^{-27}$ kg were assumed. [From Wunderlich, Baur (1970)]

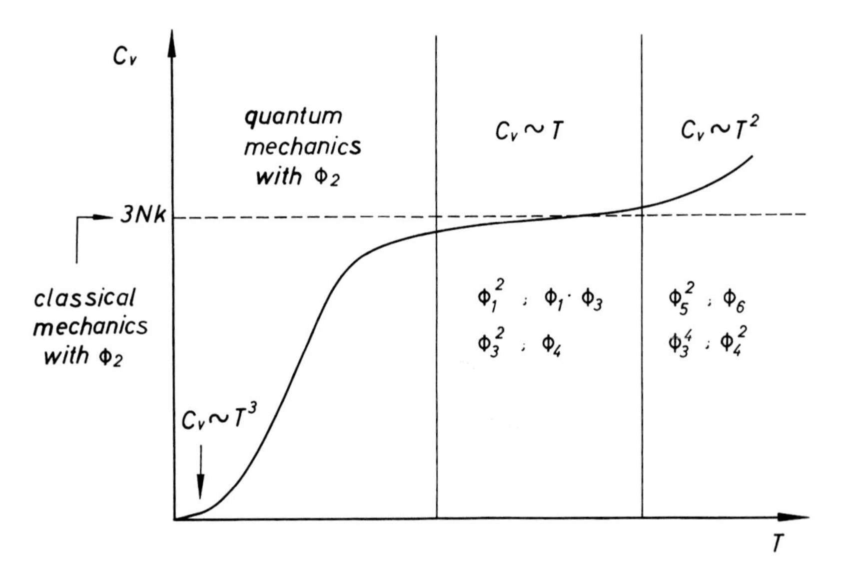

Fig. 6.30. Schematic representation of the contribution of the lattice vibrations to the heat capacity of a solid as a function of the temperature. The dashed line corresponds to the Dulong–Petit law (6.227). The solid curve agrees with the quantum mechanical treatment with a successive efficacy of the terms in (6.280). [From Wunderlich, Baur (1970)]

with increasing temperature is, of course, linked with a negative thermal expansion coefficient or a negative Grüneisen parameter. The higher perturbation terms $\Phi_1^4 \ldots \Phi_6$ cause a decrease or increase of the given quantities with the square of the temperature. There are additional perturbations in non-ideal crystals, for example, caused by the elastic or inelastic scattering of phonons at boundary layers or defects (such as impurities, occupied interstices, vacancies, or dislocations). The creation of conformational imperfections, which increases with rising temperature, is of importance in polymers (see Sect. 8.3). In addition, defects can already cause a considerable disturbance in the phonon spectrum obtained with the harmonic approximation (see the comments at the end of Sect. 6.4.2).

CHAPTER 7

Systems with Macroscopically Relevant Internal Degrees of Freedom

Equilibrium thermodynamics only retains its validity as long as the molecular internal degrees of freedom of the considered system do not attain a primary macroscopic relevance. This condition is fulfilled if the macroscopic perturbation of the system occurs so slowly (quasi-static) that the statistical equilibrium of the internal degrees of freedom is not disturbed during the perturbation. The system is then in a so-called internal equilibrium. In this case, the internal degrees of freedom are not explicitly noticeable, but only become globally apparent, for example, by a change dQ of the heat content of the system. This condition, however, is also fulfilled if the macroscopic perturbation occurs so quickly that one or more of the molecular internal degrees of freedom do not have time to react to the perturbation. The degrees of freedom concerned then appear frozen with respect to the macroscopic perturbation. Provided that the other part of the internal degrees of freedom is still able to restore equilibrium practically instantaneously during the perturbation (so that we can still talk about a thermodynamic equilibrium), the system is in a so-called arrested equilibrium with respect to the frozen degrees of freedom for the duration of the perturbation. Arrested (frozen) internal degrees of freedom do not exhibit any macroscopic effects. They can be ignored as long as different arrested states do not have to be compared with each other or with other states. How fast a system must be perturbed so that an internal degree of freedom appears arrested with respect to the perturbation naturally depends on the degree of freedom itself, but also on the physical environment in which it is embedded. Some internal degrees of freedom can, in certain temperature and pressure ranges, still be regarded as being at least approximately arrested, even in the case of a quasi-static perturbation. Two typical examples are the degrees of freedom of the diffusive translational motion of the molecules in a glass and the degree of advancement of the reaction in an oxygen and hydrogen gas mixture.

The validity of the equilibrium thermodynamics is lost if, on the other hand, the macroscopic perturbation rate reaches the order of magnitude of the rate with which any one of the molecular internal degrees of freedom strives to adjust to the statistical equilibrium. The internal or arrested equilibrium is then disturbed during the perturbation. The perturbation process turns into an irreversible sequence of non-equilibrium states whose character is determined by the primary macroscopic relevance of the internal degree of freedom. As already mentioned (Chap. 1 and Sect. 2.4), it is particularly the degrees of freedom of the conformational isomerism and the degrees of freedom of the mutual arrangement or orientation of the macromolecules that very often lead to such non-equilibrium processes in polymers.

Several theories are available for a macroscopic-phenomenological description of thermodynamic non-equilibrium processes: the now classical "thermodynamics of irreversible processes" [*e.g.*, compare Prigogine (1961); De Groot, Mazur (1962); Baur (1984)], the so-called "extended irreversible thermodynamics" [Jou, Casas-Vásquez, Lebon (1993)], the "rational thermodynamics" [*e.g.*, see Truesdell (1969)] and the so-called "entropy-free thermodynamics" [see Meixner (1969); there also a comparison of the individual theories]. In the following, we will confine ourselves to a description of the phenomena within the framework of the relatively simple classical thermodynamics of irreversible processes. With regard to the limits of validity of this theory from the standpoint of the kinetic theory of gases and statistical mechanics, for example, see Meixner, Reik (1959); De Groot, Mazur (1962); Jou, Casas-Vásquez, Lebon (1993); Kubo, Toda, Hashitsume (1985).

7.1 Preconditions, Fundamental Equations and Equations of State

In order to describe the mode of action of macroscopically relevant internal degrees of freedom, we assume a closed homogeneous system at rest which is not exposed to locally variable external force fields. "Closed" means that there should be no mass exchange between the system and its surroundings, so that the mass of the system remains constant and disappears as a variable. We refer all extensive quantities to the constant mass of the system. These specific quantities are designated with the respective small letters (see the annotation on p. 13). "Homogeneous" means above all that the system should always be in thermal and mechanical equilibrium with respect to the external variables (see Sect. 4.1). In order to maintain homogeneity, the system must be capable of adjusting its thermal and mechanical equilibrium to all changes practically instantaneously. If this is not the case, the processes, which we will consider in the following, are additionally superimposed by irreversible transport processes (energy transport, momentum transport, *etc.*). In a system at rest which is not exposed to locally variable external force fields (*e.g.*, the force of gravity), the internal energy U is identical with the total energy of the system, for which the conservation law of energy is valid. In the following, we can thus assume

$$d_i U = 0 \tag{7.1}$$

(see Sect. 2.2).

To simplify matters we will further confine ourselves to scalar mechanical deformations such as simple shearing, free uniaxial stretching, or hydrostatic deformation. Referring back to Chap. 2, we will usually use the hydrostatic pressure p or the specific volume v as representatives of this individual mechanical degree of freedom. In place of these variables, it is, of course, also possible to introduce scalar stress variables σ and scalar deformation variables ε *via*

$$p - p_0 \rightarrow -\sigma \quad \text{and} \quad \frac{v - v_0}{v_0} \rightarrow \varepsilon \tag{7.2}$$

(see Sect. 7.6). In more general cases, the static stress tensor has to be used in place of the pressure and the deformation tensor in place of the specific volume (Chap. 5). In inhomogeneous, flowing media, the static stress tensor must be replaced by the dynamic pressure or stress tensor which may still contain the frictional or viscous pressure tensor (Sect. 7.6).

In order to describe closed, homogeneous systems at rest with macroscopically relevant internal degrees of freedom, the classical thermodynamics of irreversible processes proceeds from the following assumptions:

(V1) Macroscopic so-called internal variables $\zeta_1 \dots \zeta_n$ can unambiguously be assigned to the macroscopically relevant internal degrees of freedom $1 \dots n$.

(V2) The Gibbs formalism of equilibrium thermodynamics also remains valid in non-equilibrium. The Gibbs functions, however, must be supplemented by the internal variables $\zeta_1 \dots \zeta_n$ as representatives of the additional degrees of freedom which the system possesses at non-equilibrium.

(V3) There are macroscopic dynamic laws which enable a description of the time dependence of the mutually independent internal variables.

According to (V2), the Gibbs fundamental equation, for example, is in the F-representation

$$f = f(T, v, \zeta_1 \dots \zeta_n)\,. \tag{7.3a}$$

As this equation should be valid at any time during the development of a non-equilibrium process, one can also write

$$f = f\,[T(t), v(t), \zeta_1(t) \dots \zeta_n(t)]\,. \tag{7.3b}$$

Hence, in the thermodynamics of irreversible processes one can only describe processes in which solely the independent variables directly depend on the time t. All the other thermodynamic quantities, in particular the Gibbs functions themselves, are only indirectly dependent on the time *via* the independent variables. Moreover, (V2) excludes that the Gibbs functions still depend on other variables, for example, on the rates $\dot{\zeta}_i$ of the internal variables [*e.g.*, see Meixner (1969), Jou et al. (1993)]. According to (7.3), the temporal change of the specific free energy f in non-equilibrium is given by

$$\frac{\mathrm{d}f}{\mathrm{d}t} = \left(\frac{\partial f}{\partial T}\right)_{v,\boldsymbol{\zeta}} \frac{\mathrm{d}T}{\mathrm{d}t} + \left(\frac{\partial f}{\partial v}\right)_{T,\boldsymbol{\zeta}} \frac{\mathrm{d}v}{\mathrm{d}t} + \sum_i \left(\frac{\partial f}{\partial \zeta_i}\right)_{T,\, v,\, \zeta_{k \neq i}} \frac{\mathrm{d}\zeta_i}{\mathrm{d}t}\,. \tag{7.4}$$

The index $\boldsymbol{\zeta}$ in the partial derivatives denotes the set $\{\zeta_1 \dots \zeta_n\}$ and means that all ζ_i must be kept constant when differentiating.

The fundamental question of every "non-equilibrium thermodynamics" is what should further be understood by the terms "entropy" and "temperature" which are actually only defined for the equilibrium. If in (7.3) one supposes the internal variables to be fixed at definite values $\zeta_1 = \zeta_1^0 \dots \zeta_n = \zeta_n^0$, the surface spanned by the Gibbs function

$$f = f(T, v, \zeta_1^0 \dots \zeta_n^0) \tag{7.5}$$

in the space (f, T, v) contains all possible equilibrium states of the system arrested under the condition $\zeta_1^0 \dots \zeta_n^0 = \text{const.}$ Sequences of non-equilibrium states, however, which are described by (7.3) or (7.4), successively traverse a stack of surfaces of the type (7.5) (see Fig. 7.1). Hence, (V2) also has the consequence that the non-equilibrium processes to be described can always be mapped onto a temporally ordered sequence of arrested equilibrium states. In particular, the entropy

$$s(T, v, \zeta_1 \dots \zeta_n) \equiv -\left(\frac{\partial f}{\partial T}\right)_{v, \zeta}$$

[which, according to (V2), corresponding to Eq. (2.3), is attributed to the system in non-equilibrium] is identical with the uniquely defined entropy of the equilibrium states arrested under the condition $\zeta_1 \dots \zeta_n = \text{const.}$ Transformation of (7.3) into the U-representation

$$u = u(s, v, \zeta_1 \dots \zeta_n)$$

leads to the temperature of the system [see Eq. (2.20)]

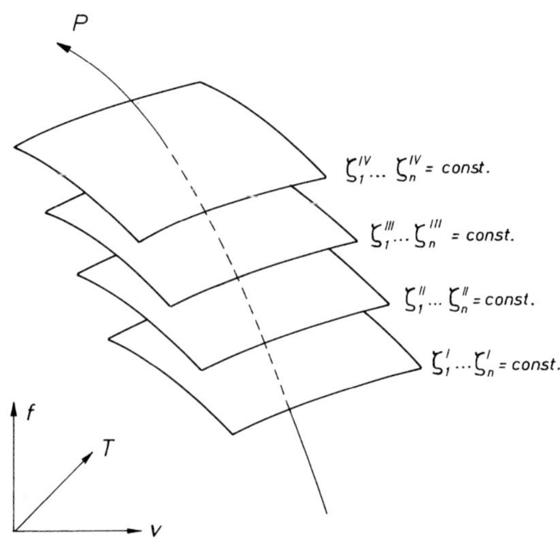

Fig. 7.1. Surfaces $f(T, v)$ of arrested equilibrium states according to (7.5). P: path of a non-equilibrium process according to (V1) and (V2).

$$T(s, v, \zeta_1 \ldots \zeta_n) \equiv \left(\frac{\partial u}{\partial s}\right)_{v,\zeta}$$

The temperature in non-equilibrium is also identified with the uniquely defined temperature of the equilibrium states arrested with respect to the relevant internal degrees of freedom. The basis of the thermodynamics of irreversible processes in homogeneous systems is thus a multiple family of equilibrium states in which the macroscopically relevant internal degrees of freedom can be ignored. The additional effect of the relevant internal degrees of freedom is then described on this basis.

The equations of state in the F-representation corresponding to the equations (2.3)–(2.5) are now

$$-\left(\frac{\partial f}{\partial T}\right)_{v,\zeta} \equiv s = s(T, v, \zeta_1 \ldots \zeta_n)\,, \tag{7.6}$$

$$-\left(\frac{\partial f}{\partial v}\right)_{T,\zeta} \equiv p = p(T, v, \zeta_1 \ldots \zeta_n)\,, \tag{7.7}$$

$$-\left(\frac{\partial f}{\partial \zeta_i}\right)_{T,v,\zeta_{k\neq i}} \equiv a_i = a_i(T, v, \zeta_1 \ldots \zeta_n)\,. \tag{7.8}$$

Apart from the specific entropy s and the hydrostatic pressure p there are additional variables of state a_i which are designated as affinities as in the theory of chemical reactions. The affinities prove to be the driving forces which set and keep irreversible processes going until the internal equilibrium is reached (Sect. 7.2). When T, $v = \text{const.}$, (2.82) leads to

$$a_i^e(T, v, \zeta_1 \ldots \zeta_n) = 0\,, \quad (i = 1 \ldots n) \tag{7.9}$$

for the internal equilibrium "e". Hence, the affinities disappear in internal equilibrium. The internal variables $\zeta_1 \ldots \zeta_n$, which are independent of the thermal and mechanical degrees of freedom in non-equilibrium, remain variable in the internal equilibrium but become dependent variables. The n functional relationships

$$\zeta_i = \zeta_i^e(T, v) \tag{7.10}$$

can be derived (at least in principle) from the n equations (7.9). Moreover, according to (2.83), the symmetric matrix

$$\begin{pmatrix} \left(\frac{\partial a_1}{\partial \zeta_1}\right)_e & \cdots & \left(\frac{\partial a_n}{\partial \zeta_1}\right)_e \\ \vdots & & \vdots \\ \left(\frac{\partial a_1}{\partial \zeta_n}\right)_e & \cdots & \left(\frac{\partial a_n}{\partial \zeta_n}\right)_e \end{pmatrix}$$

must be negative definite in a stable or metastable internal equilibrium.

With (7.6–7.8), one can write

$$\dot{f} = -s\dot{T} - p\dot{v} - \sum_i a_i \dot{\zeta}_i \tag{7.11}$$

for the temporal change (7.4) of the specific free energy. On the other hand, (2.68) with $N = \text{const.}$ leads to

$$\mathrm{d}f = -s\,\mathrm{d}T - p\,\mathrm{d}v - T\,\mathrm{d}_i s$$

or

$$\dot{f} = -s\dot{T} - p\dot{v} - T\frac{\mathrm{d}_i s}{\mathrm{d}t} \,. \tag{7.12}$$

A comparison of (7.11) with (7.12) shows that the entropy production, *i.e.*, the entropy produced during the irreversible process per time and mass unit in the interior of the system is given by

$$\frac{\mathrm{d}_i s}{\mathrm{d}t} = \frac{1}{T}\sum_i a_i \dot{\zeta}_i \,. \tag{7.13a}$$

In a homogeneous system, the entropy production during an irreversible process is thus only caused by the internal degrees of freedom. According to the second law (2.43),

$$\sum_i a_i \dot{\zeta}_i \geq 0 \tag{7.13b}$$

must always be valid. In the internal equilibrium, the entropy production disappears because of $a_i = 0$. In the arrested equilibrium, the entropy production disappears because of $\dot{\zeta}_i = 0$.

In the G-representation, Gibbs fundamental equation is

$$g = g[T(t), p(t), \zeta_1(t) \dots \zeta_n(t)] \tag{7.14}$$

and the equations of state are

$$-\left(\frac{\partial g}{\partial T}\right)_{p,\,\zeta} \equiv s = s(T, p, \zeta_1 \dots \zeta_n)\,, \tag{7.15}$$

$$\left(\frac{\partial g}{\partial p}\right)_{T,\,\zeta} \equiv v = v(T, p, \zeta_1 \dots \zeta_n)\,, \tag{7.16}$$

$$-\left(\frac{\partial g}{\partial \zeta_i}\right)_{T,\,p,\,\zeta_{k\neq i}} \equiv a_i = a_i(T, p, \zeta_1 \dots \zeta_n)\,. \tag{7.17}$$

Corresponding to (7.11), one obtains as the differential fundamental equation

$$\dot{g} = -s\dot{T} + v\dot{p} - \sum_i a_i \dot{\zeta}_i . \tag{7.18}$$

In the same way, one obtains as Gibbs fundamental equation in the U-representation

$$\dot{u} = T\dot{s} - p\dot{v} - \sum_i a_i \dot{\zeta}_i \tag{7.19}$$

and in the H-representation

$$\dot{h} = T\dot{s} + v\dot{p} - \sum_i a_i \dot{\zeta}_i . \tag{7.20}$$

Here, the affinities are defined by

$$a_i(s, v, \zeta_1 \dots \zeta_n) \equiv -\left(\frac{\partial u}{\partial \zeta_i}\right)_{s,\, v,\, \zeta_{k \neq i}} \tag{7.21}$$

or

$$a_i(s, p, \zeta_1 \dots \zeta_n) \equiv -\left(\frac{\partial h}{\partial \zeta_i}\right)_{s,\, p,\, \zeta_{k \neq i}} , \tag{7.22}$$

respectively. Resolution of (7.19) with respect to $\dot{s}$ leads to the S-representation of Gibbs fundamental equation

$$\dot{s} = \frac{1}{T}\dot{u} + \frac{p}{T}\dot{v} + \frac{1}{T}\sum_i a_i \dot{\zeta}_i . \tag{7.23}$$

In the S-representation, the affinities are given by

$$a_i(u, v, \zeta_1 \dots \zeta_n) \equiv T\left(\frac{\partial s}{\partial \zeta_i}\right)_{u,\, v,\, \zeta_{k \neq i}} . \tag{7.24}$$

The expression (7.13) for the entropy production is invariant in its form with respect to the Legendre transformations, *i.e.*, equal in all representations.

According to the first law (2.42), we have with (7.1) and $N = \text{const.}$

$$\dot{u} = \frac{\mathrm{d}_a u}{\mathrm{d}t} = \dot{q} - p\dot{v} . \tag{7.25}$$

With this, (7.23) becomes

$$\dot{s} = \frac{1}{T}\dot{q} + \frac{1}{T}\sum_i a_i \dot{\zeta}_i . \tag{7.26}$$

In the partition (2.40) of the change in entropy

$$\dot{s} = \frac{\mathrm{d}_a s}{\mathrm{d}t} + \frac{\mathrm{d}_i s}{\mathrm{d}t} , \tag{7.27}$$

the amount of entropy which has been exchanged with the surroundings of the system is given by

$$\frac{\mathrm{d}_a s}{\mathrm{d}t} = \frac{1}{T} \dot{q} . \tag{7.28}$$

Equation (7.28) demands that the temperature T defined by the arrested equilibrium states agrees with the temperature T^* of the surroundings of the system. This is by no means necessarily true [see Sect. 7.8, in particular Eq. (7.298)].

7.2 Dynamic Laws

In equilibrium thermodynamics, a system is completely and unequivocally determined by the explicit specification of Gibbs's fundamental equation. This is not true in non-equilibrium. The formalism adopted by equilibrium thermodynamics lacks information on the temporal changes of the independent variables. Although in our specific case the temporal changes of the macroscopic external variables (in the F-representation, *e.g.*, $\dot{T}$, $\dot{v}$, in the G-representation $\dot{T}$, $\dot{p}$) are given by the manner of the external perturbation, we have no information on the temporal changes $\dot{\zeta}_i$ of the internal variables. The so-called dynamic laws fill this information gap.

Dynamic laws, like Gibbs fundamental equation and the equations of state, are material-dependent. They are, therefore, together with the equations of state, often designated as the constitutive equations which classify the behaviour of the material. These are laws of experience where only experiments can give information on their form and limits of validity. In inhomogeneous media, these laws, for example, include Fourier's law of thermal heat conduction, Fick's first law of diffusion, or Ohm's law of electrical conduction, in flowing media, for example, Newton's law of viscous flow. These laws connect the fluxes equalizing the non-equilibrium (energy flow, mass flow, electrical current, momentum flow) with their driving forces (the gradients of temperature, chemical potential, electrical potential, and of the hydrodynamic velocity field). Correspondingly, we must look for an interrelation between the "scalar fluxes" $\dot{\zeta}_i$ and their driving forces, the affinities a_k.

Let us first consider a system with only one macroscopically relevant internal degree of freedom and suppose the internal variable is designated by ζ and the corresponding affinity by a. The simplest of the possible dynamic laws is then

$$\dot{\zeta} = La . \tag{7.29a}$$

In general, the quantity L, which characterizes the material, is a function of all independent variables. However, if one considers (7.29a) (corresponding to the above-mentioned dynamic laws) as a power series about the equilibrium state to be reached, broken off after the first non-vanishing term, L can only depend on the external variables [$L = L(T, v)$ in the F-representation, $L = L(T, p)$ in the G-representation]. To simplify matters, we will often also assume that $L = \text{const.}$ (strictly linear dynamic law). In any case, the second law requires with $a\dot{\zeta} > 0$ [see Eq. (7.13)]

$$L > 0 \,. \tag{7.29b}$$

For the temporal change of the affinity, the equation of state (7.8) leads to (if we only consider one relevant internal degree of freedom):

$$\dot{a} = -\left(\frac{\partial^2 f}{\partial T \partial \zeta}\right) \dot{T} - \left(\frac{\partial^2 f}{\partial v \partial \zeta}\right) \dot{v} - \left(\frac{\partial^2 f}{\partial \zeta^2}\right) \dot{\zeta} \,. \tag{7.30}$$

If we assume $L = \text{const.}$, (7.29) also results in

$$\dot{a} = \frac{1}{L} \ddot{\zeta} \,.$$

Insertion of this expression into (7.30) yields the non-linear differential equation

$$\ddot{\zeta} + L\left(\frac{\partial^2 f}{\partial \zeta^2}\right) \dot{\zeta} = -L\left[\left(\frac{\partial^2 f}{\partial T \partial \zeta}\right) \dot{T} + \left(\frac{\partial^2 f}{\partial v \partial \zeta}\right) \dot{v}\right] \tag{7.31a}$$

for the determination of the internal variables $\zeta(t)$ as a function of time. Here, $\dot{T}(t)$ and $\dot{v}(t)$ are given by the external perturbation of the system. Thus,

$$\Phi[T(t), v(t), \zeta(t), t] \equiv -L\left[\left(\frac{\partial^2 f}{\partial T \partial \zeta}\right) \dot{T} + \left(\frac{\partial^2 f}{\partial v \partial \zeta}\right) \dot{v}\right] \tag{7.31b}$$

acts as an external force. If, in addition, ζ is a dimensionless variable (*e.g.*, a concentration),

$$L\left(\frac{\partial^2 f}{\partial \zeta^2}\right)_{T,v} \equiv 1/\tau_{Tv}[T(t), v(t), \zeta(t)] \tag{7.31c}$$

has the dimension of reciprocal time. With (7.31b/c), (7.31a) can be written as

$$\ddot{\zeta} + \frac{1}{\tau_{Tv}} \dot{\zeta} = \Phi \,. \tag{7.31d}$$

This is a type of Langevin's equation. It is closely related to the differential equation of a driven damped pendulum [*e.g.*, see Goldstein (1962)] whose variety of solutions ranges from simple equilibration processes over damped oscillations to chaotic motion [see Schuster (1984)]. The role of a damping factor (frictional factor) is taken over by $1/\tau_{Tv}$. One should mention that the characteristic time τ_{T_v} is usually not a constant but depends on the instantaneous state of the system. Moreover, (7.31) is not invariant with respect to time reversal. This expresses that the equation describes an irreversible process. If one understands the coefficients in (7.31) as pure functions of time, one obtains as the first integral of the equation

$$\dot{\zeta}(t) = \mathrm{e}^{-\lambda_{Tv}(t)} \left[\dot{\zeta}(0) + \int_0^t \Phi(t')\, \mathrm{e}^{\lambda_{Tv}(t')}\, \mathrm{d}t' \right] \tag{7.32a}$$

with the dimensionless relative time variable

$$\lambda_{Tv}(t) \equiv \int_0^t \frac{\mathrm{d}t'}{\tau_{Tv}} . \tag{7.32b}$$

Equations (7.32) reveal one of the most important features of the macroscopic relevance of an internal degree of freedom: According to (7.32), the instantaneous value of $\dot{\zeta}(t)$ at the time t not only depends on the instantaneous stress $\Phi(t)$, but also on the stress $\Phi(t')$ at all previous times $t' < t$. The previous history of the system influences the instantaneous state of the system *via* the internal degree of freedom. The internal degree of freedom provides the system with a kind of "memory". Among other things, this leads (see Sect. 7.4 and 7.6) to the so-called phenomena of aftereffects.

If the non-equilibrium states of the system deviate only to a small extent from an internal equilibrium state "e" during an external perturbation (*e.g.*, a periodic perturbation with small amplitudes), one can approximately refer the coefficients in (7.31) to this fiducial state. With (7.6), (7.7), and (7.31 c), one then obtains

$$\left(\frac{\partial^2 f}{\partial T \partial \zeta}\right)_e = \left(\frac{\partial^2 f}{\partial \zeta \partial T}\right)_e = -\left(\frac{\partial s}{\partial \zeta}\right)^e_{T,\,v} \equiv -\,\sigma^e_{Tv}(T,\, v)\,, \tag{7.33}$$

$$\left(\frac{\partial^2 f}{\partial v \partial \zeta}\right)_e = \left(\frac{\partial^2 f}{\partial \zeta \partial v}\right)_e = -\left(\frac{\partial p}{\partial \zeta}\right)^e_{T,\,v} \equiv -\,\pi^e_{Tv}(T,\, v)\,, \tag{7.34}$$

$$L\left(\frac{\partial^2 f}{\partial \zeta^2}\right)^e_{T,\,v} \equiv 1/\tau^e_{Tv}(T,\, v) > 0\,, \tag{7.35}$$

where σ^e_{Tv} is the partial specific entropy of the internal degree of freedom in the internal equilibrium, and π^e_{Tv} is a measure of the contribution of the internal degree of freedom to the pressure of the system in internal equilibrium [*cf.* Sect. 7.3, Eq. (7.68)]. If the fiducial state is a stable or metastable equilibrium state, τ^e_{Tv} is, according to (2.78) and (7.29b), a positive quantity. In a system left to itself in non-equilibrium under the condition T, $v =$ const., τ^e_{Tv} becomes identical with the so-called Debye relaxation time (Sect. 7.4). In place of (7.31), one then obtains the linear differential equation

$$\ddot{\zeta} + \frac{1}{\tau^e_{Tv}} \dot{\zeta} = L\,[\sigma^e_{Tv}\,\dot{T} + \pi^e_{Tv}\,\dot{v}]\,. \tag{7.36}$$

In the G-representation, the temporal change of the affinity is, according to (7.18), given by

$$\dot{a} = -\left(\frac{\partial^2 g}{\partial T \partial \zeta}\right) \dot{T} - \left(\frac{\partial^2 g}{\partial p \partial \zeta}\right) \dot{p} - \left(\frac{\partial^2 g}{\partial \zeta^2}\right) \dot{\zeta}. \tag{7.37}$$

Together with the dynamic law (7.29), this leads to the non-linear differential equation

$$\ddot{\zeta} + \frac{1}{\tau_{Tp}} \dot{\zeta} = \Gamma \tag{7.38a}$$

for the determination of the internal variable $\zeta(t)$ as a function of time with the external driving force

$$\Gamma[T(t), p(t), \zeta(t), t] \equiv -L\left[\left(\frac{\partial^2 g}{\partial T \partial \zeta}\right) \dot{T} + \left(\frac{\partial^2 g}{\partial p \partial \zeta}\right) \dot{p}\right] \tag{7.38b}$$

and the reciprocal characteristic time

$$L\left(\frac{\partial^2 g}{\partial \zeta^2}\right)_{T,p} \equiv 1/\tau_{Tp}[T(t), p(t), \zeta(t)]. \tag{7.38c}$$

The first integral of (7.38) is

$$\dot{\zeta}(t) = \mathrm{e}^{-\lambda_{Tp}(t)}\left[\dot{\zeta}(0) + \int_0^t \Gamma(t')\,\mathrm{e}^{\lambda_{Tp}(t')}\,\mathrm{d}t'\right] \tag{7.39a}$$

with the dimensionless relative time variable

$$\lambda_{Tp}(t) \equiv \int_0^t \frac{\mathrm{d}t'}{\tau_{Tp}}\,. \tag{7.39b}$$

If one refers the coefficients in (7.38) approximately to a given internal equilibrium state "e", (7.16), (7.17), and (7.38c) lead to

$$\left(\frac{\partial^2 g}{\partial T \partial \zeta}\right)_e = \left(\frac{\partial^2 g}{\partial \zeta \partial T}\right)_e = -\left(\frac{\partial s}{\partial \zeta}\right)^e_{T,p} \equiv -\sigma^e_{Tp}(T,p)\,, \tag{7.40}$$

$$\left(\frac{\partial^2 g}{\partial p\,\partial\zeta}\right)_e = \left(\frac{\partial^2 g}{\partial\zeta\,\partial p}\right)_e = \left(\frac{\partial v}{\partial\zeta}\right)^e_{T,p} \equiv \varphi^e_{Tp}(T,p), \tag{7.41}$$

$$L\left(\frac{\partial^2 g}{\partial\zeta^2}\right)^e_{T,p} \equiv 1/\tau^e_{Tp}(T,p)\,. \tag{7.42}$$

In place of (7.38), we then obtain the linear differential equation

$$\ddot{\zeta} + \frac{1}{\tau^e_{Tp}}\dot{\zeta} = L\,[\sigma^e_{Tp}\,\dot{T} - \varphi^e_{Tp}\,\dot{v}]\,, \tag{7.43}$$

where σ^e_{Tp} is the partial specific entropy, and φ^e_{Tp}, the partial specific volume of the internal degree of freedom in the internal equilibrium state "e". If this is a stable or metastable state, (2.88) leads to

$$\left(\frac{\partial^2 g}{\partial\zeta^2}\right)^e_{T,p} > 0 \tag{7.44}$$

and with (7.29b) to

$$\tau^e_{Tp} > 0\,. \tag{7.45}$$

In this case, τ^e_{Tp} is the Debye relaxation time in the G-representation (Sect. 7.4).

If several internal degrees of freedom are macroscopically relevant, one can postulate the dynamic laws

$$\dot{\zeta}_i = L_i\,a_i\,, \qquad (i = 1\,\dots\,n) \tag{7.46}$$

in analogy to (7.29). As the a_i generally depend on all internal variables $\zeta_1 \dots \zeta_n$, (7.46) represents a system of n coupled differential equations. Equation (7.46), however, does not express that the internal degrees of freedom can also be dynamically interlinked. Experience teaches us that, for example, an energy or heat flow can also be started by the gradient of the chemical potential (Dufour effect) and in the reverse case that mass flow can be induced by a temperature gradient (Soret effect); the gradient of an electrical potential can cause a heat flow (Peltier effect), and the temperature gradient an electrical current (Thomson–Seebeck effects) [*e.g.*, see De Groot (1963)]. In general, one can assume that all possible fluxes are driven by all forces that are present. In our specific case, it can thus be expected that the temporal change of the internal variables ζ_i is not only caused by the affinity a_i, but also by the affinities $a_{k \neq i}$. As the simplest dynamic law, one then obtains

$$\dot{\zeta}_i = \sum_k L_{ik}\,a_k\,. \tag{7.47a}$$

The coefficients L_{ik} form a symmetric matrix

$$L_{ik} = L_{ki} \tag{7.47b}$$

[the so-called Onsager reciprocal relations; see Onsager (1931), Casimir (1945), De Groot, Mazur (1962)]. Insertion of (7.47) into (7.13) yields for the entropy production the expression

$$\frac{\mathrm{d}_i s}{\mathrm{d}t} = \frac{1}{T} \sum_i \sum_k L_{ik}\, a_i\, a_k > 0\,. \tag{7.48}$$

Hence, the second law (2.43) requires that the matrix of the coefficients L_{ik} be positive definite.

According to (7.18), the temporal change of the affinities, for example, in the G-representation is given by

$$\dot{a}_k = -\left(\frac{\partial^2 g}{\partial T\, \partial \zeta_k}\right) \dot{T} - \left(\frac{\partial^2 g}{\partial p\, \partial \zeta_k}\right) \dot{p} - \sum_\ell \left(\frac{\partial^2 g}{\partial \zeta_\ell\, \partial \zeta_k}\right) \dot{\zeta}_\ell\,. \tag{7.49}$$

If the L_{ik} in (7.47) are constant, *i.e.*, if (7.47a) is a true linear law, we have, on the other hand,

$$\ddot{\zeta}_i = \sum_k L_{ik}\, \dot{a}_k\,.$$

Insertion of (7.49) results in

$$\ddot{\zeta}_i = -\sum_k L_{ik} \left[\left(\frac{\partial^2 g}{\partial T\, \partial \zeta_k}\right) \dot{T} + \left(\frac{\partial^2 g}{\partial p\, \partial \zeta_k}\right) \dot{p}\right] - \sum_k \sum_\ell L_{ik} \left(\frac{\partial^2 g}{\partial \zeta_\ell\, \partial \zeta_k}\right) \dot{\zeta}_\ell\,.$$

With

$$\Gamma_i \equiv -\sum_k L_{ik} \left[\left(\frac{\partial^2 g}{\partial T\, \partial \zeta_k}\right) \dot{T} + \left(\frac{\partial^2 g}{\partial p\, \partial \zeta_k}\right) \dot{p}\right] \tag{7.50a}$$

and

$$M_{i\ell} \equiv \sum_k L_{ik} \left(\frac{\partial^2 g}{\partial \zeta_\ell\, \partial \zeta_k}\right), \tag{7.50b}$$

one obtains the system of n coupled non-linear differential equations

$$\ddot{\zeta}_i + \sum_\ell M_{i\ell}\, \dot{\zeta}_\ell = \Gamma_i \tag{7.50c}$$

for the determination of the internal variables $\zeta_i(t)$ as a function of time. The Γ_i are understood to be external driving forces acting on the internal variables ζ_i. The $M_{i\ell}$ act as damping or frictional factors which, if the ζ_i are dimensionless, have the dimension of reciprocal time.

7.3 Dynamic Response Functions

To simplify matters, we will in the following sections once again consider a system with only one macroscopically relevant internal degree of freedom. In Sect. 7.7, we deal with the particularities which occur in systems with several simultaneously relevant internal degrees of freedom. In Sect. 7.3, we examine the effects of a relevant internal degree of freedom on the response functions of a system.

According to (2.44), the isothermal bulk modulus is a relative measure of the change in pressure with the change in volume at constant temperature. The mechanical equation of state (7.7) results in

$$dp = \left(\frac{\partial p}{\partial T}\right)_{v,\zeta} dT + \left(\frac{\partial p}{\partial v}\right)_{T,\zeta} dv + \left(\frac{\partial p}{\partial \zeta}\right)_{T,v} d\zeta \tag{7.51}$$

for the change in pressure of a system with a relevant internal degree of freedom. As the isothermal bulk modulus, one thus obtains

$$k_T \equiv -v\left(\frac{dp}{dv}\right)_T = -v\left(\frac{\partial p}{\partial v}\right)_{T,\zeta} - v\left(\frac{\partial p}{\partial \zeta}\right)_{T,v}\left(\frac{d\zeta}{dv}\right)_T . \tag{7.52a}$$

As ζ and v are mutually independent, but time-dependent variables in non-equilibrium,

$$\left(\frac{d\zeta}{dv}\right)_T = \dot{\zeta}_T/\dot{v}_T \tag{7.52b}$$

must be interpreted as the ratio of the temporal change $\dot{\zeta}_T$ of the internal variables at constant temperature to the temporal change $\dot{v}_T$ of the specific volume at constant temperature. If one considers the coefficient of thermal pressure (2.45) as the relative measure of the change in pressure with the temperature at constant volume, (7.51) leads in the same way to

$$\beta_v \equiv \frac{1}{p}\left(\frac{dp}{dT}\right)_v = \frac{1}{p}\left(\frac{\partial p}{\partial T}\right)_{v,\zeta} + \frac{1}{p}\left(\frac{\partial p}{\partial \zeta}\right)_{T,v} \dot{\zeta}_v/\dot{T}_v . \tag{7.53}$$

According to (2.45), the coefficient of thermal pressure can also be understood as a relative measure of the change in entropy with volume at constant temperature. The differential form of the thermal equation of state (7.6)

$$ds = \left(\frac{\partial s}{\partial T}\right)_{v,\zeta} dT + \left(\frac{\partial s}{\partial v}\right)_{T,\zeta} dv + \left(\frac{\partial s}{\partial \zeta}\right)_{T,v} d\zeta \tag{7.54}$$

then leads to

$$\beta_T \equiv \frac{1}{p}\left(\frac{\mathrm{d}s}{\mathrm{d}v}\right)_T = \frac{1}{p}\left(\frac{\partial s}{\partial v}\right)_{T,\zeta} + \frac{1}{p}\left(\frac{\partial s}{\partial \zeta}\right)_{T,v} \dot{\zeta}_T / \dot{v}_T . \tag{7.55}$$

We will see further below that $\beta_T = \beta_v = \beta$ is valid in an arrested, as well as in an internal equilibrium. On the other hand, $\beta_T \neq \beta_v$ generally holds in non-equilibrium, so that two different coefficients of thermal pressure must be distinguished depending on the perturbation of the system. Equation (7.54) further leads to

$$c_v \equiv T\left(\frac{\mathrm{d}s}{\mathrm{d}T}\right)_v = T\left(\frac{\partial s}{\partial T}\right)_{v,\zeta} + T\left(\frac{\partial s}{\partial \zeta}\right)_{T,v} \dot{\zeta}_v / \dot{T}_v \tag{7.56}$$

as the isochoric specific heat capacity. The equations (7.52/7.53) and (7.55/7.56) reveal an additional important feature of the macroscopic relevance of an internal degree of freedom: At internal equilibrium, the response functions of the F-representation at constant temperature and volume are matter constants. In the case of variable temperature and volume, the response functions are matter functions. This is no longer the case when the internal degree of freedom is disturbed. The response functions depend on the ratio of the rate $\dot{\zeta}_T$ or $\dot{\zeta}_v$ of the internal variables to the rate of external perturbation $\dot{v}_T$ or $\dot{T}_v$. In order to characterize the material, the rate of perturbation must be given in addition to the response functions. Two limiting cases can be distinguished:

In the first limiting case, $\dot{\zeta}_T \ll \dot{v}_T$ or $\dot{\zeta}_v \ll \dot{T}_v$ holds. The external perturbation occurs much faster than the change in the internal variables. During the perturbation, the system is then practically in an arrested equilibrium with respect to the internal degree of freedom. The response functions of the arrested equilibrium are measured:

$$k_T = k_{T,\zeta} \equiv -\, v\left(\frac{\partial p}{\partial v}\right)_{T,\zeta} , \tag{7.57}$$

$$\beta = \beta_\zeta = \beta_{v,\zeta} \equiv \frac{1}{p}\left(\frac{\partial p}{\partial T}\right)_{v,\zeta} = \frac{1}{p}\left(\frac{\partial s}{\partial v}\right)_{T,\zeta} \equiv \beta_{T,\zeta} , \tag{7.58}$$

$$c_v = c_{v,\zeta} \equiv T\left(\frac{\partial s}{\partial T}\right)_{v,\zeta} . \tag{7.59}$$

The second limiting case occurs if $\dot{\zeta}_T \gg \dot{v}_T$ or $\dot{\zeta}_v \gg \dot{T}_v$. The external perturbation occurs now considerably more slowly than the change in the internal variables. In this case, the internal variable is at any time capable of reaching its equilibrium value practically instantaneously. The perturbation process becomes a sequence of internal equilibrium states in which

$$a(T, v, \zeta) = 0 \quad \text{and} \quad \zeta = \zeta_e(T, v)$$

hold at any time, and ζ becomes a dependent variable which with $\zeta = \zeta_e(T, v)$ directly follows the external changes. According to (7.30), we obtain for this process

$$da = -\left(\frac{\partial^2 f}{\partial T \partial \zeta}\right)_e dT - \left(\frac{\partial^2 f}{\partial v \partial \zeta}\right)_e dv - \left(\frac{\partial^2 f}{\partial \zeta^2}\right)_e d\zeta = 0$$

or with the abbreviations (7.33–7.35)

$$\sigma^e_{Tv}\, dT + \pi^e_{Tv}\, dp - \frac{1}{L\tau^e_{Tv}}\, d\zeta_e = 0 \,.$$

Accordingly, the dependence of the internal variables on the external variables is described by

$$d\zeta_e = L\tau^e_{Tv}\, \sigma^e_{Tv}\, dT + L\tau^e_{Tv}\, \pi^e_{Tv}\, dv \tag{7.60a}$$

with

$$\left(\frac{\partial \zeta_e}{\partial T}\right)_v = L\tau^e_{Tv}\, \sigma^e_{Tv} \tag{7.60b}$$

and

$$\left(\frac{\partial \zeta_e}{\partial v}\right)_T = L\tau^e_{Tv}\, \pi^e_{Tv} \,. \tag{7.60c}$$

As response functions which can be measured in internal equilibrium, (7.52), (7.53), (7.55), and (7.56) lead to

$$k^e_T = -\left(\frac{\partial p}{\partial v}\right)_{T,\zeta_e} - v\left(\frac{\partial p}{\partial \zeta}\right)^e_{T,v}\left(\frac{\partial \zeta_e}{\partial v}\right)_T$$

$$= k_{T,\zeta_e} - v\, L\tau^e_{Tv}\, (\pi^e_{Tv})^2 \,, \tag{7.61}$$

$$\beta_e = \beta^e_T = \beta^e_v = \beta_{\zeta_e} + \frac{1}{p} L\tau^e_{Tv}\pi^e_{Tv}\sigma^e_{Tv} \,, \tag{7.62}$$

$$c^e_v = c_{v,\zeta_e} + T\, L\tau^e_{Tv}\, (\sigma^e_{Tv})^2 \,. \tag{7.63}$$

In (7.61) to (7.63), k_{T,ζ_e}, β_{ζ_e} and c_{v,ζ_e} are those response functions which are measured if the system in the internal equilibrium state $Z[T, v, \zeta_e(T, v)]$ is perturbed so rapidly that the internal variable ζ_e appears arrested with respect to this perturbation. The response functions of the arrested equilibrium contain the contributions of all internal degrees of freedom except for the contribution of ζ. Hence, the contribution of the macroscopically relevant internal degree of freedom to the response functions which can be measured in internal equilibrium is given by the differences

$$\Delta k_T \equiv k_T^e - k_{T,\zeta_e} = -vL\tau_{Tv}^e(\pi_{Tv}^e)^2 < 0, \tag{7.64}$$

$$\Delta\beta \equiv \beta_e - \beta_{\zeta_e} = \frac{1}{p} L\tau_{Tv}^e \pi_{Tv}^e \sigma_{Tv}^e , \tag{7.65}$$

$$\Delta c_v \equiv c_v^e - c_{v,\zeta_e} = TL\tau_{Tv}^e (\sigma_{Tv}^e)^2 > 0 . \tag{7.66}$$

According to (7.29b) and (7.35), the smaller and greater sign hold if the internal equilibrium is a stable or metastable state. In this case, the bulk modulus measured during a very slow perturbation is always smaller than the bulk modulus measured during a rapid perturbation. The internal degree of freedom contributes to a "softening" of the material. On the other hand, the heat capacity measured during a very slow perturbation is larger than the heat capacity measured during a rapid perturbation. The sign of $\Delta\beta$ remains open as long as the internal degree of freedom is not explicitly specified. Insertion of (7.64–7.66) easily confirms the Davies relation

$$-\Delta c_v \Delta k_T = Tv p^2 (\Delta\beta)^2 \tag{7.67}$$

[Davies (1952); Davies, Jones (1953); see also Prigogine, Defay (1950)].

The retention of the internal variable in internal equilibrium may be somewhat irritating, as the internal variables are usually eliminated in equilibrium thermodynamics. However, it becomes necessary as soon as a comparison of non-equilibrium states, arrested equilibrium states, and internal equilibrium states is required. The consistency of the equations (7.60–7.66) with equilibrium thermodynamics may be demonstrated, for example, on the basis of the mechanical equation of state (7.51): in the internal equilibrium, (7.51) is

$$\mathrm{d}p = \left(\frac{\partial p}{\partial T}\right)_{v,\zeta_e} \mathrm{d}T + \left(\frac{\partial p}{\partial v}\right)_{T,\zeta_e} \mathrm{d}v + \left(\frac{\partial p}{\partial \zeta}\right)_{T,v}^{e} \mathrm{d}\zeta_e$$

or with (7.57), (7.58), and (7.34)

$$\mathrm{d}p = p\,\beta_{\zeta_e}\,\mathrm{d}T - \frac{1}{v} k_{T,\zeta_e}\,\mathrm{d}v + \pi_{Tv}^e\,\mathrm{d}\zeta_e \tag{7.68}$$

and with (7.60) and (7.64–7.66)

$$\begin{aligned}\mathrm{d}p &= p\beta_{\zeta_e}\mathrm{d}T - \frac{1}{v} k_{T,\zeta_e}\,\mathrm{d}v + L\tau_{Tv}^e\,\sigma_{Tv}^e\,\pi_{Tv}^e\mathrm{d}T + L\tau_{Tv}^e(\pi_{Tv}^e)^2\mathrm{d}v \\ &= p\,(\beta_{\zeta_e} + \Delta\beta)\,\mathrm{d}T - \frac{1}{v}\,(k_{T,\zeta_e} + \Delta k_T)\,\mathrm{d}v \\ &= p\beta_e\mathrm{d}T - \frac{1}{v} k_T^e\,\mathrm{d}v .\end{aligned}$$

Because of $N = \text{const.}$, the last line obviously corresponds to the differential equation of state (2.56) of equilibrium thermodynamics according to (2.45) and (2.44).

In order to obtain information on the development of the response functions between the two limiting cases, *i.e.*, in non-equilibrium, the external perturbation $\dot{v}_T$ or $\dot{T}_v$ must, of course, be known explicitly. Examples can be found in the two following sections and in Sect. 7.8.

In the G-representation, the mechanical and thermal equations of state (7.17) and (7.16) are in their differential form

$$dv = \left(\frac{\partial v}{\partial T}\right)_{p,\zeta} dT + \left(\frac{\partial v}{\partial p}\right)_{T,\zeta} dp + \left(\frac{\partial v}{\partial \zeta}\right)_{T,p} d\zeta\,, \tag{7.69}$$

$$ds = \left(\frac{\partial s}{\partial T}\right)_{p,\zeta} dT + \left(\frac{\partial s}{\partial p}\right)_{T,\zeta} dp + \left(\frac{\partial s}{\partial \zeta}\right)_{T,p} d\zeta\,. \tag{7.70}$$

This leads with

$$\left(\frac{d\zeta}{dp}\right)_T = \dot{\zeta}_T/\dot{p}_T \quad \text{and} \quad \left(\frac{d\zeta}{dT}\right)_p = \dot{\zeta}_p/\dot{T}_p$$

to the isothermal compressibility

$$\kappa_T \equiv -\frac{1}{v}\left(\frac{dv}{dp}\right)_T = -\frac{1}{v}\left(\frac{\partial v}{\partial p}\right)_{T,\zeta} - \frac{1}{v}\left(\frac{\partial v}{\partial \zeta}\right)_{T,p} \dot{\zeta}_T/\dot{p}_T\,, \tag{7.71}$$

the coefficient of thermal expansion at constant pressure

$$\alpha_p \equiv \frac{1}{v}\left(\frac{dv}{dT}\right)_p = \frac{1}{v}\left(\frac{\partial v}{\partial T}\right)_{p,\zeta} + \frac{1}{v}\left(\frac{\partial v}{\partial \zeta}\right)_{T,p} \dot{\zeta}_p/\dot{T}_p\,, \tag{7.72}$$

the coefficient of thermal expansion at constant temperature

$$\alpha_T \equiv -\frac{1}{v}\left(\frac{ds}{dp}\right)_T = -\frac{1}{v}\left(\frac{\partial s}{\partial p}\right)_{T,\zeta} - \frac{1}{v}\left(\frac{\partial s}{\partial \zeta}\right)_{T,p} \dot{\zeta}_T/\dot{p}_T\,, \tag{7.73}$$

and the isobaric specific heat capacity

$$c_p \equiv T\left(\frac{ds}{dT}\right)_p = T\left(\frac{\partial s}{\partial T}\right)_{p,\zeta} + T\left(\frac{\partial s}{\partial \zeta}\right)_{T,p} \dot{\zeta}_p/\dot{T}_p \tag{7.74}$$

[see Eqs. (2.47–2.49)]. In an arrested equilibrium, the response functions

$$\kappa_{T,\zeta} \equiv -\frac{1}{v}\left(\frac{\partial v}{\partial p}\right)_{T,\zeta}\,, \tag{7.75}$$

$$\alpha_\zeta = \alpha_{T,\zeta} \equiv -\frac{1}{v}\left(\frac{\partial s}{\partial p}\right)_{T,\zeta} = \frac{1}{v}\left(\frac{\partial v}{\partial T}\right)_{p,\zeta} \equiv \alpha_{p,\zeta}\,, \tag{7.76}$$

$$c_{p,\zeta} \equiv T\left(\frac{\partial s}{\partial T}\right)_{p,\zeta} \tag{7.77}$$

are measured with $\dot{\zeta}_T \ll \dot{p}_T$ or $\dot{\zeta}_p \ll \dot{T}_p$, respectively. In the case of an internal equilibrium [$\dot{\zeta}_T \gg \dot{p}_T$ or $\dot{\zeta}_p \gg \dot{T}_p$, respectively, and $a = 0$, $\zeta = \zeta_e(T, p)$], (7.37) with the abbreviations (7.40–7.42) results in

$$\mathrm{d}a = \sigma^e_{Tp}\,\mathrm{d}T - \varphi^e_{Tp}\,\mathrm{d}p - \frac{1}{L\tau^e_{Tp}}\,\mathrm{d}\zeta_e = 0\,,$$

$$\mathrm{d}\zeta_e = L\tau^e_{Tp}\sigma^e_{Tp}\,\mathrm{d}T - L\tau^e_{Tp}\,\varphi^e_{Tp}\,\mathrm{d}p\,, \tag{7.78a}$$

i.e.,

$$\left(\frac{\partial \zeta_e}{\partial T}\right)_p = L\tau^e_{Tp}\sigma^e_{Tp}\,, \tag{7.78b}$$

$$\left(\frac{\partial \zeta_e}{\partial p}\right)_T = -\,L\tau^e_{Tp}\,\varphi^e_{Tp}\,. \tag{7.78c}$$

One thus obtains for the response functions to be measured in internal equilibrium

$$\kappa^e_T = \kappa_{T,\zeta_e} + \frac{1}{v}L\tau^e_{Tp}(\varphi^e_{Tp})^2\,, \tag{7.79}$$

$$\alpha_e = \alpha^e_p = \alpha^e_T = \alpha_{\zeta_e} + \frac{1}{v}L\tau^e_{Tp}\varphi^e_{Tp}\sigma^e_{Tp}\,, \tag{7.80}$$

$$c^e_p = c_{p,\zeta_e} + TL\tau^e_{Tp}(\sigma^e_{Tp})^2\,. \tag{7.81}$$

The contribution of the internal degree of freedom to the response functions of the internal equilibrium is given by the differences

$$\Delta\kappa_T \equiv \kappa^e_T - \kappa_{T,\zeta_e} = \frac{1}{v}L\tau^e_{Tp}(\varphi^e_{Tp})^2 > 0\,, \tag{7.82}$$

$$\Delta\alpha \equiv \alpha_e - \alpha_{\zeta_e} = \frac{1}{v}L\tau^e_{Tp}\varphi^e_{Tp}\sigma^e_{Tp}\,, \tag{7.83}$$

$$\Delta c_p \equiv c^e_p - c_{p,\zeta_e} = TL\tau^e_{Tp}(\sigma^e_{Tp})^2 > 0\,. \tag{7.84}$$

The greater-than sign is valid if the internal equilibrium state $Z[T, p, \zeta_e(T, p)]$ is a stable or metastable state, so that $\tau^e_{Tp} > 0$. The contributions (7.82–7.84) fulfil the Davies relation

$$\Delta c_p \Delta \kappa_T = T v (\Delta \alpha)^2 \tag{7.85}$$

[see Eq. (7.67)]. The Davies relation (7.85) is formally identical with the so-called Ehrenfest relation (8.8), which holds for second order phase transformations (Sect. 8.1). However, the meaning of the two relations differs considerably. The Ehrenfest relation is a purely thermostatic relation in which the differences Δ refer to two different states $Z[T, p, x_I(T, p)]$ and $Z[T, p, x_{II}(T, p)]$ of the internal equilibrium. The Davies relation, on the other hand, is a dynamic relation in which the differences Δ refer to one and the same state $Z(T, p, \zeta_e)$ which is stressed once very rapidly ($\zeta_e = \text{const.}$) and once quasi-statically [$\zeta_e = \zeta_e(T, p)$].

There are some purely mathematical relations between the response functions in equilibrium thermodynamics (Sect. 2.3). These relations naturally also exist among the response functions of an arrested equilibrium or among the response functions of an internal equilibrium. In addition, there are still very similar relations between the equilibrium coefficients τ_e, φ_e π_e and σ_e. As a result, the response functions of the arrested equilibrium and the response functions of the internal equilibrium are not independent of each other.

For example,

$$a(T, v, \zeta) = a[T, p(T, v, \zeta), \zeta]$$

holds for the transition from the G-representation to the F-representation, *i.e.*,

$$\left(\frac{\partial a}{\partial \zeta}\right)_{T,v} = \left(\frac{\partial a}{\partial p}\right)_{T,\zeta} \left(\frac{\partial p}{\partial \zeta}\right)_{T,v} + \left(\frac{\partial a}{\partial \zeta}\right)_{T,p}$$

If one relates this equation to the internal equilibrium, one gets for the individual terms according to (7.8) and (7.35),

$$\left(\frac{\partial a}{\partial \zeta}\right)^e_{T,v} = -\left(\frac{\partial^2 f}{\partial \zeta^2}\right)^e_{T,v} = -1/L\tau^e_{Tv},$$

according to (7.17) and (7.42),

$$\left(\frac{\partial a}{\partial \zeta}\right)^e_{T,p} = -\left(\frac{\partial^2 g}{\partial \zeta^2}\right)^e_{T,p} = -1/\tau^e_{Tp},$$

according to (7.34),

$$\left(\frac{\partial p}{\partial \zeta}\right)^e_{T,v} = \pi^e_{Tv},$$

and according to (7.17), (7.16), and (7.41),

$$\left(\frac{\partial a}{\partial p}\right)^e_{T,\zeta} = -\left(\frac{\partial^2 g}{\partial p\,\partial \zeta}\right)_e = -\left(\frac{\partial^2 g}{\partial \zeta\,\partial p}\right)_e = -\left(\frac{\partial v}{\partial \zeta}\right)^e_{T,p} = -\,\varphi^e_{Tp}\,.$$

This leads to the relation

$$\frac{1}{\tau^e_{Tv}} = \frac{1}{\tau^e_{Tp}} + L\,\varphi^e_{Tp}\,\pi^e_{Tv}\,. \tag{7.86}$$

With $T, p =$ const., the F-representation (7.51) of the mechanical equation of state results in

$$0 = \left(\frac{\partial p}{\partial v}\right)_{T,\zeta} \mathrm{d}v + \left(\frac{\partial p}{\partial \zeta}\right)_{T,v} \mathrm{d}\zeta$$

or, since we have $T, p =$ const. as a secondary condition so that

$$\frac{\mathrm{d}v}{\mathrm{d}\zeta} = \left(\frac{\partial v}{\partial \zeta}\right)_{T,p},$$

in

$$\left(\frac{\partial p}{\partial \zeta}\right)_{T,v} = -\left(\frac{\partial p}{\partial v}\right)_{T,\zeta}\left(\frac{\partial v}{\partial \zeta}\right)_{T,p}\,.$$

If one relates this equation again to an internal equilibrium, one obtains, according to (7.34), (7.57), and (7.41), the relation

$$\pi^e_{Tv} = \frac{1}{v}\,k_{T,\zeta_e}\,\varphi^e_{Tp}\,, \tag{7.87 a}$$

or, in addition, because of

$$k_{T,\zeta_e} = 1/\kappa_{T,\zeta_e}\,, \tag{7.88}$$

$$v\kappa_{T,\zeta_e}(\pi^e_{Tv})^2 = \frac{1}{v}\,k_{T,\zeta_e}(\varphi^e_{Tp})^2\,. \tag{7.87 b}$$

Insertion of (7.87 a) into (7.86) yields the relation

$$\frac{1}{\tau^e_{Tv}} = \frac{1}{\tau^e_{Tp}} + \frac{L}{v}\,k_{T,\zeta_e}(\varphi^e_{Tp})^2 \tag{7.89 a}$$

and with (7.87b) the alternative relation

$$\frac{1}{\tau_{Tv}^{e}} = \frac{1}{\tau_{Tp}^{e}} + vL\,\kappa_{T,\zeta_e}(\pi_{Tv}^{e})^2 \,. \tag{7.89b}$$

If the arrested equilibrium state is mechanically stable, we have $k_{T,\zeta_e} > 0$, $\kappa_{T,\zeta_e} > 0$ [see (4.22)]. The relation (7.89) then expresses that

$$\frac{1}{\tau_{Tv}^{e}} \geq \frac{1}{\tau_{Tp}^{e}} \quad \text{or} \quad \tau_{Tp}^{e} \geq \tau_{Tv}^{e} \,, \tag{7.90}$$

respectively, must always be valid. Hence, the processes to be described with the differential equation (7.36) are generally subject to a stronger damping than the processes to be described with the differential equation (7.43). The equal sign only holds in (7.90) if $\pi_{Tv}^{e} = \varphi_{Tp}^{e} = 0$ is valid, *i.e.*, if the pressure or the volume do not depend at all on the internal degree of freedom (see below).

Equation (7.89 a) with (7.88) and (7.82) further leads to

$$\frac{\tau_{Tp}^{e}}{\tau_{Tv}^{e}} = 1 + \frac{1}{v\kappa_{T,\zeta_e}} L\,\tau_{Tp}^{e}(\varphi_{Tp}^{e})^2 = 1 + \frac{\Delta\kappa_T}{\kappa_{T,\zeta_e}} \,,$$

i.e.,

$$\Delta\kappa_T = \left(\frac{\tau_{Tp}^{e}}{\tau_{Tv}^{e}} - 1\right)\kappa_{T,\zeta_e} \tag{7.91a}$$

and

$$\kappa_T^{e} = \frac{\tau_{Tp}^{e}}{\tau_{Tv}^{e}}\,\kappa_{T,\zeta_e} \,. \tag{7.91b}$$

The compressibility to be measured in an internal equilibrium is linked with the compressibility to be measured in an arrested equilibrium *via* the ratio $\tau_{Tp}^{e}/\tau_{Tv}^{e}$ of the damping factors. In the same way, (7.89b) with (7.88) and (7.64) results in

$$\Delta k_T = \left(\frac{\tau_{Tv}^{e}}{\tau_{Tp}^{e}} - 1\right)k_{T,\zeta_e} \tag{7.92a}$$

and

$$k_T^{e} = \frac{\tau_{Tv}^{e}}{\tau_{Tp}^{e}}\,k_{T,\zeta_e} \,. \tag{7.92b}$$

The validity of (7.88) is obvious. (7.91b) and (7.92b) now clearly show that

$$k_T^{e} = 1/\kappa_T^{e} \tag{7.93}$$

is also valid.

Similar relations can be obtained from other representations. One thus obtains, for example, for the damping factors of the U- and H-representation

$$L\left(\frac{\partial^2 u}{\partial \zeta^2}\right)^e_{s,\,v} \equiv 1/\tau^e_{sv} \qquad \text{or} \qquad L\left(\frac{\partial^2 h}{\partial \zeta^2}\right)^e_{s,\,p} \equiv 1/\tau^e_{sp}\,,$$

respectively, the relations

$$\frac{1}{\tau^e_{sv}} = \frac{1}{\tau^e_{Tv}} + \frac{LT}{c_{v,\,\zeta_e}}\,(\sigma^e_{Tv})^2\,, \tag{7.94a}$$

$$\frac{1}{\tau^e_{sp}} = \frac{1}{\tau^e_{Tp}} + \frac{LT}{c_{p,\,\zeta_e}}\,(\sigma^e_{Tp})^2\,, \tag{7.94b}$$

$$\frac{1}{\tau^e_{sv}} = \frac{1}{\tau^e_{sp}} + \frac{L}{v}\,k_{s,\,\zeta_e}(\varphi_{sp})^2\,, \tag{7.94c}$$

where $k_{s,\,\zeta_e}$ is the isentropic bulk modulus (2.50) which is measured in an arrested equilibrium. The partial specific volume φ^e_{sp} is defined by

$$\varphi^e_{sp} \equiv \left(\frac{\partial v}{\partial \zeta}\right)^e_{s,\,p}\,.$$

If the reference state "e" is a stable or metastable equilibrium state, we have the time hierarchy

$$\tau^e_{Tp} \geq \tau^e_{Tv} \geq \tau^e_{sv} > 0\,, \tag{7.95a}$$

$$\tau^e_{Tp} \geq \tau^e_{sp} \geq \tau^e_{sv} > 0\,. \tag{7.95b}$$

Processes in which T, p occur as free variables are damped the least. Processes in which s, v occur as free variables are damped the most. Equation (7.94a) with (7.66) leads to

$$c^e_v = \frac{\tau^e_{Tv}}{\tau^e_{sv}}\,c_{v,\,\zeta_e} \tag{7.96}$$

and (7.94b) with (7.84) to

$$c^e_p = \frac{\tau^e_{Tp}}{\tau^e_{sp}}\,c_{p,\,\zeta_e}\,. \tag{7.97}$$

The observation that $\tau^e_{Tv} = \tau^e_{Tp}$ by no means entails $\tau^e_{sp} = \tau^e_{sv}$ is finally of significance. According to (7.89), $\tau^e_{Tv} = \tau^e_{Tp}$ leads to $\varphi^e_{Tp} = \pi^e_{Tv} = 0$ and, according to (7.91) and (7.92), to $\kappa^e_T = \kappa_{T,\,\zeta_e}$, $k^e_T = k_{T,\,\zeta_e}$. In this case, the internal degree of freedom does not have any influence on the isothermal response functions. The internal degree of freedom is irrelevant with respect to the isothermal mechanical processes. With $\tau^e_{sp} \neq \tau^e_{sv} \neq \tau^e_{Tv}$, the same internal degree of freedom can still become noticeable in the heat capacity or during an isentropic mechanical perturbation. In the reverse case, the isothermal mechanical relevance of an internal degree of freedom by no means necessarily entails a relevance of this degree of freedom with respect to the heat capacity. The relevance of an internal degree of freedom or its strength thus depends on the external boundary conditions which the system is subject to.

The statements we have obtained about the mode of action of a macroscopically relevant molecular internal degree of freedom can easily be applied to more complicated systems. In general, the internal degree of freedom provides the system with a "memory" which, among other things, leads to the so-called phenomena of aftereffects. The response functions become dependent on the rate with which the system is perturbed (the mechanical moduli or compliances, the coefficients of thermal expansion and stress, the heat capacities, and also, for example, the so-called dielectric constants and magnetic permeabilities, which are not treated further here). One measures the response functions of an arrested equilibrium with a very fast perturbation and those of an internal equilibrium with a quasi-static perturbation. The strength of reaction of an internal degree of freedom depends on the external boundary conditions which the system is subject to. The reaction of an internal degree of freedom which is particularly pronounced during a certain manner of perturbation may disappear completely in the case of another manner of perturbation.

7.4 Relaxation Phenomena

If a system is brought into a non-equilibrium state $Z[T, v, \zeta(0)]$ and then left to itself under the condition T, $v = \text{const.}$, it will, with increasing time, usually strive to achieve an internal equilibrium state $Z[T, v, \zeta_e(T, v)]$. This process is called relaxation. In order to describe such a relaxation process, (7.31) or (7.32), respectively, lead to

$$\ddot{\zeta} = -\frac{1}{\tau_{Tv}}\dot{\zeta} \tag{7.98a}$$

or

$$\dot{\zeta}(t) = \dot{\zeta}(0)\,\mathrm{e}^{-\lambda_{Tv}(t)} \quad \text{with} \quad \lambda_{Tv}(t) \equiv \int_0^t \frac{\mathrm{d}t'}{\tau_{Tv}}. \tag{7.98b}$$

If the process should come to a standstill at a time $t = t_e$ after reaching the internal equilibrium state,

$$\lim_{t \to t_e} \lambda_{Tv}(t) = +\infty \tag{7.98c}$$

must necessarily be valid. If the attained equilibrium state is a stable or metastable state, it follows further that [see Eq. (7.35)]:

$$\lim_{t \to t_e} \tau_{Tv}(t) = \tau_{Tv}^e > 0 \,. \tag{7.98d}$$

Moreover, in the case of an ideal, non-singular continuous function $\tau_{Tv}(t)$, $t_e = +\infty$ can theoretically be expected according to (7.98b).

The relaxation processes to be described by (7.98) are generally non-linear processes, as the characteristic time

$$\tau_{Tv} = \tau_{Tv}[T, v, \zeta(t)] \tag{7.99}$$

depends on the instantaneous state of the system. However, if the initial state is not too far from the final state, one can approximatively assume according to (7.98d):

$$\tau_{Tv}(t) \approx \tau_{Tv}^e = \text{const.} > 0 \,. \tag{7.100}$$

Equation (7.98a) thus becomes a linear differential equation. The relaxation process is then only determined by the data of the initial and the final state, in particular by the time constant $\tau_{Tv}^e(T, v)$ which is determined by the final state.

If one inserts (7.100) into (7.98a), one obtains as the first integral of the differential equation

$$\dot{\zeta}(t) = \dot{\zeta}(0)\, e^{-t/\tau_{Tv}^e} \tag{7.101}$$

and as the second integral

$$\zeta(t) = [\zeta(0) - \zeta_e]\, e^{-t/\tau_{Tv}^e} + \zeta_e \,. \tag{7.102}$$

The initial rate of the process is given by

$$\dot{\zeta}(0) = -\frac{1}{\tau_{Tv}^e}[\zeta(0) - \zeta_e] \,. \tag{7.103}$$

With this, one can also replace (7.101) by

$$\dot{\zeta}(t) = -\frac{1}{\tau_{Tv}^e}[\zeta(t) - \zeta_e] \,. \tag{7.104}$$

The relaxation process becomes a monotonous exponential equilibration process. Equation (7.104) has the form of a "decay law", as is valid for many "naturally" proceeding processes (occurring without external disturbance). The constant τ_{Tv}^e is often designated as the Debye relaxation time.

In the neighbourhood of the final state, the equations of state (7.6–7.8) can be expanded in Taylor series and the series broken off after the linear terms.

Under the condition $T, v = \text{const.}$, one thus obtains with (7.33–7.35) and (7.9)

$$s = s_e + \left(\frac{\partial s}{\partial \zeta}\right)^e_{T,v} (\zeta - \zeta_e) = s_e + \sigma^e_{Tv}(\zeta - \zeta_e) , \tag{7.105}$$

$$p = p_e + \left(\frac{\partial p}{\partial \zeta}\right)^e_{T,v} (\zeta - \zeta_e) = p_e + \pi^e_{Tv}(\zeta - \zeta_e) , \tag{7.106}$$

$$a = a_e + \left(\frac{\partial a}{\partial \zeta}\right)^e_{T,v} (\zeta - \zeta_e) = -\frac{1}{L\tau^e_{Tv}} (\zeta - \zeta_e) . \tag{7.107}$$

Insertion of (7.102) results in

$$s(t) = [s(0) - s_e]\, \mathrm{e}^{-t/\tau^e_{Tv}} + s_e , \tag{7.108a}$$

$$p(t) = [p(0) - p_e]\, \mathrm{e}^{-t/\tau^e_{Tv}} + p_e , \tag{7.109a}$$

$$a(t) = a(0)\, \mathrm{e}^{-t/\tau^e_{Tv}} , \tag{7.110a}$$

with

$$s(0) - s_e = \sigma^e_{Tv}[\zeta(0) - \zeta_e] , \tag{7.108b}$$

$$p(0) - p_e = \pi^e_{Tv}[\zeta(0) - \zeta_e] , \tag{7.109b}$$

$$a(0) = -[\zeta(0) - \zeta_e]/L\tau^e_{Tv} . \tag{7.110b}$$

The entropy, the pressure, and the affinity decrease exponentially from their initial value to the final value. The rates of change of these quantities are given with (7.29) by

$$\dot{s} = \left(\frac{\partial s}{\partial \zeta}\right)_{T,v} \dot{\zeta} = L\, \sigma^e_{Tv}\, a(t) , \tag{7.111}$$

$$\dot{p} = \left(\frac{\partial p}{\partial \zeta}\right)_{T,v} \dot{\zeta} = L \pi^e_{Tv} a(t) , \tag{7.112}$$

$$\dot{a} = \left(\frac{\partial a}{\partial \zeta}\right)_{T,v} \dot{\zeta} = -\frac{1}{\tau^e_{Tv}} a(t) . \tag{7.113}$$

According to (2.18), we get for the internal energy of the relaxing system

$$\left(\frac{\partial u}{\partial \zeta}\right)_{T,v} = \left(\frac{\partial f}{\partial \zeta}\right)_{T,v} + T\left(\frac{\partial s}{\partial \zeta}\right)_{T,v} = -a + T\sigma^e_{Tv} ,$$

i.e., with (7.29)

$$\dot{u} = \left(\frac{\partial u}{\partial \zeta}\right)_{T,\,v} \dot{\zeta} = - La^2 + TL\sigma^e_{Tv}\, a\,. \tag{7.114a}$$

According to the first law (2.42) or (7.25), the change in the internal energy $\dot{u}$ under the condition T, v, $N =$ const. is identical with the heat $\dot{q}$ which must be exchanged within the time unit with the surroundings of the system. We also have

$$\frac{\dot{q}}{T} = L\,\sigma^e_{Tv}\, a - \frac{L}{T}\, a^2\,. \tag{7.114b}$$

Here, the first term on the right describes the change $\dot{s}$ in total entropy [according to (7.111)]. From this, one substracts a term which, according to (7.13) and (7.29), is identical to the entropy produced during the relaxation per time and mass unit in the interior of the system.

One should mention that an internal excess temperature

$$\Theta_{Tv}(t) \equiv - \frac{a(t)}{\sigma^e_{Tv}} = \left(\frac{\partial T}{\partial \zeta_e}\right)_v [\zeta(t) - \zeta_e] \tag{7.115}$$

can also be introduced in place of the internal variable $\zeta(t)$ in the description of the relaxation process [see Eqs. (7.107) and (7.60b)]. According to (7.113), this temperature follows the differential equation

$$\dot{\Theta}_{Tv} = - \frac{1}{\tau^e_{Tv}}\, \Theta_{Tv}(t)\,. \tag{7.116}$$

It vanishes when the internal equilibrium state is reached. With (7.65) and (7.66), one then obtains in place of (7.108) and (7.109)

$$s(t) = s_e + \frac{\Delta c_v}{T}\, \Theta_{Tv}(t)\,, \tag{7.117}$$

$$p(t) = p_e\,[1 + \Delta\beta\, \Theta_{Tv}(t)]\,, \tag{7.118}$$

i.e., equations which directly contain the contribution of the internal degree of freedom to the response functions. One obtains in place of (7.111) and (7.112)

$$\dot{s} = - \frac{1}{\tau^e_{Tv}} \frac{\Delta c_v}{T}\, \Theta_{Tv}(t)\,, \tag{7.119}$$

$$\dot{p} = - \frac{1}{\tau^e_{Tv}}\, p_e\, \Delta\beta\, \Theta_{Tv}(t)\,. \tag{7.120}$$

For the change in internal energy (7.114), one finds

$$\dot{u} = \dot{q} = - \frac{\Delta c_v}{\tau^e_{Tv}} \left(\Theta_{Tv} + \frac{1}{T} \Theta^2_{Tv} \right) . \tag{7.121}$$

When $T, p = \text{const.}$, relaxation processes have to be described, according to (7.38) or (7.39), by:

$$\ddot{\zeta} = - \frac{1}{\tau_{Tp}} \dot{\zeta} \tag{7.122a}$$

or

$$\dot{\zeta}(t) = \dot{\zeta}(0)\, \mathrm{e}^{-\lambda_{Tp}(t)} \qquad \text{with} \qquad \lambda_{Tp}(t) \equiv \int_0^t \frac{\mathrm{d}t'}{\tau_{Tp}} . \tag{7.122b}$$

The limiting conditions

$$\lim_{t \to t_e} \lambda_{Tp}(t) = + \infty \tag{7.122c}$$

and

$$\lim_{t \to t_e} \tau_{Tp}(t) = \tau^e_{Tp} > 0 \tag{7.122d}$$

are valid, and τ^e_{Tp} is the Debye relaxation time of the *G*-representation. Linearization of Eq. (7.122a) with

$$\tau_{Tp}(t) \approx \tau^e_{Tp} = \text{const.} > 0 \tag{7.123}$$

leads to the equation

$$\dot{\zeta}(t) = - \frac{1}{\tau^e_{Tp}} [\zeta(t) - \zeta_e] \tag{7.124}$$

with the solution

$$\zeta(t) = [\zeta(0) - \zeta_e]\, \mathrm{e}^{-t/\tau^e_{Tp}} + \zeta_e . \tag{7.125}$$

According to (7.90) or (7.95), the relaxation of an internal degree of freedom under the condition $T, p = \text{const.}$ is always slower than under the condition $T, v = \text{const.}$

Linearization of the equations of state (7.15) – (7.17) leads to

$$s = s_e + \sigma^e_{Tp} (\zeta - \zeta_e) , \tag{7.126}$$

$$v = v_e + \varphi^e_{Tp} (\zeta - \zeta_e) , \tag{7.127}$$

$$a = - \frac{1}{L\tau^e_{Tp}} (\zeta - \zeta_e) \tag{7.128}$$

or

$$s(t) = [s(0) - s_e] \, e^{-t/\tau^e_{Tp}} + s_e , \tag{7.129 a}$$

$$v(t) = [v(0) - v_e] \, e^{-t/\tau^e_{Tp}} + v_e , \tag{7.130 a}$$

$$a(t) = a(0) \, e^{-t/\tau^e_{Tp}} , \tag{7.131 a}$$

with

$$s(0) - s_e = \sigma^e_{Tp} [\zeta(0) - \zeta_e] , \tag{7.129 b}$$

$$v(0) - v_e = \varphi^e_{Tp} [\zeta(0) - \zeta_e] , \tag{7.130 b}$$

$$a(0) = - [\zeta(0) - \zeta_e]/L\tau^e_{Tp} \tag{7.131 b}$$

[see Eqs. (7.40–7.42)]. With (7.29), we further obtain

$$\dot{s} = \left(\frac{\partial s}{\partial \zeta}\right)_{T,p} \dot{\zeta} = L\sigma^e_{Tp} a(t) , \tag{7.132}$$

$$\dot{v} = \left(\frac{\partial v}{\partial \zeta}\right)_{T,p} \dot{\zeta} = L\varphi^e_{Tp} a(t) , \tag{7.133}$$

$$\dot{a} = \left(\frac{\partial a}{\partial \zeta}\right)_{T,p} \dot{\zeta} = - \frac{1}{\tau^e_{Tp}} a(t) . \tag{7.134}$$

According to the first law (7.25) or (2.42) with $T, p, N = \text{const.}$, the amount of heat to be exchanged with the surroundings of the system is given by

$$\dot{q} = \dot{u} + p\dot{v} = \dot{h}$$

[see the equations (2.29) and (2.18)]. As $\mu \equiv g$ holds with $N = \text{const.}$, (2.30) results in

$$\left(\frac{\partial h}{\partial \zeta}\right)_{T,p} = \left(\frac{\partial g}{\partial \zeta}\right)_{T,p} + T\left(\frac{\partial s}{\partial \zeta}\right)_{T,p}$$

and thus, according to (7.17), (7.29), and (7.126), in

$$\dot{q} = \dot{h} = \left(\frac{\partial h}{\partial \zeta}\right)_{T,p} \dot{\zeta} = - La^2 + TL\sigma^e_{Tp} a . \tag{7.135}$$

In addition,

$$\Theta_{Tp}(t) = -\frac{a(t)}{\sigma_{Tp}^{e}} = \left(\frac{\partial T}{\partial \zeta_e}\right)_p [\zeta(t) - \zeta_e] \tag{7.136}$$

can now be defined as the internal excess temperature [see the equations (7.128) and (7.78b)]. It satisfies the differential equation

$$\dot{\Theta}_{Tp} = -\frac{1}{\tau_{Tp}^{e}} \Theta_{Tp}(t) . \tag{7.137}$$

According to (7.83) and (7.84), one obtains

$$s(t) = s_e + \frac{\Delta c_p}{T} \Theta_{Tp}(t) , \tag{7.138}$$

$$v(t) = v_e[1 + \Delta a \Theta_{Tp}(t)] \tag{7.139}$$

and

$$\dot{s} = -\frac{1}{\tau_{Tp}^{e}} \frac{\Delta c_p}{T} \Theta_{Tp}(t) , \tag{7.140}$$

$$\dot{v} = -\frac{1}{\tau_{Tp}^{e}} v_e \Delta a \Theta_{Tp}(t) , \tag{7.141}$$

$$\dot{q} = \dot{h} = -\frac{\Delta c_p}{\tau_{Tp}^{e}} \left(\Theta_{Tp} + \frac{1}{T} \Theta_{Tp}^{2}\right) . \tag{7.142}$$

Similar "exponential laws" can easily be derived for more complicated systems [*e.g.*, for the deformation and stress variables of elastic systems or for the dielectric displacement (electric induction) and the electric field strength in a dielectric]. However, only in specific cases will we deal with true laws with a larger region of validity. For example, three presumptive approximations were necessary in order to obtain an exponential decrease or increase in the pressure (7.109) or in the volume (7.130): 1. the presumption of a linear dynamic law (7.29) with $L(T, v) = \text{const.}$ or $L(T, p) = \text{const.}$, 2. the linearization of the differential equations (7.98) or (7.122) with the presumption $\tau_{Tv} = \text{const.} > 0$ or $\tau_{Tp} = \text{const.} > 0$, respectively, and 3. the linearization of the mechanical equations of state. If the linear dynamic law (7.29) is valid, an exponential attenuation of the internal variable at a larger distance from the final state can only be expected if the Gibbs potentials, *e.g.*, the free energy, have the quadratic form

$$f = f_0 + f_1(\zeta - \zeta_e) + f_2(\zeta - \zeta_e)^2 .$$

For example, this is the case when the internal energy in the free energy outweighs the entropy, and the internal energy, similar to Eq. (6.109), is a quadratic function of the internal variable. According to (7.31 c), we then obtain τ_{Tv} = const. Such a constancy, however, cannot be expected in the case of a more complicated intermolecular interaction.

If the relaxation represents an attenuation process during which the internal variable monotonously drops from the initial value $\zeta(0) > \zeta_e$ to the equilibrium value, ζ_e, $\dot{\zeta} < 0$ is always valid. Moreover, if $\tau_{Tv} > 0$, (7.98) necessarily results in $\ddot{\zeta} > 0$. The attenuation curve $\zeta(t)$ then always has, as has the exponential function (7.102), a convex curvature *versus* the time axis. Due to (7.98 d), such a curvature is essential in the vicinity of the final state. With a larger distance from the final state, on the other hand, $(\partial^2 f/\partial\zeta^2)_{T,\,v}$ and thus τ_{Tv} can definitely also assume negative values. If $\tau_{Tv} < 0$ holds together with $\dot{\zeta} < 0$ at the beginning of the process, we get $\ddot{\zeta} < 0$. The monotonous attenuation curve is then first, as in an autocatalytic reaction, concavely curved *versus* the time axis. According to (7.98 a), a singularity occurs with $\ddot{\zeta} = 0$ following the transition from the concave curvature ($\ddot{\zeta} < 0$) to the convex curvature ($\ddot{\zeta} > 0$). With $\dot{\zeta} < 0$, either $\tau_{Tv} = 0$, $\dot{\zeta} = -\infty$ or $\tau_{Tv} = \pm\infty$, $\dot{\zeta}$ = finite is valid at this point. In the first case, the affinity also runs through a singularity with $a = -\infty$. In the second case, the affinity $a \sim \dot{\zeta}$ (because of $\ddot{\zeta} = 0$) passes through a finite minimum value. Two simple examples are shown in Figs. 7.2 and 7.3.

If the entropic part predominates in the free energy, one can expect a proportionality $f \sim \ln\zeta$ which leads to $\tau_{Tv} \sim -\zeta^2$. With the formulation

$$\tau_{Tv} = \tau_{Tv}^e - \tau_2[\zeta(t) - \zeta_e]^2, \quad \tau_2 = \text{const.} > 0\,, \tag{7.143}$$

two cases must be distinguished: If $\tau_{Tv}^e > \tau_2[\zeta(0) - \zeta_e]^2$ is valid, $\tau_{Tv}^e > \tau_{Tv} > 0$ always holds for all $\zeta(t) > \zeta_e$. The attenuation curve, like the exponential curve, is

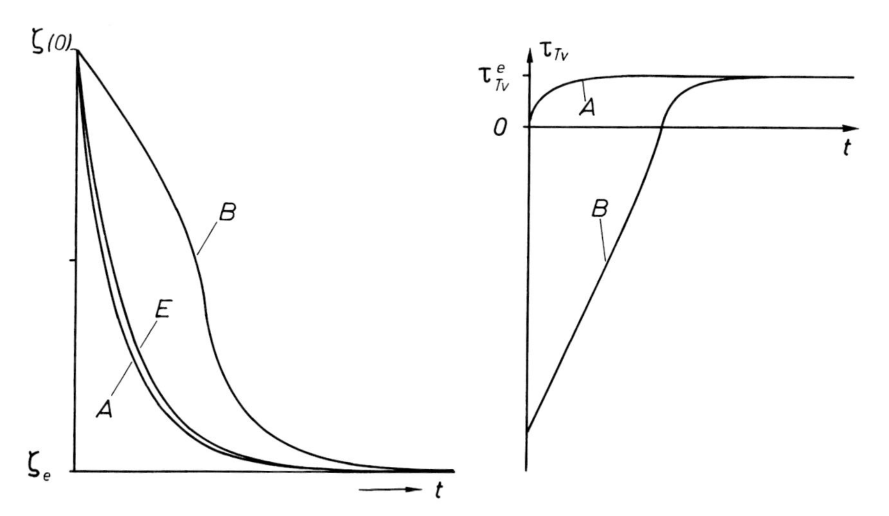

Fig. 7.2. Non-linear relaxation of the internal variable ζ according to (7.98) and (7.143). *A*: $0 < \tau_{Tv}(0) < \tau_{Tv}^e$; B: $_{Tv}(0) < 0 < \tau_{Tv}^e$. *E*: in comparison, linear relaxation according to (7.104). *t*: time

convexly curved *versus* the time axis. Relaxation, however, occurs – especially in the first process intervals – faster than in an exponential relaxation (Fig. 7.2 A). If, on the other hand, $\tau_{Tv}^e < \tau_2[\zeta(0) - \zeta_e]^2$ is valid, the process starts with $\tau_{Tv} < 0$. The attenuation curve is at first concavely curved *versus* the time axis. $\tau_{Tv} = 0$ and $\dot{\zeta} = -\infty$ result during the transition from the concave curvature to the convex curvature (Fig. 7.2 B). A singularity of the second case, for example, occurs if

$$\tau_{Tv} = \tau_0 - \frac{\tau_1}{\zeta - \zeta_s}, \quad \zeta_e < \zeta_s < \zeta(0) \tag{7.144a}$$

holds. In order to fulfil the condition (7.98 d),

$$\tau_0 = \tau_{Tv}^e - \frac{\tau_1}{\zeta_s - \zeta_e} \tag{7.144b}$$

must be valid, so that it follows that

$$\tau_{Tv} = \tau_{Tv}^e - \tau_1' \frac{\zeta - \zeta_e}{\zeta - \zeta_s}, \quad \tau_1' \equiv \tau_1/(\zeta_s - \zeta_e) \tag{7.144c}$$

(see Fig. 7.3).

Apart from exponential relaxation processes, hyperbolic relaxation processes with

$$[\zeta(t) - \zeta_e] \sim t^{-\beta}, \quad \beta > 0 \tag{7.145}$$

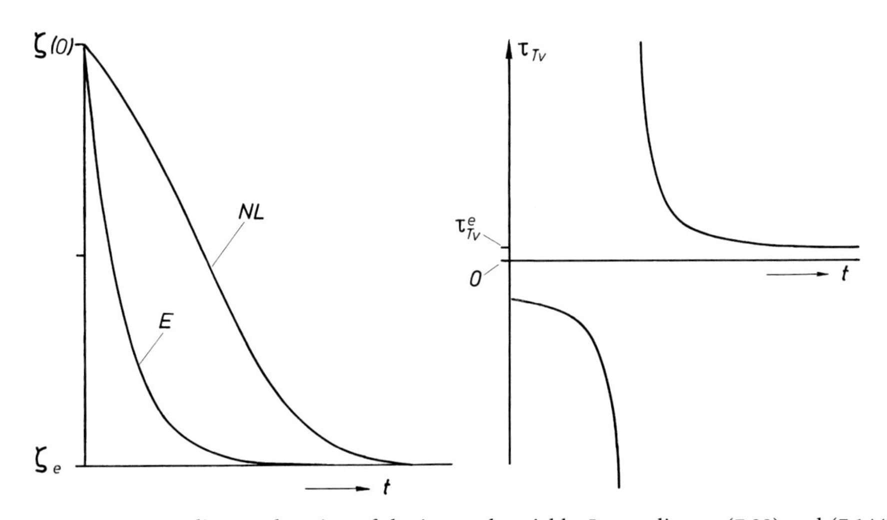

Fig. 7.3. *NL*: non-linear relaxation of the internal variable ζ according to (7.98) and (7.144); *E*: linear relaxation according to (7.104). Right: relaxation time in the case *NL*. *t*: time

are also observed (*e.g.*, when a Leyden jar is discharged). Hyperbolic relaxation phenomena are invariant under change of scale, *i.e.*, a change in the time scale $t \to t' = \gamma t$ leads to

$$\zeta(\gamma t) - \zeta_e = \gamma^{-\beta}[\zeta(t) - \zeta_e] .$$

An arbitrary stretching of the time scale can always be corrected by a corresponding compression of the curve $\zeta(t) - \zeta_e$. The appearance of $\zeta(t) - \zeta_e$ is similar in all time scales. Equation (7.145) is also possible as the solution of the differential equation (7.98a). Insertion of (7.145) into (7.98a) yields

$$\tau_{Tv} = \frac{t}{\beta + 1} \sim [\zeta(t) - \zeta_e]^{-1/\beta} . \tag{7.146}$$

With $\beta = 1/2$, we, for example, presuppose a free energy of the form

$$f = f_0 + f_1(\zeta - \zeta_e) + f_4(\zeta - \zeta_e)^4 .$$

Equation (7.146), however, only satisfies the subsidiary condition (7.98d) if

$$\tau^e_{Tv} = +\infty , \qquad i.e., \qquad \left(\frac{\partial^2 f}{\partial \zeta^2}\right)^e_{T,v} = 0 \qquad \text{or} \qquad L \to 0 \tag{7.147}$$

is valid, *i.e.*, if the final state with respect to the internal variable is a neutral or arrested internal equilibrium state (see further below). Proceeding from a finite initial value $\zeta(0)$ (7.145) must be normalized to

$$\zeta(t) = \left[\frac{t}{\tau} + [\zeta(0) - \zeta_e]^{-1/\beta}\right]^{-\beta} + \zeta_e . \tag{7.148}$$

This solution is scaling-invariant for all times

$$t \gg \tau[\zeta(0) - \zeta_e]^{-1/\beta} .$$

A characteristic of the hyperbolic relaxation is that it occurs considerably faster than a comparable exponential relaxation at the beginning of the process, but considerably slower at the end of the process (see Fig. 7.4 H).

A similar behaviour is observed during the so-called Kohlrausch–Williams–Watts relaxation [see Kohlrausch (1847); Williams, Watts (1970)], where it is assumed that the relaxing quantity follows the proportionality

$$[\zeta(t) - \zeta_e] \sim \exp\left[-\left(\frac{t}{\tau}\right)^\beta\right] , \quad 0 < \beta \le 1 \tag{7.149}$$

(Fig. 7.4, KWW). If one considers (7.149), as is common in the literature, as the

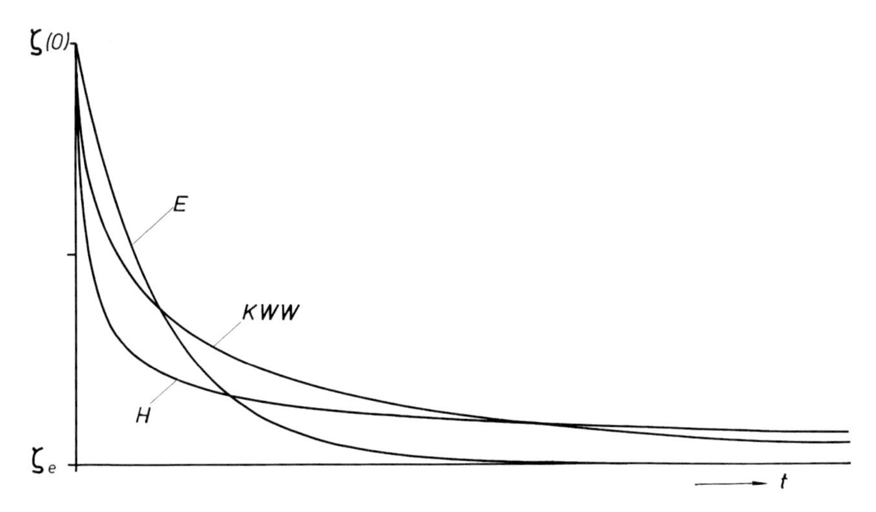

Fig. 7.4. *H*: hyperbolic relaxation according to (7.148) with $\beta = 1/2$; KWW: Kohlrausch–Williams–Watts relaxation according to (7.149) with $\beta = 1/2$; *E*: exponential relaxation according to (7.102)

solution of a differential equation of the form (7.104), one obtains

$$\tau_{Tv} = \frac{1}{\beta}\,\tau^{\beta} t^{1-\beta}\,. \tag{7.150}$$

As the solution of the somewhat more general equation (7.98 a), (7.149), on the other hand, leads to

$$\tau_{Tv} = \frac{t}{(1-\beta) + \beta\left(\dfrac{t}{\tau}\right)^{\beta}}\,. \tag{7.151}$$

The expressions (7.150) and (7.151) become identical for large values of time *t*. The proportionality (7.149) also requires infinitely large Debye relaxation times if it is supposed to satisfy the equations (7.98) or (7.122) [see (7.147)]. The Kohlrausch–Williams–Watts relaxation, therefore, only complies with the assumption of a linear dynamic law of the form (7.29) if the final state of the relaxation with respect to the internal degree of freedom is a neutral equilibrium state, or if the internal degree of freedom increasingly freezes-in during the relaxation and finally appears completely arrested in the final state. Freezing processes, however, generally require a variable coefficient of coupling *L* (see Sect. 7.8), while $L = \text{const.}$ was assumed in (7.98).

The *crux* of the phenomenological description of the relaxation processes is that the non-linear relaxation of a single internal degree of freedom can often equally well be described by the additive superimposition of the exponential (*i. e.*, linear) relaxation of many internal degrees of freedom with a spectrum of Debye relaxation times (see Sect. 7.7). Reasons for the assumption of a relaxation time spectrum can almost always be found (*e.g.*, in the always present fluctuations of the mass density). On the other hand, linear relaxation equations can only be expected, even in the relatively simple thermodynamics of irreversible processes, if the coupling coefficient L is constant and the Gibbs potentials can be represented in a simple quadratic form. In the case of non-exponentially proceeding relaxation processes, one, therefore, often concerns oneself with a superimposition of coupled non-linear processes with varying damping. On the diversity of relaxation phenomena, for example, see Campbell, Giovannella (1990).

Another kind of relaxation (or retardation) occurs if one of the external variables is suddenly brought to another value. As an example, we have chosen an instantaneous isothermal change in volume at the time t_1 from a value v_0 to a value v_1. If we fix the zero point of the time to $t = -\infty$,

$$-\infty \leq t \leq t_1 : v = v_0 \,; \quad t_1 \leq t \leq +\infty : v = v_1 > v_0 \,; \quad -\infty \leq t \leq +\infty : T = \text{const.}$$

should be valid in the indicated time intervals. We assume the initial state to be an internal equilibrium state $Z[T, v_0, \zeta_0(T, v_0)]$, so that $\dot{\zeta}(-\infty) = 0$ results. The change in volume at the time t_1 occurs so rapidly that the internal degree of freedom is not capable of following this change. At time t_1 the system changes into the arrested equilibrium state $Z[T, v_1, \zeta = \zeta_0(T, v_0)]$. The arrest, however, only exists during the instantaneous change in volume from v_0 to v_1. The arrest is removed with $v_1 = \text{const.}$ during the rest of the perturbation process. The state arrested at $t = t_1$ turns into an unarrested non-equilibrium state from which the system relaxes into the internal equilibrium state $Z[T, v_1, \zeta_1(T, v_1)]$. In the following, we will confine ourselves again to the strictly linear case and suppose that all coefficients of the determinant equations can be replaced by the constant coefficients of the initial state. For the description of the process, (7.36), corresponding to (7.32), leads as the first integral to

$$\dot{\zeta}(t) = L\pi^0_{Tv}\, \mathrm{e}^{-t/\tau^0_{Tv}} \int_{-\infty}^{t} \mathrm{e}^{t'/\tau^0_{Tv}}\, \dot{v}_T(t')\, \mathrm{d}t' \,. \tag{7.152}$$

The isothermal volume change is given by the Dirac delta function

$$\dot{v}_T(t) = (v_1 - v_0)\, \delta(t - t_1) \,.$$

With this, (7.152) further leads for times $t \geq t_1$ to

$$\dot{\zeta}(t) = L\pi^0_{Tv}(v_1 - v_0)\, \mathrm{e}^{-(t - t_1)/\tau^0_{Tv}} \,. \tag{7.153}$$

In this type of relaxation, the driving force of the process, the affinity $a(t) = \dot{\zeta}/L$, is not, as in the preceding cases, determined by the difference $\zeta(0) - \zeta_e$ of the internal variable, but by the volume difference $v_1 - v_0$. At the time t_1, it starts with its maximum value

$$a(t_1) = \pi^0_{Tv}\,(v_1 - v_0)$$

in order to exponentially fade out in the course of time. The integration of (7.153) then yields the equation

$$\zeta(t) = \zeta_0 + L\tau^0_{Tv}\pi^0_{Tv}(v_1 - v_0)\,[1 - \mathrm{e}^{-(t-t_1)/\tau^0_{Tv}}] \tag{7.154}$$

for the change in the internal variable with respect to time. The internal variable increases or decreases exponentially from the initial value ζ_0 to the new equilibrium value

$$\zeta_1 = \zeta_0 + L\tau^0_{Tv}\pi^0_{Tv}(v_1 - v_0) = \zeta_0 + \left(\frac{\partial \zeta_0}{\partial v}\right)_T (v_1 - v_0)$$

[see Eq. (7.60c)]. If ζ_1 is smaller or greater than ζ_0, of course, depends on whether the concentration ζ_0 of the internal degree of freedom in the internal equilibrium increases or decreases with the volume.

Linearization of the mechanical equation of state yields with (7.57) and (7.34)

$$p = p_0 + \left(\frac{\partial p}{\partial v}\right)_{T,\,\zeta_0} (v_1 - v_0) + \left(\frac{\partial p}{\partial \zeta}\right)^0_{T,\,v} (\zeta - \zeta_0)$$

$$= p_0 - \frac{1}{v_0} k_{T,\,\zeta_0} (v_1 - v_0) + \pi^0_{Tv} (\zeta - \zeta_0)\,.$$

If one inserts (7.154), one obtains with (7.64) for the change in pressure during the relaxation of the internal degree of freedom

$$p(t) - p_0 = -\,[k_{T,\,\zeta_0} + \Delta k_T\,[1 - \mathrm{e}^{-(t-t_1)/\tau^0_{Tv}}]]\,\frac{v_1 - v_0}{v_0}\,. \tag{7.155a}$$

The pressure reacts to the instantaneous volume change at the time t_1 with the instantaneous Hookean change of the arrested equilibrium

$$p(t_1) - p_0 = -\,k_{T,\,\zeta_0}\,\frac{v_1 - v_0}{v_0}$$

[see Eqs. (5.34/5.35)]. However, the pressure still continues to react. The relaxation of the internal degree of freedom induces an aftereffect. With increasing time, the contribution Δk_T of the internal degree of freedom becomes more and more important until the Hookean pressure of the internal equilibrium is finally reached

$$p_1 - p_0 \equiv p(\infty) - p_0 = k^0_T\,\frac{v_1 - v_0}{v_0}\,.$$

Aftereffect phenomena are naturally atypical for Hookean bodies which follow a linear mechanical equation of state with constant coefficients. However, after elimination of the internal variables in the mechanical equation of state, the relaxation process can be described by a pseudo-Hookean law

$$p(t) - p_0 = - k_T(t) \frac{v_1 - v_0}{v_0} \tag{7.155b}$$

in which the so-called relaxation modulus

$$k_T(t) \equiv k_{T,\zeta_0} + \Delta k_T [1 - e^{-(t - t_1)/\tau^0_{Tv}}] , \quad t \geq t_1 \tag{7.155c}$$

is a time-dependent response function.

In the same way, (7.43) yields in the G-representation in place of (7.152)

$$\dot{\zeta}(t) = - L\, \varphi^0_{Tp}\, e^{-t/T^0_{Tp}} \int_{-\infty}^{t} e^{t'/\tau^0_{Tp}}\, \dot{p}_T(t')\, dt' , \tag{7.156}$$

in the case of an instantaneous isothermal change in pressure at the time t_1

$$\dot{p}_T(t) = (p_1 - p_0)\, \delta(t - t_1) ,$$

$$\dot{\zeta}(t) = - L\varphi^0_{Tp}(p_1 - p_0)\, e^{-(t - t_1)/\tau^0_{Tp}} , \quad t \geq t_1 \tag{7.157}$$

and

$$\zeta(t) = \zeta_0 + L\tau^0_{Tp}\varphi^0_{Tp}(p_1 - p_0)\, [e^{-(t - t_1)/\tau^0_{Tp}} - 1] . \tag{7.158}$$

The new equilibrium value, which the internal variable strives to achieve after the pressure change, corresponds to

$$\zeta_1 = \zeta_0 - L\tau^0_{Tp}\varphi^0_{Tp}(p_1 - p_0) = \zeta_0 + \left(\frac{\partial \zeta_0}{\partial p}\right)_T (p_1 - p_0)$$

[see Eq. (7.78c)]. Linearization of the mechanical equation of state with (7.75) and (7.41) leads to

$$v = v_0 + \left(\frac{\partial v}{\partial p}\right)_{T,\zeta_0} (p_1 - p_0) + \left(\frac{\partial v}{\partial \zeta}\right)^0_{T,p} (\zeta - \zeta_0)$$

$$= v_0 - v_0 \kappa_{T,\zeta_0}(p_1 - p_0) + \varphi^0_{Tp}(\zeta - \zeta_0) .$$

Elimination of the internal variables *via* (7.158) yields with (7.82) the equation

$$v(t) - v_0 = - v_0 \kappa_{T,\zeta_0}(p_1 - p_0) + v_0 \Delta\kappa_T(p_1 - p_0)\, [e^{-(t - t_1)/\tau^0_{Tp}} - 1] ,$$

i.e., the pseudo-Hookean law

$$\frac{v(t) - v_0}{v_0} = - \kappa_T(t)\, (p_1 - p_0) \tag{7.159a}$$

with the time-dependent compressibility

$$\kappa_T(t) = \kappa_{T,\,\zeta_0} + \Delta\kappa_T[1 - \mathrm{e}^{-(t-t_1)/\tau^0_{Tp}}]\,, \quad t \geq t_1\,. \tag{7.159b}$$

Upon a sudden change in the value of a macroscopic variable, the macroscopic relevance of an internal degree of freedom generally has the effect that the adequate response function first assumes the arrested equilibrium value and then relaxes to the internal equilibrium value.

7.5 Linear Response to Periodic Perturbation

An aid which is often used in practice to obtain information on the macroscopic relevance of internal degrees of freedom is the isothermal periodic perturbation of a material. We will first consider the example of an isothermal periodic change in volume. The initial state is once again an internal equilibrium state $Z[T, v, \zeta_e(T, v)]$, so that $\dot{\zeta} = 0$ holds at time zero. The maximum deviation Δv from this equilibrium state is assumed to be so small that the coefficients in Eqs. (7.31) and (7.52) can be replaced by coefficients referring to the initial state [see Eqs. (7.33)–(7.36)]. With $\dot{T} = 0$, we then obtain from (7.36), as in the case of (7.152),

$$\dot{\zeta}_T(t) = L\pi^e_{Tv}\,\mathrm{e}^{-t/\tau^e_{Tv}} \int_{-\infty}^{t} \mathrm{e}^{t'/\tau^e_{Tv}}\, \dot{v}_T(t')\,\mathrm{d}t'\,. \tag{7.160}$$

One took into account here that the zero point of the time scale has to be fixed at $t = -\infty$ in a periodic process with a sharply defined frequency. Periodic perturbation of the volume with the frequency $\nu = \omega/2\pi$ means in a complex description

$$v(t) = v_e + \Delta v\,\mathrm{e}^{i\omega t} \tag{7.161}$$

and

$$\dot{v}_T = i\omega\,\Delta v\,\mathrm{e}^{i\omega t} = i\omega[v(t) - v_e]\,. \tag{7.162}$$

Insertion of (7.162) into (7.160) leads to

$$\dot{\zeta}(t) = L\pi^e_{Tv}\, i\omega\,\Delta v\,\mathrm{e}^{-t/\tau^e_{Tv}} \int_{-\infty}^{t} \mathrm{e}^{(i\omega + 1/\tau^e_{Tv})t'}\,\mathrm{d}t'\,,$$

i.e.,

$$\dot{\zeta}(t) = L\tau^e_{Tv}\pi^e_{Tv}\,\frac{i\omega\Delta v}{1 + i\omega\tau^e_{Tv}}\,\mathrm{e}^{i\omega t}\,. \tag{7.163}$$

Further integration results in

$$\zeta(t) = \zeta_e + L\tau_{Tv}^e \pi_{Tv}^e \frac{[v(t) - v_e]}{1 + i\,\omega\,\tau_{Tv}^e} \,. \tag{7.164}$$

Hence, the internal variable is also subject to a periodic change. Because of [7]

$$\frac{1}{1 + i\,\omega\,\tau_{Tv}^e} \equiv \frac{1}{\sqrt{1 + \omega^2 (\tau_{Tv}^e)^2}}\, e^{-i\gamma} \quad \text{with} \quad \tan\gamma = \omega\,\tau_{Tv}^e \,,$$

however, there is a phase shift γ between the external perturbation $v(t) - v_e$ and the oscillations of the internal variable, which is dependent on the frequency ω. In the range $\omega\,\tau_{Tv}^e \ll 1$, the internal variable $\zeta(t)$ follows the external perturbation practically instantaneously with $\gamma \approx 0$. This corresponds to a quasi-static perturbation. In the range $\omega\,\tau_{Tv}^e \gg 1$, on the other hand, $\zeta(t) - \zeta_e$ lags behind the external perturbation $v(t) - v_e$ by $\gamma \approx 90°$. The amplitude of $\zeta(t) - \zeta_e$ then becomes ever smaller with increasing frequency until in the limit $\omega \to \infty$, one obtains $\zeta(t) = \zeta_e = \text{const}$. This corresponds to a rapid perturbation with an arrested internal variable.

Equations (7.163), (7.162), and (7.60 c) lead to

$$\left(\frac{d\zeta}{dv}\right)_T = \frac{\dot{\zeta}_T}{\dot{v}_T} = L\tau_{Tv}^e \pi_{Tv}^e \frac{1}{1 + i\omega\tau_{Tv}^e} = \left(\frac{\partial \zeta_e}{\partial v}\right)_T \frac{1}{1 + i\,\omega\,\tau_{Tv}^e} \,.$$

Insertion of this expression into (7.52) yields with (7.57) and (7.64) the complex frequency-dependent bulk modulus

$$k_T(\omega) = k_{T,\zeta_e} + \frac{\Delta k_T}{1 + i\omega\tau_{Tv}^e} \tag{7.165 a}$$

with the real part

$$k_T'(\omega) = k_{T,\zeta_e} + \frac{\Delta k_T}{1 + \omega^2 (\tau_{Tv}^e)^2} \tag{7.165 b}$$

and the imaginary part

$$k_T''(\omega) = -\frac{\omega\,\tau_{Tv}^e\,\Delta k_T}{1 + \omega^2 (\tau_{Tv}^e)^2} \,. \tag{7.165 c}$$

7 One should remember that every complex number $z = a + ib$ can also be represented by

$$z = |z|\,e^{i\varphi} = |z|\,(\cos\varphi + i\sin\varphi)$$

with

$$|z| = \sqrt{a^2 + b^2}, \quad \cos\varphi = \frac{a}{|z|}, \quad \sin\varphi = \frac{b}{|z|} \,.$$

Alternatively, one can also write in place of (7.165b):

$$k_T'(\omega) = k_T^e - \frac{\omega^2 (\tau_{Tv}^e)^2 \, \Delta k_T}{1 + \omega^2 \, (\tau_{Tv}^e)^2} \, . \tag{7.165d}$$

The relations

$$k_T'(\omega) = k_T^e + \omega \, \tau_{Tv}^e \, k_T''(\omega) \tag{7.166a}$$

and

$$k_T''(\omega) = \omega \, \tau_{Tv}^e \, [k_{T,\zeta_e} - k_T'(\omega)] \tag{7.166b}$$

obviously exist between the real part and the imaginary part.

The real part $k_T'(\omega)$, the so-called storage modulus, is, according to (2.44), the dynamic analogue to the equilibrium modulus k_T. Hence,

$$\omega \, \tau_{Tv}^e \gg 1 \quad : \quad k_T(\omega) = k_T'(\omega) = k_{T,\zeta_e} \, ,$$

$$\omega \, \tau_{Tv}^e \ll 1 \quad : \quad k_T(\omega) = k_T'(\omega) = k_T^e$$

follow in the limiting cases of the arrested and the internal equilibrium. The internal degree of freedom yields the total contribution Δk_T to the bulk modulus only when the frequency is very low. This contribution becomes ever smaller with increasing frequency.

The imaginary part $k_T''(\omega)$, the so-called loss modulus, on the other hand, does not have an analogue in the equilibrium thermodynamics. For the limiting conditions, one finds

$$\omega \, \tau_{Tv}^e \gg 1 \quad : \quad k_T''(\omega) = -\Delta k_T / \omega \, \tau_{Tv}^e \, ,$$

$$\omega \, \tau_{Tv}^e \ll 1 \quad : \quad k_T''(\omega) = - \, \Delta k_T \omega \, \tau_{Tv}^e \, ,$$

so that the loss modulus of the arrested equilibrium ($\omega \to \infty$) and the internal equilibrium ($\omega \to 0$) vanishes. The loss modulus is a measure of the entropy produced in non-equilibrium per half-period of the oscillation $v(t) - v_e$, *i.e.*, a measure of the mechanical energy dissipated per half-period. If we denote the conjugated complex quantity with "*", (7.13), (7.29), and (7.163) lead for the entropy production of the process to

$$\frac{\mathrm{d}_i s}{\mathrm{d}t} = \frac{1}{LT} \dot{\zeta}_T^* \, \dot{\zeta}_T = \frac{L}{T} \, (\pi_{Tv}^e \Delta v)^2 \, \frac{\omega^2 (\tau_{Tv}^e)^2}{1 + \omega^2 (\tau_{Tv}^e)^2} \, .$$

The entropy produced per half-period of the oscillation amounts to

$$\Delta_i s \equiv \int_{\pi n/\omega}^{\pi(n+1)/\omega} \frac{\mathrm{d}_i s}{\mathrm{d}t}\,\mathrm{d}t = \frac{\pi}{T} L \tau_{Tv}^e (\pi_{Tv}^e)^2 (\Delta v)^2 \frac{\omega \tau_{Tv}^e}{1+\omega^2(\tau_{Tv}^e)^2}$$

$$= -\frac{\pi}{T}\frac{(\Delta v)^2}{v_e}\Delta k_T \frac{\omega \tau_{Tv}^e}{1+\omega^2(\tau_{Tv}^e)^2}$$

[see Eq. (7.64)]. According to (7.165 c), we thus have

$$k_T''(\omega) = \frac{v_e}{\pi(\Delta v)^2} T \Delta_i s \,. \tag{7.167}$$

In Fig. 7.5, the storage and loss moduli are represented as the function of $\ln \omega \tau_{Tv}^e$. The bell-shaped loss curve between the internal and the arrested equilibrium has a half width of $\Delta(\ln \omega \tau_{Tv}^e) = 2.634$ or $\Delta(\log \omega \tau_{Tv}^e) = 1.44$, *i. e.*, comprises more than a decade. There can be two causes for an experimental determination of a greater half width: Eqs. (7.31) are either non-linear, *i. e.*, the coefficients depend on $\zeta(t)$, or several internal degrees of freedom are effective whose Debye relaxation times τ_{Tv}^e do not differ greatly (see Fig. 7.14). Moreover, the storage and loss moduli are not independent of each other. When the equilibrium moduli k_T^e, k_{T,ζ_e} are known, one modulus can be determined from the other *via* (7.166). Equations (7.166) are a special case of the so-called Kramers–

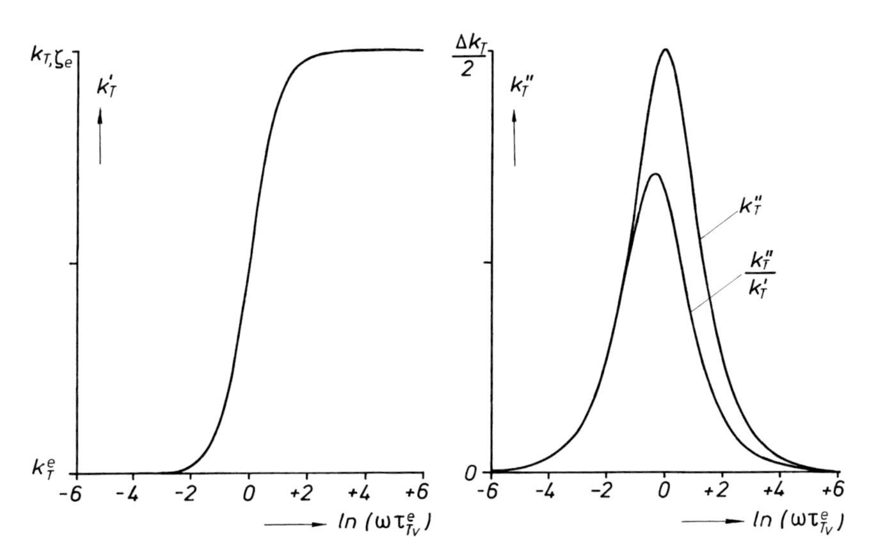

Fig. 7.5. Dynamic modulus (storage modulus) k_T' according to (7.165 b), loss modulus k_T'' according to (7.165 c) and loss factor $\tan\delta = k_T''/k_T'$ according to (7.169) as a function of $\ln(\omega\tau_{Tv}^e)$. Frequency of the external perturbation: ω; Debye's relaxation time in the F-representation: τ_{Tv}^e (Correction: On the ordinate Δk_T should be replaced by $-\Delta k_T > 0$)

Kronig relations which, when certain preconditions are satisfied, generally link the real part $\chi'(\omega)$ with the imaginary part $\chi''(\omega)$ of the Fourier components of complex response functions $\chi(t)$ [*e.g.*, see De Groot, Mazur (1962) or Isihara (1971)].

The assumed vicinity to the initial state also allows a linearization of the mechanical equations of state. Between the pressure and volume, we then have the pseudo-Hookean law

$$p(t) - p_e = - k_T(\omega) \frac{v(t) - v_e}{v_e} \tag{7.168}$$

in which the response function depends on the frequency [see Sect. 7.4]. Because of

$$k_T(\omega) = \sqrt{(k_T')^2 + (k_T'')^2}\, e^{i\delta} \quad \text{with} \quad \tan\delta = \frac{k_T''}{k_T'}, \tag{7.169}$$

a phase difference δ develops between the pressure and the volume which only disappears in the limiting cases of the internal and the arrested equilibrium ($k_T'' \to 0$). Angle δ is designated as the loss angle, tan δ as the loss factor. One should bear in mind, however, that the loss angle and loss factor only provide a relative measurement of the actual losses (the entropy produced or the energy dissipated per half-period of the oscillation). The maximum of tan δ as a function of ω or ln $\omega\tau_{Tv}^e$ by no means coincides with the maximum of the energy loss (Fig. 7.5). The curve tan $\delta(\omega)$ always appears deformed as compared to the true loss curve $k_T''(\omega)$, because the denominator $k_T'(\omega)$ changes greatly, especially in the range of greater losses.

For an isothermal, periodic pressure loading, (7.43) leads to

$$\dot{\zeta}_T(t) = - L\varphi_{Tp}^e\, e^{-t/\tau_{Tp}^e} \int_{-\infty}^{t} e^{t'/\tau_{Tp}^e}\, \dot{p}_T(t')\, dt' \tag{7.170}$$

and from this with

$$p(t) = p_e + \Delta p\, e^{i\omega t}, \tag{7.171}$$

$$\dot{p}_T(t) = i\omega\Delta p\, e^{i\omega t} = i\omega[p(t) - p_e], \tag{7.172}$$

corresponding to (7.163) and (7.164), to

$$\dot{\zeta}_T(t) = - L\tau_{Tp}^e\, \varphi_{Tp}^e \frac{i\omega\Delta p}{1 + i\omega\tau_{Tp}^e}\, e^{i\omega t}, \tag{7.173}$$

$$\zeta(t) = \zeta_e - L\tau_{Tp}^e\, \varphi_{Tp}^e \frac{[p(t) - p_e]}{1 + i\omega\tau_{Tp}^e}\,. \tag{7.174}$$

Insertion of (7.173) and (7.172) into (7.71) yields with (7.82) the complex frequency-dependent compressibility

$$\kappa_T(\omega) = \kappa_{T,\zeta_e} + \frac{\Delta\kappa_T}{1 + i\omega\tau^e_{Tp}} \ . \tag{7.175a}$$

When splitting the compressibility into a reactive and a dissipative part, corresponding to the partition (7.165), it is useful to select the dissipative part in such a way that it is always positive [compare the signs in (7.13), (7.64), and (7.82)]. When defining the storage and loss quantities, one should then proceed from

$$\kappa_T(\omega) = \kappa'_T(\omega) - i\kappa''(\omega) \ . \tag{7.175b}$$

The dynamic compressibility is given by

$$\kappa'_T(\omega) = \kappa_{T,\zeta_e} + \frac{\Delta\kappa_T}{1 + \omega^2(\tau^e_{Tp})^2} \tag{7.175 c}$$

and the loss compressibility by

$$\kappa''_T(\omega) = \frac{\omega\tau^e_{Tp}\Delta\kappa_T}{1 + \omega^2(\tau^e_{Tp})^2} \tag{7.175d}$$

(see Fig. 7.6).

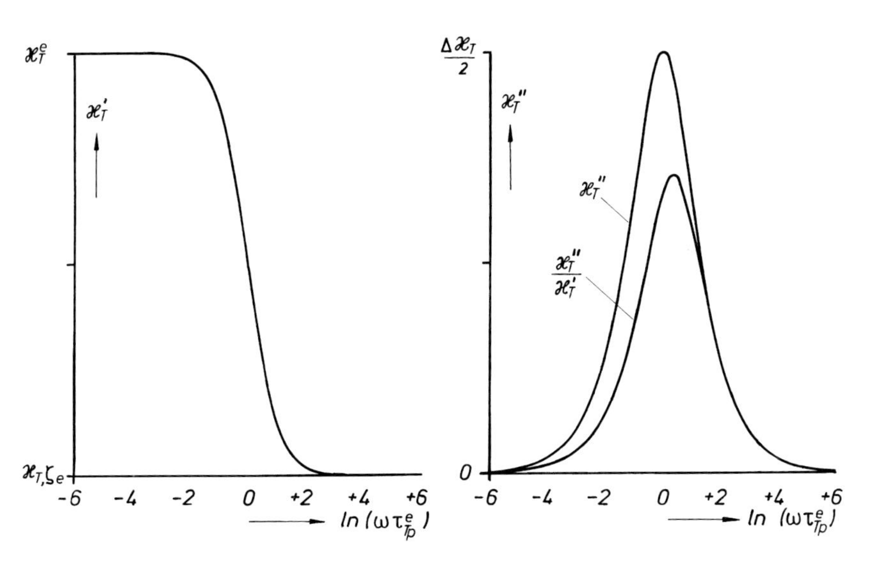

Fig. 7.6. Dynamic compressibility κ'_T according to (7.175c), loss compressibility κ''_T according to (7.175d) and loss factor $\tan\delta = \kappa''_T/\kappa'_T$ according to (7.180) as a function of $\ln(\omega\tau^e_{Tp})$. Frequency of the external perturbation: ω; Debye's relaxation time in the G-representation: τ^e_{Tp}

Insertion of (7.91) into (7.175 a) yields

$$\kappa_T(\omega) = \kappa_{T,\zeta_e} \frac{\tau^e_{Tp}(1 + i\omega\tau^e_{Tv})}{\tau^e_{Tv}(1 + i\omega\tau^e_{Tp})} = \kappa^e_T \frac{1 + i\omega\tau^e_{Tv}}{1 + i\omega\tau^e_{Tp}} . \tag{7.176}$$

In the same way, (7.92) and (7.165 a) lead to

$$k_T(\omega) = k_{T,\zeta_e} \frac{\tau^e_{Tv}(1 + i\omega\tau^e_{Tp})}{\tau^e_{Tp}(1 + i\omega\tau^e_{Tv})} = k^e_T \frac{1 + i\omega\tau^e_{Tp}}{1 + i\omega\tau^e_{Tv}} . \tag{7.177}$$

Hence, with (7.88) or (7.93), we have for the complex mechanical response functions

$$k_T(\omega) = 1/\kappa_T(\omega) , \tag{7.178}$$

which is as it has to be if the pseudo-Hookean law (7.168) should be invertible to

$$\frac{v(t) - v_e}{v_e} = - \kappa_T(\omega)\,[p(t) - p_e] . \tag{7.179}$$

Equations (7.171), (7.179), and (7.175b) lead to a phase difference δ between the pressure and the volume, which is given by

$$\kappa_T(\omega) = \sqrt{(\kappa'_T)^2 + (\kappa''_T)^2}\, e^{i\delta}, \quad \tan\delta = \frac{\kappa''_T}{\kappa'_T} . \tag{7.180}$$

Further, (7.176) and (7.175 b) result in

$$\kappa'_T(\omega) = \kappa^e_T \frac{1 + \omega^2\tau^e_{Tv}\tau^e_{Tp}}{1 + \omega^2(\tau^e_{Tp})^2} , \quad \kappa''_T(\omega) = \kappa^e_T \frac{\omega(\tau^e_{Tp} - \tau^e_{Tv})}{1 + \omega^2(\tau^e_{Tp})^2} \tag{7.181}$$

and (7.177) in

$$k'_T(\omega) = k^e_T \frac{1 + \omega^2\tau^e_{Tp}\tau^e_{Tv}}{1 + \omega^2(\tau^e_{Tv})^2} ; \quad k''_T(\omega) = k^e_T \frac{\omega(\tau^e_{Tp} - \tau^e_{Tv})}{1 + \omega^2(\tau^e_{Tv})^2} . \tag{7.182}$$

We thus obtain

$$\frac{k''_T}{k'_T} = \frac{\kappa''_T}{\kappa'_T} = \frac{\omega(\tau^e_{Tp} - \tau^e_{Tv})}{1 + \omega^2\tau^e_{Tp}\tau^e_{Tv}} . \tag{7.183}$$

The loss factor and the loss angle are equal on periodic perturbation of the pressure or the volume. The true loss curves, however, are different. The maximum of the loss quantity of the modulus is always at higher frequencies than the maximum of the loss quantity of the compliance because of $\tau^e_{Tp} > \tau^e_{Tv}$ [see

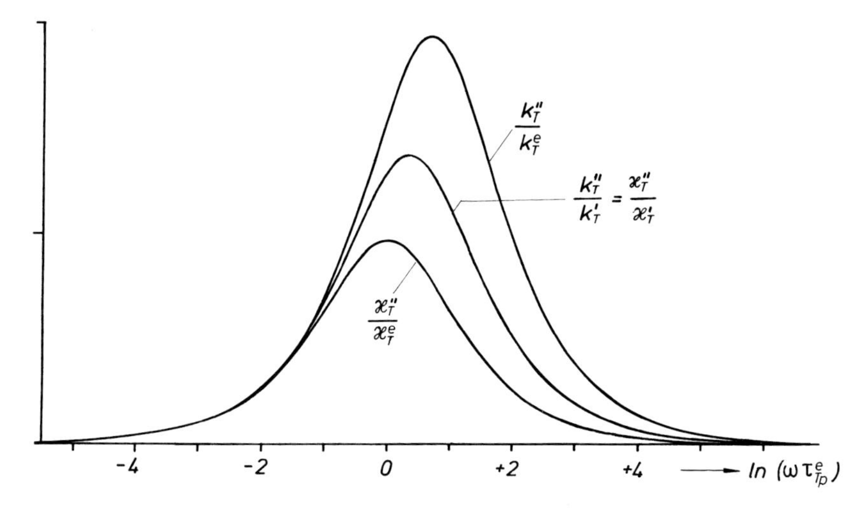

Fig. 7.7. The different loss quantities in the case of a periodic change in volume or pressure according to (7.181), (7.182), and (7.183) with $\tau_{Tv}^e = \tau_{Tp}^e/2$

(7.90) or (7.95)] (Fig. 7.7). After a long but simple calculation, we finally obtain from (7.181) and (7.182)

$$|k_T(\omega)|\,|\kappa_T(\omega)| = [(k_T')^2 + (k_T'')^2]^{1/2}\,[(\kappa_T')^2 + (\kappa_T'')^2]^{1/2} = 1\,,$$

i.e.,

$$|k_T(\omega)| = 1/|\kappa_T(\omega)| \tag{7.184}$$

and thus

$$\kappa_T'(\omega) = \frac{k_T'(\omega)}{|k_T(\omega)|^2}\,;\quad \kappa_T''(\omega) = \frac{k_T''(\omega)}{|k_T(\omega)|^2}\,. \tag{7.185}$$

As can be seen in Figs. 7.5 to 7.7, the dispersion ranges of the mechanical response functions (the steps in the moduli and compliances, the bell-shaped loss curves) can possibly extend over several decades of the perturbation frequency even if only one internal degree of freedom is relevant. The variability of the frequency, however, is usually very limited in mechanically vibrating devices. In order to generate the complete dispersion curve of mechanical response functions one has, therefore, to depend on a series of different testing methods [*e.g.*, testing with low-frequency forced vibrations, with sound and ultrasonic waves and Brillouin scattering; *e.g.*, compare McCrum, Read, Williams (1967); Truell, Elbaum, Chick (1969); Novick, Berry (1972)]. A much simpler alternative arises when the temperature dependence of the Debye relaxation times $\tau^e_{Tv}(T)$, $\tau^e_{Tp}(T)$ is known. It is then possible, in order, for example, to cover all of the abscissas in Figs. 7.5 to 7.7, to vary the relaxation times by means of the temperature at constant frequency ω. Nevertheless, this procedure is a viable substitute only if one can be sure that the molecular structure of the test sample does not change over the required temperature range.

If the rate-limiting step for the equilibration of the internal degree of freedom is a movement over an energy barrier (*e.g.*, see Fig. 6.1), one can suppose according to Arrhenius

$$\tau_e = \tau_\infty \, e^{\Delta\varepsilon/RT} \tag{7.186}$$

[τ_e: Debye relaxation time; $\Delta\varepsilon$: activation energy; R: gas constant; see Eq. (6.44)]. This leads to

$$\ln \omega \, \tau_e = \frac{\Delta\varepsilon}{RT} + \ln \omega \tau_\infty \, .$$

If $\Delta\varepsilon$ and τ_∞ are temperature-independent quantities, there is a linear relation between $\ln \omega\tau_e$ and $1/T$. Then a plot of the dynamic response functions as a function of $\ln \omega \, \tau_e$ (Figs. 7.5–7.7) only differs from a plot of the same functions as a function of $1/T$ by a fixed parallel translation of the abscissas and a change in the scale of the abscissas by a constant factor $R/\Delta\varepsilon$. The measurements at different frequencies and constant temperature are under these circumstances completely equivalent to (isothermal) measurements at different temperatures and constant frequency.

If the relaxation time decreases with increasing temperature, the dispersion ranges of the response functions shift to higher frequencies with increasing temperature. The form of the dispersion curves is then, at least approximatively, often retained [*e.g.*, see Ferry (1980), Schwarzl (1990)]. With an increase in temperature from T_0 to T, the curves are only shifted parallel to the abscissa with:

$$\ln \omega\tau_e(T) = \ln \omega \, \tau_e(T_0) + \ln a(T, T_0) \, . \tag{7.187a}$$

The shift factor a is given by the ratio

$$a(T, T_0) = \frac{\tau_e(T)}{\tau_e(T_0)} \, . \tag{7.187b}$$

Systems which satisfy this reduction are referred to as thermorheologically simple systems [Schwarzl, Staverman (1952)].

7.6 Connection with the Linear Theory of Aftereffects

The statements we have obtained in Sects. 7.4 and 7.5 from the linearized equations of thermodynamics on the external variables, especially on the variation of the pressure, the volume, and the mechanical response functions during an isothermal relaxation or an isothermal periodic external perturbation, agree with the corresponding statements of the so-called linear theory of aftereffects. Therefore, the question suggests itself whether and under which conditions the linear theory of aftereffects is contained in the thermodynamics of irreversible processes [Meixner (1953)].

The linear theory of aftereffects [*e.g.*, see Gross (1953); Christensen (1971); Pipkin (1972)] proceeds from the following premises:

(P1) If the perturbation $P_1(t)$ of a system leads to the response $R_1(t)$ and the perturbation $P_2(t)$ to the response $R_2(t)$, the perturbation $P_1(t) + P_2(t)$ leads to the response $R_1(t) + R_2(t)$ (Boltzmann superposition principle).

(P2) The response $R(t)$ depends on the present and the previous perturbation $P(t - t')$ with $0 \leq t' \leq +\infty$ (principle of causality).

(P3) If $P(t)$ leads to $R(t)$, $P(t + t'')$ leads to $R(t + t'')$ with arbitrary t'' (principle of the homogeneity of time).

(P4) If $P(t)$ is a smooth function of time, $R(t)$ is also a smooth function of time (principle of stability).

According to these principles, the interrelation

$$R(t) = M(0)\,P(t) + \int_0^\infty \dot{M}(t')\,P(t - t')\,\mathrm{d}t' \tag{7.188a}$$

exists between an isothermal perturbation $P(t)$ and the response $R(t)$, whereby $M(t)$ represents an adequate time-dependent response function with respect to the perturbation $P(t)$. If $P(-\infty) = 0$ is valid, a partial integration of (7.188a) also leads to

$$R(t) = \int_{-\infty}^{t} M(t - t')\,\dot{P}(t')\,\mathrm{d}t' \,. \tag{7.188b}$$

In the case of a mechanical perturbation, (7.188a) resembles a generalization of Hooke's law $R(t) = MP(t)$. Superimposed on the instantaneous Hookean reaction $M(0)\,P(t)$, however, is, in addition, an aftereffect which is dependent on all the previous times $-\infty \leq t - t' \leq t$. One designates, therefore,

$$N(t) \equiv \dot{M}(t)\,, \quad t \geq 0 \tag{7.188c}$$

as an aftereffect function or a memory function. The form of (7.188b), on the other hand, reminds us more of Newton's friction law $R(t) = M\dot{P}(t)$ for viscous fluids. The elastic behaviour of the systems obeying Eqs. (7.188) also seems to be associated with a certain viscous aspect (see further below).

The validity of Boltzmann's superposition principle (P1) presupposes that the equations controlling the system are linear. In order to be able to establish a connection between thermodynamics and the linear theory of aftereffects, one has to proceed from the linearized equations of thermodynamics. Moreover, in the linear theory of aftereffects every macroscopic perturbation $P(t)$ is, according to (7.188), directly linked with the macroscopic response:

$$P(t) \rightarrow R(t) .$$

The internal degrees of freedom remain concealed in the theory of aftereffects. In the thermodynamics of irreversible processes, however, one explicitly takes into account that the macroscopic perturbation also acts on the internal molecular degrees of freedom, and these exert a determinative influence on the macroscopic response:

$$P(t) \rightarrow \{\zeta_1(t) \ldots \zeta_n(t)\} \rightarrow R(t) .$$

In order to be able to establish a connection between thermodynamics and the theory of aftereffects, the internal variables must be eliminated in the thermodynamics.

In the following, we will replace *via* (7.2) the pressure p by a stress variable σ and the specific volume v by a strain variable ε to obtain somewhat more general formulations. The equations of state (7.6)–(7.8) of the F-representation are then

$$-\left(\frac{\partial f}{\partial T}\right)_{\varepsilon,\zeta} \equiv s = s(T, v_e\varepsilon, \zeta) , \tag{7.189}$$

$$\frac{1}{v_e}\left(\frac{\partial f}{\partial \varepsilon}\right)_{T,\zeta} \equiv \sigma = \sigma(T, v_e\, \varepsilon, \zeta) , \tag{7.190}$$

$$-\left(\frac{\partial f}{\partial \zeta}\right)_{T,\varepsilon} \equiv a = a(T, v_e\varepsilon, \zeta) \tag{7.191}$$

[see also the Eqs. (5.5) and (5.6)]. As a reference state, we will choose a fixed internal equilibrium state "e". Under the condition $T = \text{const.}$, the Taylor series about this state yields the linear equations of state

$$\sigma = \sigma_e + \left(\frac{\partial \sigma}{\partial \varepsilon}\right)_{T,\zeta_e} (\varepsilon - \varepsilon_e) + \left(\frac{\partial \sigma}{\partial \zeta}\right)^e_{T,\varepsilon} (\zeta - \zeta_e) \tag{7.192}$$

$$a = a_e + \left(\frac{\partial a}{\partial \varepsilon}\right)_{T,\zeta_e} (\varepsilon - \varepsilon_e) + \left(\frac{\partial a}{\partial \zeta}\right)^e_{T,\varepsilon} (\zeta - \zeta_e) . \tag{7.193}$$

In the non-deformed reference state we have $\sigma_e = \varepsilon_e = a_e = 0$. According to (7.190) and (5.17),

$$\left(\frac{\partial\sigma}{\partial\varepsilon}\right)_{T,\,\zeta_e} = \frac{1}{v_e}\left(\frac{\partial^2 f}{\partial\varepsilon^2}\right)_{T,\,\zeta_e} \equiv M_{T,\,\zeta_e} \tag{7.194}$$

is the isothermal modulus of the arrested internal equilibrium state "e". Depending on the kind of stress ε, we have Young's modulus, the torsion modulus, or the bulk modulus [see Sect. 5.2 and Eq. (7.57) in Sect. 7.3]. Corresponding to (7.34) and (7.35), one can further set

$$\left(\frac{\partial\sigma}{\partial\zeta}\right)^e_{T,\,\varepsilon} \equiv \pi^e_{T\varepsilon}\,, \tag{7.195}$$

$$\left(\frac{\partial a}{\partial\zeta}\right)^e_{T,\,\varepsilon} = -\left(\frac{\partial^2 f}{\partial\zeta^2}\right)^e_{T,\,\varepsilon} \equiv -\,1/L\,\tau^e_{T\varepsilon} < 0\,. \tag{7.196}$$

$\pi^e_{T\varepsilon}$ is a measure of the contribution of the internal degree of freedom to the stress. $\tau^e_{T\varepsilon}$ is the Debye relaxation time under the condition $T = \text{const.}$, $\varepsilon = \text{const.}$ With (7.190), (7.191), and (7.195), one finally obtains

$$\left(\frac{\partial a}{\partial\varepsilon}\right)_{T,\,\zeta_e} = -\left(\frac{\partial^2 f}{\partial\varepsilon\partial\zeta}\right)_e = -\left(\frac{\partial^2 f}{\partial\zeta\partial\varepsilon}\right)_e = -\,v_e\left(\frac{\partial\sigma}{\partial\zeta}\right)^e_{T,\,\varepsilon} = -\,v_e\,\pi^e_{T\varepsilon}\,. \tag{7.197}$$

Hence, the linear equations of state (7.192) and (7.193) also read as follows

$$\sigma = M_{T,\,\zeta_e}\,\varepsilon + \pi^e_{T\varepsilon}\,(\zeta - \zeta_e)\,, \tag{7.198}$$

$$a = -\,v_e\,\pi^e_{T\varepsilon}\,\varepsilon - \frac{1}{L\tau^e_{T\varepsilon}}\,(\zeta - \zeta_e)\,. \tag{7.199}$$

Resolution of (7.198) with respect to $(\zeta - \zeta_e)$ leads to

$$\zeta - \zeta_e = \frac{1}{\pi^e_{T\varepsilon}}\,(\sigma - M_{T,\,\zeta_e}\,\varepsilon)\,. \tag{7.200}$$

With this, the internal variable in (7.199) can be eliminated. Insertion of (7.200) into (7.199) yields for the affinity

$$a = -\,v_e\,\pi^e_{T\varepsilon}\,\varepsilon - \frac{1}{L\tau^e_{T\varepsilon}\pi^e_{T\varepsilon}}\,(\sigma - M_{T,\,\zeta_e}\,\varepsilon)\,. \tag{7.201}$$

Differentiation of (7.200) with respect to time further leads to

$$\dot{\zeta} = \frac{1}{\pi^e_{T\varepsilon}}\,(\dot{\sigma} - M_{T,\,\zeta_e}\,\dot{\varepsilon})\,. \tag{7.202}$$

If one inserts (7.202) and (7.201) into the dynamic law (7.29), one obtains for the linear dynamic equation of state

$$\tau^e_{T\varepsilon}\dot{\sigma} - \tau^e_{T\varepsilon} M_{T,\zeta_e}\dot{\varepsilon} = - \upsilon_e L \tau^e_{T\varepsilon}(\pi^e_{T\varepsilon})^2 \varepsilon - \sigma + M_{T,\zeta_e}\varepsilon ,$$

which is valid for all isothermal processes and free from the internal variable. Here, corresponding to the equation (7.64),

$$- \upsilon_e L \tau^e_{T\varepsilon}(\pi^e_{T\varepsilon})^2 = \Delta M_T \equiv M^e_T - M_{T,\zeta_e} < 0 \tag{7.203}$$

is the contribution of the internal degree of freedom to the modulus M^e_T of the reference state "*e*". Thus, the dynamic equation of state results in

$$\tau^e_{T\varepsilon}\dot{\sigma} + \sigma = \tau^e_{T\varepsilon} M_{T,\zeta_e}\dot{\varepsilon} + M^e_T\varepsilon . \tag{7.204}$$

This equation is now identical with the rheological equation of state of the so-called standard anelastic body in the linear theory of aftereffects.

As in electrodynamics, one can also describe systems by equivalent-circuit diagrams in mechanics. Fundamental elements of the mechanical "equivalent-circuits" are:

1. The elastic spring (Fig. 7.8 A). Within the framework of the linear theory, its behaviour is determined by Hooke's law

   $$\sigma(t) = M\varepsilon(t) . \tag{7.205}$$

   Its reaction is instantaneous. It is capable of storing and releasing elastic energy without loss (the electrodynamic analogue is the condenser).

2. The dashpot (Fig. 7.8 B). This is a cylinder filled with a viscous liquid with a moving piston in its interior. Within the framework of the linear theory, its

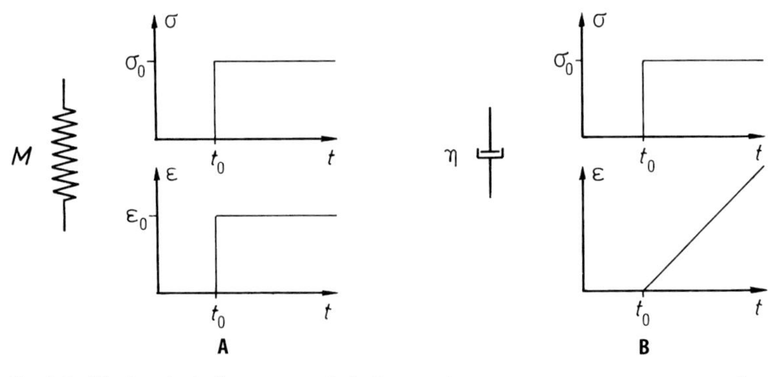

Fig. 7.8. Mechanical elements and their reaction ε to constant stress σ; t: time. **A**: Hookean spring with modulus M; **B**: Newtonian dashpot with viscosity η

behaviour is determined by Newton's viscous law

$$\sigma(t) = \eta \dot{\varepsilon}(t) \tag{7.206}$$

(η: depending on the kind of deformation, with the exception of a numerical factor, Newton's shear viscosity, Trouton's viscosity, or volume viscosity). The dashpot induces a retardation of the motional actions by viscous flow linked with the dissipation of mechanical energy by friction. If the dashpot is loaded with a constant stress σ_0 from the time t_0, it initially shows only an infinitesimal instantaneous reaction at t_0. The subsequent deformation ε is proportional to time (the electrodynamic analogue is an ohmic resistor).

3. The inert mass. We do not need this element in the following. It is only mentioned for the sake of completeness. For example, it serves for the construction of mechanical "oscillatory circuits" (the electrodynamic analogue is the induction coil).

The following combination rules apply for springs and dashpots: In a parallel combination of two elements (Fig. 7.9 A), the stresses are additive while the strains are equal,

$$\sigma = \sigma_1 + \sigma_2 \, ; \quad \varepsilon = \varepsilon_1 = \varepsilon_2 \, . \tag{7.207a}$$

When two elements are combined in series (Fig. 7.9 B), the strains are additive while the stresses are equal,

$$\varepsilon = \varepsilon_1 + \varepsilon_2 \, ; \quad \sigma = \sigma_1 = \sigma_2 \, . \tag{7.207b}$$

The equivalent combination for the standard anelastic body requires three fundamental elements, two springs and one dashpot. A possible combination is shown in Fig. 7.10 A, the so-called Maxwell three-parameter model for a standard anelastic body. According to (7.205) and (7.206),

$$\sigma_1 = M_1 \varepsilon_1 \, ; \quad \sigma_2 = M_2 \varepsilon_2 \, ; \quad \sigma_3 = \eta_F \dot{\varepsilon}_3$$

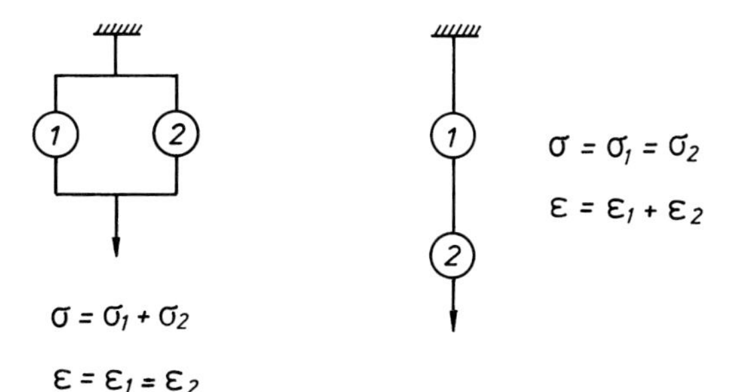

Fig. 7.9. Mechanical elements combined in parallel and in series. Stress: σ; measure of the deformation: ε

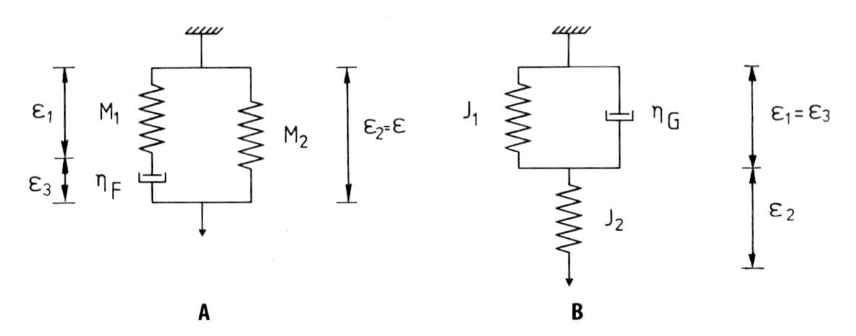

Fig. 7.10. Three-parameter models of the linear theory of aftereffects. **A**: Maxwell model; **B**: Kelvin–Voigt model. Measure of the deformation: ε (see text)

hold for the individual elements of this model. According to the combination rules (7.207), we further obtain

$$\varepsilon = \varepsilon_1 + \varepsilon_3 = \varepsilon_2 \quad \text{and} \quad \sigma = \sigma_1 + \sigma_2 \, ; \quad \sigma_1 = \sigma_3 \, ,$$

that is

$$\sigma = \eta_F \, \dot{\varepsilon}_3 + M_2 \, \varepsilon_2 = \eta_F \, \dot{\varepsilon} - \eta_F \, \dot{\varepsilon}_1 + M_2 \, \varepsilon$$

and in here

$$\begin{aligned} \eta_F \, \dot{\varepsilon}_1 &= \eta_F \, \dot{\varepsilon}_1 + \frac{M_2}{M_1} \, \eta_F \, \dot{\varepsilon}_2 - \frac{M_2}{M_1} \, \eta_F \, \dot{\varepsilon}_2 \\ &= \frac{\eta_F}{M_1} \, (M_1 \, \dot{\varepsilon}_1 + M_2 \, \dot{\varepsilon}_2) - \frac{M_2}{M_1} \, \eta_F \, \dot{\varepsilon}_2 \\ &= \frac{\eta_F}{M_1} \, \dot{\sigma} - \frac{M_2}{M_1} \, \eta_F \, \dot{\varepsilon} \, , \end{aligned}$$

i. e.,

$$\sigma = \eta_F \, \dot{\varepsilon} - \frac{\eta_F}{M_1} \, \dot{\sigma} + \frac{\eta_F}{M_1} \, M_2 \, \dot{\varepsilon} + M_2 \, \varepsilon$$

or

$$\frac{\eta_F}{M_1} \, \dot{\sigma} + \sigma = \frac{\eta_F}{M_1} \, (M_1 + M_2) \, \dot{\varepsilon} + M_2 \, \varepsilon \, . \tag{7.208}$$

This is the so-called rheological equation of state in the Maxwell three-parameter model.

The mechanical equation (7.208) and the thermodynamic equation (7.204) become identical with the identifications

$$\frac{\eta_F}{M_1} \equiv \tau^e_{T\varepsilon} , \tag{7.209a}$$

$$M_1 + M_2 \equiv M_{T,\zeta_e} , \tag{7.209b}$$

$$M_2 \equiv M^e_T . \tag{7.209c}$$

The identity (7.209b) is plausible, as the dashpot does not react in the case of a very rapid deformation in the Maxwell model ($\varepsilon_3 = 0$), so that only the two parallelly combined springs are strained and the model responds with a stress

$$\sigma = M_1 \varepsilon_1 + M_2 \varepsilon_2 = (M_1 + M_2)\, \varepsilon .$$

The identity (7.209c) is equally plausible: In the case of a very slow (quasi-static) deformation of the Maxwell model, the piston within the dashpot can practically instantaneously follow the deformation at any time, so that only the spring M_2 is strained. The identification (7.209a) is possible, as η_F/M_1 has the dimension of time.

As a viscous element is of importance in the Maxwell three-parameter model (Fig. 7.10A), the behaviour of systems following the equation (7.208) is often classified as being viscoelastic (in general, the theory of the mechanical phenomena of aftereffects is often also designated as the theory of viscoelasticity). According to the identification (7.209), however, it becomes clear that this designation does not quite agree with the physical facts. Viscosity is a phenomenon of transport which requires a flowing medium controlled by a velocity gradient. In the usual sense, the cause of the viscosity can be seen in the dissipative interaction between the mass elements of a medium inhomogeneous with respect to the velocity $\vec{v}(x_1 x_2 x_3)$ of the mass elements. For the deduction of the dynamic equation of state (7.204), however, we explicitly assumed a homogeneous system at rest. According to the identification (7.209), the individual elements of the Maxwell three-parameter model have the following meaning: $M_2 = M^e_T$ is the equilibrium modulus of a Hookean body; M_2 should thus be assigned to the macroscopic degree of freedom of the system represented by ε. The relation

$$M_1 \equiv M_{T,\zeta_e} - M^e_T = -\Delta M_T > 0 , \tag{7.209d}$$

on the other hand, corresponds to the contribution (7.203) of the internal molecular degree of freedom to the equilibrium modulus M^e_T of the system. The dashpot connected in series with the spring M_1 has the effect that the maximum contribution ΔM_T of the internal degree of freedom manifests itself depending on the rate of perturbation with differing importance, possibly delayed and linked with a varying entropy production (energy dissipation). Hence, the so-called Maxwell element consisting of a spring with the modulus M_1 and a dashpot with the viscosity η_F connected to it in series has to be assigned as a whole to the internal degree of freedom. It characterizes the dissipative elastic reaction of a disturbed internal degree of freedom in every individual mass element of a homogeneous Hookean body. The behaviour of systems following Eqs. (7.204) or (7.208) is designated as being anelastic in the solid state physics [Zener (1948)]. The Maxwell three-parameter model specifically corresponds to the simplest of the possible anelastic systems, the linear standard anelastic body [Nowick, Berry (1972)].

One obtains as the integral of the differential equation (7.204)

$$\sigma(t) = \mathrm{e}^{-t/\tau_{T\varepsilon}^e} \left[\text{const.} + \int_{-\infty}^{t} \mathrm{e}^{t'/\tau_{T\varepsilon}^e} \left[M_{T,\zeta_e}\, \dot{\varepsilon}(t') + \frac{M_T^e}{\tau_{T\varepsilon}^e}\, \varepsilon(t') \right] \mathrm{d}t' \right].$$

The term containing the constant of integration only describes an initial effect which practically vanishes for times $t \gg \tau_{T\varepsilon}^e$. If the initial state is an internal equilibrium state with $\sigma(-\infty) = 0$, this initial effect does not occur at all. A partial integration of the term containing $\dot{\varepsilon}(t')$ then leads to

$$\sigma(t) = M_{T,\zeta_e}\varepsilon(t) + \frac{\Delta M_T}{\tau_{T\varepsilon}^e} \int_{-\infty}^{t} \mathrm{e}^{(t'-t)/\tau_{T\varepsilon}^e}\, \varepsilon(t')\, \mathrm{d}t' \,. \tag{7.210}$$

If the system is exposed to a constant deformation $\varepsilon(t) = \varepsilon = \text{const.}$ starting from a time $t = 0$, (7.210) yields the pseudo-Hookean law

$$\sigma(t) = M(t)\,\varepsilon\,, \quad t \geq 0 \tag{7.211 a}$$

with the time-dependent relaxation modulus

$$M(t) = M_{T,\zeta_e} + \Delta M_T (1 - \mathrm{e}^{-t/\tau_{T\varepsilon}^e}) = M_T^e - \Delta M_T\, \mathrm{e}^{t'/\tau_{T\varepsilon}^e}\,. \tag{7.211 b}$$

This corresponds to the relaxation experiment of the second kind in Sect. 7.4, Eqs. (7.153)–(7.155). The memory function (7.188c) of the standard anelastic body is given by

$$N(t) = \dot{M}(t) = \frac{\Delta M_T}{\tau_{T\varepsilon}^e}\, \mathrm{e}^{-t/\tau_{T\varepsilon}^e}\,, \quad t \geq 0\,. \tag{7.212}$$

Because of $t - t' \geq 0$, (7.212) can be substituted in (7.210). With (7.211), one then obtains

$$\sigma(t) = M(0)\,\varepsilon(t) + \int_{-\infty}^{t} \dot{M}(t-t')\,\varepsilon(t')\,\mathrm{d}t'. \tag{7.213 a}$$

When one counts the time backwards starting from the present time t with $t'' \equiv t - t'$, one gets

$$\sigma(t) = M(0)\,\varepsilon(t) + \int_{0}^{\infty} \dot{M}(t'')\,\varepsilon(t-t'')\,\mathrm{d}t''\,. \tag{7.213 b}$$

This corresponds to the general interrelation (7.188) of the linear theory of aftereffects. Besides, the standard anelastic body, according to (7.212), possesses a fading memory. The longer ago the deformation $\varepsilon(t - t'')$ occurred (the larger t'' is), the lower the weight with which it enters the momentaneous stress $\sigma(t)$.

The equations of state (7.15)–(7.17) of the G-representation transformed with (7.2) in the same way lead, after linearization and elimination of the internal variables, to the dynamic equation of state

$$\tau^e_{T\sigma}\dot{\varepsilon} + \varepsilon = \tau^e_{T\sigma} J_{T,\zeta_e}\dot{\sigma} + J^e_T\,\sigma\,, \tag{7.214}$$

where $\tau^e_{T\sigma}$ is the Debye relaxation time

$$\tau^e_{T\sigma} \equiv 1/L\left(\frac{\partial^2 g}{\partial\zeta^2}\right)^e_{T,\sigma} > 0$$

of the G-representation,

$$J_{T,\zeta_e} = 1/M_{T,\zeta_e} \quad \text{and} \quad J^e_T = 1/M^e_T \tag{7.215}$$

are the compliances of the arrested and internal equilibria in the reference state "e" [see Eqs. (7.88) and (7.93)]. Equations (7.204) and (7.214) must, of course, be identical if they refer to one and the same system. With (7.215), identification of the coefficients of the two equations results in

$$J^e_T = \frac{\tau^e_{T\sigma}}{\tau^e_{T\varepsilon}} J_{T,\zeta_e} \quad \text{and} \quad M^e_T = \frac{\tau^e_{T\varepsilon}}{\tau^e_{T\sigma}} M_{T,\zeta_e}\,, \tag{7.216}$$

i.e., relations whose validity we have already proved (with p, v as external variables) with (7.91) and (7.92).

The mechanical equivalent-combination of the dynamic equation of state (7.214), however, is not Maxwell's three-parameter model (Fig. 7.10 A), but the so-called three-parameter model of Kelvin–Voigt or also the Poynting–Thomson model (Fig. 7.10 B). In thermodynamics, the Maxwell model thus corresponds to the F-representation, the Kelvin–Voigt model to the G-representation. According to (7.205) and (7.206),

$$\varepsilon_1 = J_1\sigma_1\,; \quad \sigma_2 = J_2\sigma_2\,; \quad \sigma_3 = \eta_G\,\dot{\varepsilon}_3$$

hold for the Kelvin–Voigt model (Fig. 7.10 B) and according to the combination rules (7.207)

$$\varepsilon = \varepsilon_1 + \varepsilon_2\,, \quad \varepsilon_1 = \varepsilon_3 \quad \text{and} \quad \sigma = \sigma_1 + \sigma_3 = \sigma_2\,.$$

This yields

$$\begin{aligned}\varepsilon &= J_1\sigma_1 + J_2\sigma_2 = J_2\sigma + J_1(\sigma - \sigma_3)\\ &= (J_1 + J_2)\,\sigma - \eta_G J_1\,\dot{\varepsilon}_3\end{aligned}$$

and here because of

$$\dot{\varepsilon}_3 = \dot{\varepsilon}_1 = \dot{\varepsilon} - \dot{\varepsilon}_2\,, \quad \sigma_2 = \sigma$$

$$\eta_G J_1 \dot{\varepsilon}_3 = \eta_G J_1 \dot{\varepsilon} - \eta_G J_1 J_2 \dot{\sigma}\,,$$

i.e.,

$$\eta_G J_1 \dot{\varepsilon} + \varepsilon = \eta_G J_1 J_2 \dot{\sigma} + (J_1 + J_2)\,\sigma\,. \tag{7.217}$$

Identification of the thermodynamic equation of state (7.214) with the rheological equation of state (7.217) of the Kelvin–Voigt three-parameter model leads to the identities

$$\eta_G J_1 \equiv \tau^e_{T\sigma}\,, \tag{7.218a}$$

$$J_1 + J_2 \equiv J^e_T\,, \tag{7.218b}$$

$$J_2 \equiv J_{T,\,\zeta_e}\,. \tag{7.218c}$$

The spring with the compliance J_2 corresponds to the Hookean compliance of the arrested equilibrium and must be assigned to the macroscopic degree of freedom represented by σ. The so-called Kelvin–Voigt element consisting of the spring with the compliance

$$J_1 \equiv J^e_T - J_{T,\,\zeta_e} = \Delta J_T > 0 \tag{7.218d}$$

and the parallel-connected dashpot with the viscosity η_G must be assigned to the internal degree of freedom.

From the standpoint of thermodynamics, the models Fig. 7.10 thus describe a homogeneous Hookean body with a mechanically relevant internal degree of freedom. The dashpot in these models imitates the retardations and the entropy production connected with the elastic reaction of the disturbed internal degree of freedom. In contrast to that, the dashpots may naturally also correspond to a true viscosity in the equivalent-combinations of viscous liquids. The equivalent-combinations of the simplest Newtonian liquid with phenomena of after-effects are depicted in Fig. 7.11. These are the three-parameter models of Jeffrey and Letherisch. One obtains as the rheological equation of state in the Jeffrey model

$$\frac{\eta_1}{\Delta M}\,\dot{\sigma} + \sigma = (\eta_1 + \eta_2)\,\dot{\varepsilon} + \frac{\eta_1}{\Delta M}\,\eta_2 \ddot{\varepsilon} \tag{7.219a}$$

[this equation is derived in the same way as Eq. (7.208)]. If one disregards initial effects (see above) and introduces the relaxation time τ *via*

$$\frac{\eta_1}{\Delta M} \equiv \tau\,, \tag{7.219b}$$

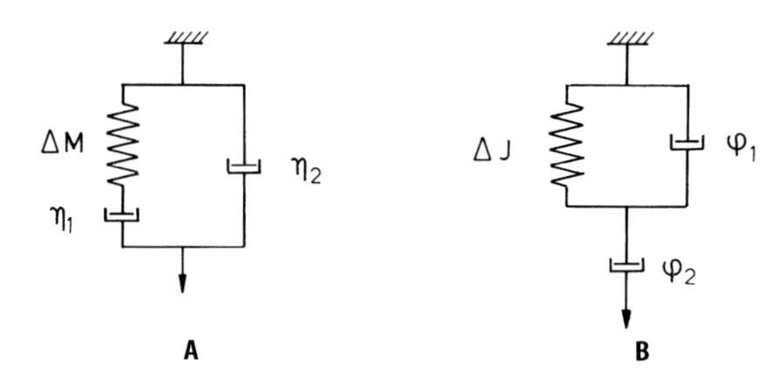

Fig. 7.11. Three-parameter models of the linear theory of aftereffects. **A**: Jeffrey model; **B**: Letherisch model. Fluidity of the dashpots: $\varphi = 1/\eta$

the integral of this differential equation with respect to $\sigma(t)$ is

$$\sigma(t) = \mathrm{e}^{-t/\tau} \int_{-\infty}^{t} \mathrm{e}^{t'/\tau} \left[\frac{1}{\tau} (\eta_1 + \eta_2)\, \dot{\varepsilon}(t') + \eta_2\, \ddot{\varepsilon}(t') \right] \mathrm{d}t' .$$

A partial integration over the term containing $\ddot{\varepsilon}(t')$ yields

$$\sigma(t) = \eta_2\, \dot{\varepsilon}(t) + \frac{\eta_1}{\tau} \int_{-\infty}^{t} \mathrm{e}^{(t'-t)/\tau}\, \dot{\varepsilon}(t')\, \mathrm{d}t'. \tag{7.220}$$

If one exposes the Jeffrey body starting at the time $t' = 0$ to a constant rate of deformation $\dot{\varepsilon}(t') = \dot{\varepsilon} = \text{const.}$ ($\dot{\varepsilon}(t') = 0$ should hold for $t' < 0$), and (7.220) leads to the pseudo-Newtonian law

$$\sigma(t) = \eta(t)\, \dot{\varepsilon}, \quad t \geq 0 \tag{7.221a}$$

with the time-dependent viscosity

$$\eta(t) = \eta_2 + \eta_1 (1 - \mathrm{e}^{-t/\tau}) . \tag{7.221b}$$

At the beginning of the perturbation, the system reacts with a viscosity η_2 which in time increases up to the viscosity $\eta_1 + \eta_2$. The Jeffrey body apparently becomes more viscous during the perturbation (see further below and Fig. 7.12).

The Jeffrey body, however, undoubtedly also acquires elastic properties *via* the spring with the modulus ΔM. With the relaxation modulus

$$M(t) = \Delta M\, \mathrm{e}^{-t/\tau}, \quad t \geq 0 \tag{7.222}$$

and Eq. (7.219b), one can also write in place of (7.220)

$$\sigma(t) = \eta_2\, \dot{\varepsilon}(t) + \int_{-\infty}^{t} M(t - t')\, \dot{\varepsilon}(t')\, \mathrm{d}t' . \tag{7.223a}$$

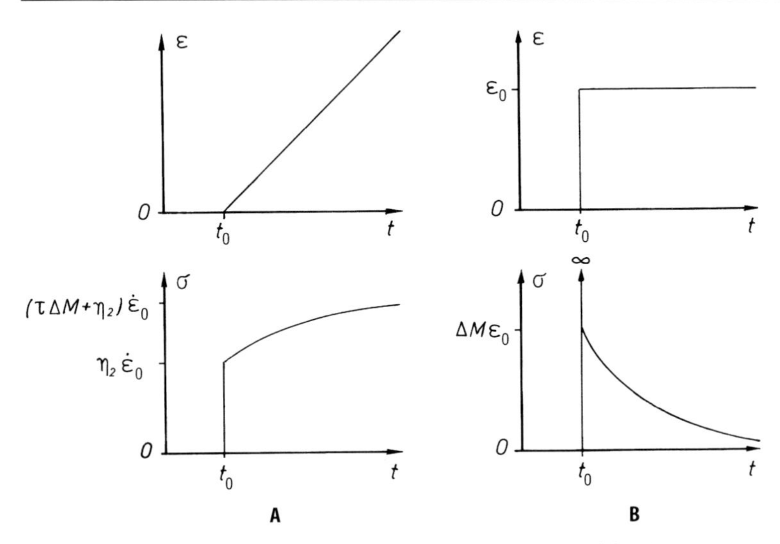

Fig. 7.12. Dissipative elastic reaction of an internal degree of freedom in an incompressible Newtonian liquid. **A:** with a constant velocity of deformation $\dot{\varepsilon}_0$ according to (7.221); **B:** with a constant deformation ε_0 according to (7.224). Given deformation: ε; resulting stress: σ

If $\varepsilon(-\infty) = 0$ is valid and if one counts the time backwards with $t'' \equiv t - t'$, a partial integration of (7.223 a) leads to

$$\sigma(t) = \eta_2\,\dot{\varepsilon}(t) + M(0)\,\varepsilon(t) + \int\limits_0^\infty \dot{M}(t'')\,\varepsilon(t - t'')\,\mathrm{d}t''. \tag{7.223b}$$

This form of the integrated rheological equation of state (7.219) suggests that one can interpret the Jeffrey body as an elasto-viscous liquid in which the Newtonian flow $\eta_2\dot{\varepsilon}(t)$ is superimposed by a Hookean elasticity with aftereffects [compare Eq. (7.223b) with Eq. (7.213b)]. As an equilibrium modulus M_T^e is missing in (7.222) [compare Eq. (7.222) with Eq. (7.211)], the Jeffrey-body, like every Newtonian liquid, does not acquire elastic properties in internal equilibrium.

From the standpoint of thermodynamics, the models of Fig. 7.11 describe a Newtonian liquid with a macroscopically relevant internal degree of freedom. In the Jeffrey model, as in the Maxwell model, the Maxwell element consisting of a spring with the modulus ΔM and a dashpot with the viscosity η_1 connected in series has to be assigned to an internal degree of freedom concealed in the theory of aftereffects. The dashpot with the viscosity η_2 has to be attributed to the external degree of freedom represented by the variable $\varepsilon(t)$. The true Newtonian viscosity is η_2 which, however, only occurs in a pure form if the internal degree of freedom is arrested or in internal equilibrium. The internal degree of freedom first appears arrested at $t = 0$ with respect to the suddenly starting deformation if the Jeffrey body is stressed from time $t = 0$ onwards with a rate of deformation $\dot{\varepsilon} = \text{const.}$ [Eq. (7.221)]. The viscosity η_2 is measured.

In the course of time, however, the internal degree of freedom is capable of adjusting to the new situation. The viscosity $\eta_2 + \eta_1$ is then measured at times $t \gg \tau$ (Fig. 7.12 A). Here, the first part η_2 is due to the mutual dissipative interaction of the mass elements, the second part η_1, on the other hand, is due to the dissipative reaction of the internal degree of freedom in every individual mass element. That $\eta_1 = \tau \Delta M$ does not correspond to a true viscosity but to a dissipative elastic reaction, which only appears in a thermodynamic non-equilibrium, becomes particularly evident when stressing the Jeffrey body with a constant deformation ε from a time $t = 0$ (Fig. 7.12 B). Every purely viscous substance merely reacts to such a perturbation with a stress peak $\sigma(t) = \eta \varepsilon \delta(t)$ at $t = 0$, or otherwise not at all [$\delta(t)$: Dirac delta function]. With

$$-\infty \leq t \leq 0 : \varepsilon(t) = 0 \,; \quad 0 \leq t \leq +\infty : \varepsilon(t) = \varepsilon = \text{const.},$$

i.e.,

$$\dot{\varepsilon}(t) = \varepsilon \delta(t) \,,$$

(7.223) then yields for the Jeffrey body

$$\sigma(t) = [\eta_2 \delta(t) + \Delta M \, \mathrm{e}^{-t/\tau}] \, \varepsilon \,. \tag{7.224}$$

The purely viscous reaction at $t = 0$ is followed by a purely elastic reaction in the course of time which, however, disappears completely as soon as the internal degree of freedom has reached internal equilibrium.

In the hydromechanics of viscous media, the pressure tensor $\boldsymbol{P} = -\boldsymbol{\sigma}$ has to be split into two parts

$$\boldsymbol{P} = \boldsymbol{P}_e + \boldsymbol{\Pi} \,,$$

where $\boldsymbol{P}_e$ is the tensor of hydrostatic pressure, and $\boldsymbol{\Pi}$, the tensor of frictional pressure which is dependent on the velocity of mass flow $\vec{v}$ [*e.g.*, see Landau, Lifshitz (1987)]. The tensor $\boldsymbol{P}_e$ is given by the mechanical equations of state. For $\boldsymbol{\Pi}$, on the other hand, if the system is supposed to be completely determinable, a dynamic law must additionally be formulated which connects $\boldsymbol{\Pi}$ with the gradient $\boldsymbol{Grad}\,\vec{v}$ of the velocity field (this is Newton's viscous law in the simplest case). At the same time, $\boldsymbol{P}$ has the meaning of a relative density of momentum flow. The tensor $\boldsymbol{\Pi}$ is the dissipative part of the density of momentum flow which yields the contribution

$$\frac{\mathrm{d}_i s}{\mathrm{d}t} = -\frac{1}{T} \boldsymbol{\Pi} : \boldsymbol{Grad}\,\vec{v}$$

to the entropy production in the interior of the system (":" designates the scalar product of the two tensors). This so-called Rayleigh dissipation function describes the consumption of kinetic energy by friction which simultaneously appears as frictional heat in the internal energy of the system. In contrast to our pre-condition Eq. (7.1), $\mathrm{d}_i V \neq 0$ is given in viscous media.

Two dynamic laws have to be formulated for viscous systems with relevant internal degrees of freedom, one for the dissipative momentum flow $\boldsymbol{\Pi}$ and one for the equally dissipative "scalar flow" $\dot{\zeta}_i$ of the internal variables. One generally has to assume here that $\boldsymbol{\Pi}$ as well as $\dot{\zeta}_i$ depend on all of the driving forces present, *i.e.*, on the components of the velocity gradient and the affinities:

$$\boldsymbol{\Pi} = \boldsymbol{\Pi}\left(\ldots \frac{\partial v_\ell}{\partial x_k} \ldots ; \ldots a_k \ldots\right),$$

$$\dot{\zeta}_i = \dot{\zeta}_i\left(\ldots \frac{\partial v_\ell}{\partial x_k} \ldots ; \ldots a_k \ldots\right)$$

[v_ℓ ($\ell = 1, 2, 3$): components of the velocity of mass flow $\vec{v}$; x_k ($k = 1, 2, 3$): components of a Cartesian coordinate system]. In contrast to homogeneous elastic media, one can find a dynamic coupling between the internal and the external degrees of freedom in viscous media. As the behaviour of tensors and scalars with respect to spatial transformations is completely different, only the tensorial or scalar parts of the fluxes and forces can be interdependent in an isotropic medium. The $\dot{\zeta}_i$ can then only depend on the scalar part of the velocity gradient, *i.e.*, on the divergence $\operatorname{div} \vec{v}$ of the velocity field. Finally, $\operatorname{div} \vec{v} = 0$ holds in incompressible media. There is no dynamic coupling between the internal and external degrees of freedom in isotropic incompressible media.

The models of Jeffrey and Letherisch suppose isotropic incompressible Newtonian liquids. The confinement to only one external degree of freedom in these models further entails a confinement to simple velocity profile developments. In the case of a laminar flow $\vec{v}$ in the x_1-direction with a gradient in the x_2-direction, for example, one gets

$$\pi_{12} = \pi_{21} = -\eta \frac{\partial v_1}{\partial x_2} = -2\eta\dot{\varepsilon} \equiv -\eta\dot{\gamma}_{12}\,; \quad \pi_{ik} = 0 \quad \text{otherwise}$$

[η: Newton's shear viscosity; γ_{12}: shearing (see Sect. 5.1); π_{ik}: components of the tensor $\boldsymbol{\Pi}$]. After elimination of the internal degree of freedom, the dissipative elastic contribution of the internal degree of freedom appears as an additional term in p_{12}, so that, corresponding to (7.223), it follows that

$$p_{12} = -\eta\dot{\gamma}_{12} - \int_{-\infty}^{t} M(t-t')\,\dot{\gamma}_{12}(t')\,\mathrm{d}t'.$$

7.7 Systems with Several Intercoupled Internal Degrees of Freedom

In this section, we will point out that if several internal degrees of freedom, which are intercoupled *via* the dynamic law (7.47), are macroscopically relevant, one can find some differences when compared to systems with only one relevant internal degree of freedom (Sect. 7.3–7.6).

In order to describe the relaxation of a system with n relevant internal degrees of freedom when $T, p = \text{const.}$ or $T, \sigma = \text{const.}$, (7.50) leads to the system of n non-linear differential equations

$$\ddot{\zeta}_i + \sum_{\ell} M_{i\ell}\, \dot{\zeta}_\ell = 0 \tag{7.225}$$

On the other hand, if one assumes that one is not too far from the final state of relaxation, an internal equilibrium state "e" (see Sect. 7.4), so that the coefficients can be referred to the final state with

$$M_{i\ell} \approx M^e_{i\ell} \equiv \sum_k L_{ik} \left(\frac{\partial^2 g}{\partial\zeta_k \partial\zeta_\ell}\right)_e = \text{const.}\,, \tag{7.226a}$$

one obtains the system of linear equations

$$\ddot{\zeta}_i = -\sum_{\ell} M^e_{i\ell}\, \dot{\zeta}_\ell$$

or

$$\dot{\zeta}_i = -\sum_{\ell} M^e_{i\ell}\, (\zeta_\ell - \zeta^e_\ell). \tag{7.226b}$$

If one combines the ζ_ℓ ($\ell = 1 \ldots n$) into "vectors" $\boldsymbol{\zeta}$ and designates the matrices of the coefficients $M^e_{i\ell}$, L_{ik} and

$$G_{k\ell} = G_{\ell k} \equiv \left(\frac{\partial^2 g}{\partial\zeta_k \partial\zeta_\ell}\right)_e$$

with $\boldsymbol{M}$, $\boldsymbol{L}$ and $\boldsymbol{G}$, one can also write in place of (7.226a/b) as an abbreviation

$$\dot{\boldsymbol{\zeta}} = -\boldsymbol{M}(\boldsymbol{\zeta} - \boldsymbol{\zeta}_e)\,; \quad \boldsymbol{M} = \boldsymbol{L}\boldsymbol{G} \tag{7.226c}$$

According to (2.88), the matrix $\boldsymbol{G}$ is positive definite if the final state is stable or metastable. Moreover, the matrix $\boldsymbol{L}$ is also positive definite according to the second law [see (7.48)]. All the eigenvalues μ_i of the matrix $\boldsymbol{M}$ are then real and positive. With dimensionless internal variables ζ_i (*e.g.*, concentrations), the eigenvalues $\mu_i > 0$, like the matrix elements $M^e_{i\ell}$, have the dimension of reciprocal time. One can introduce the time constants τ^G_i *via*

$$\mu_i(T, \sigma) \equiv 1/\tau^G_i(T, \sigma) > 0\,, \tag{7.227}$$

which, like the μ_i and $M^e_{i\ell}$, refer to the final state. In the following, we will confine ourselves to the case that all the eigenvalues μ_i of $\boldsymbol{M}$ are different from each other.[8)]

8 See any textbook on systems of linear equations or Margenau, Murphy (1956) also for the case that some eigenvalues are equal.

We then have a matrix $\boldsymbol{R}$ with which $\boldsymbol{M}$ can be brought into the diagonal form

$$\boldsymbol{R}\boldsymbol{M}\boldsymbol{R}^{-1} = \boldsymbol{D} \equiv \begin{pmatrix} \mu_1 & 0 & 0 \\ 0 & \mu_2 & 0 \\ 0 & 0 & \mu_3 \ddots \end{pmatrix} \tag{7.228}$$

(so-called principal axis transformation). In this case, the matrix of time constants τ_i^G is given by

$$\begin{pmatrix} \tau_1^G & 0 & 0 \\ 0 & \tau_2^G & 0 \\ 0 & 0 & \tau_3^G \ddots \end{pmatrix} = \boldsymbol{D}^{-1} = (\boldsymbol{R}\boldsymbol{M}\boldsymbol{R}^{-1})^{-1} = \boldsymbol{R}\boldsymbol{G}^{-1}\,\boldsymbol{L}^{-1}\,\boldsymbol{R}^{-1}\,. \tag{7.229}$$

The transformation

$$\boldsymbol{R}\,(\boldsymbol{\zeta} - \boldsymbol{\zeta}_e) = \bar{\boldsymbol{\zeta}} - \bar{\boldsymbol{\zeta}}_e \tag{7.230 a}$$

changes the differences $\zeta_k - \zeta_k^e$ of the internal variables into the differences of the so-called normal coordinates (normal modes)

$$\bar{\zeta}_i - \bar{\zeta}_i^e = \sum_k R_{ik}\,(\zeta_k - \zeta_k^e)\,. \tag{7.230 b}$$

One should take into account that $\boldsymbol{R}$, like $\boldsymbol{M}$, depends on the values of the external variables. If one inserts (7.230) into (7.226), one obtains with (7.228)

$$\boldsymbol{R}^{-1}\,\dot{\bar{\boldsymbol{\zeta}}} = -\,\boldsymbol{M}\boldsymbol{R}^{-1}\,(\bar{\boldsymbol{\zeta}} - \bar{\boldsymbol{\zeta}}_e)$$

$$\boldsymbol{R}\boldsymbol{R}^{-1}\,\dot{\bar{\boldsymbol{\zeta}}} = -\,\boldsymbol{R}\boldsymbol{M}\boldsymbol{R}^{-1}\,(\bar{\boldsymbol{\zeta}} - \bar{\boldsymbol{\zeta}}_e)$$

$$\dot{\bar{\boldsymbol{\zeta}}} = -\,\boldsymbol{D}\,(\bar{\boldsymbol{\zeta}} - \bar{\boldsymbol{\zeta}}_e) \tag{7.231 a}$$

or in the component notation

$$\dot{\bar{\zeta}}_i = -\,\mu_i(\bar{\zeta}_i - \bar{\zeta}_i^e)\,. \tag{7.231 b}$$

With (7.227), the solutions of the mutually independent differential equations are

$$\bar{\zeta}_i(t) = [\bar{\zeta}_i(0) - \bar{\zeta}_i^e]\,\mathrm{e}^{-t/\tau_i^G} + \bar{\zeta}_i^e\,. \tag{7.232}$$

For the internal variables assigned to the internal degree of freedom, one thus obtains *via* (7.230) the solutions

$$\zeta_i(t) = \sum_k (\boldsymbol{R}^{-1})_{ik}\,[\bar{\zeta}_k(0) - \bar{\zeta}_k^e]\,\mathrm{e}^{-t/\tau_k^G} + \zeta_i^e\,. \tag{7.233}$$

The following conclusions can be drawn from these equations: If several internal degrees of freedom are macroscopically relevant and if these degrees of freedom are coupled *via* (7.47), the relaxation times (the Debye relaxation times in the sense of the linear theory of aftereffects) can no longer be assigned to the individual internal degrees of freedom. Rather, the relaxation time τ_i^G is attributed to the normal mode $\bar{\zeta}_i$ which generally includes all the internal variables $\zeta_1 \ldots \zeta_n$ [Eq. (7.230)]. In this case, the weight with which the internal variables are included in the normal modes depends on the thermostatic state of the system, as the R_{ik} are dependent on this state. Hence, the composition of the normal modes of the internal variables varies depending on the temperature, the pressure *etc.* of the system. According to (7.233), the temporal behaviour of the individual degrees of freedom includes all the relaxation times $\tau_1^G \ldots \tau_n^G$, and these also with a weight which is dependent on the thermostatic state of the system. The spectrum of the relaxation times is a collective property of the internal degrees of freedom. Moreover, according to (7.233), the relaxation of the individual internal degree of freedom is by no means necessarily exponential. It can proceed in several steps or also take an oscillating course (Fig. 7.13). If n internal variables are intercoupled *via* (7.47), the individual internal variable $\zeta_i(t)$ as a function of time can pass through $n-1$ stopping points (steps or extreme values) and intersect the equilibrium value ζ_i^e with $n-1$ "zero passages" [Meixner (1949)].

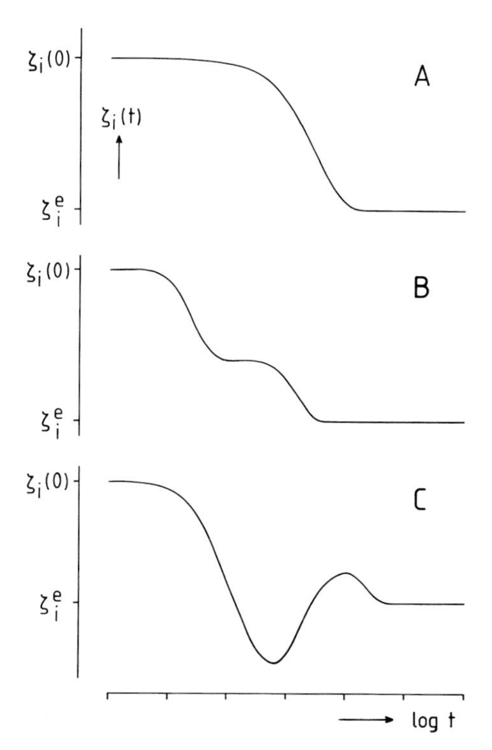

Fig. 7.13. Relaxation of an internal variable ζ_i, intercoupled *via* (7.47), according to (7.233) with $n = 3$. **A:** monotonous relaxation; **B:** monotonous relaxation with stopping points; **C:** oscillating relaxation

In addition to the $\zeta_i(t)$ and $\bar{\zeta}_i(t)$, every arbitrary linear combination of these variables is also a solution of the differential equations (7.226). This is a result of the linearity of these equations, but does not mean that the problem is not uniquely solvable. If the relevant internal degrees of freedom are explicitly known and if macroscopic internal variables can, corresponding to the assumption (V1), be unambiguously assigned to these degrees of freedom, the solution of the problem is also unique. An ambiguity only arises if, for example, in order to explain experimental findings, one postulates *ad hoc* the relevance of several internal degrees of freedom, but is not able to specify them individually.

One obtains from the differential form of the equations of state (15–17) of the G-representation

$$\mathrm{d}s = \left(\frac{\partial s}{\partial T}\right)_{\sigma,\,\zeta} \mathrm{d}T + \left(\frac{\partial s}{\partial \sigma}\right)_{T,\,\zeta} \mathrm{d}\sigma + \sum_i \left(\frac{\partial s}{\partial \zeta_i}\right)_{T,\,\sigma,\,\zeta_{k\neq i}} \mathrm{d}\zeta_i\,, \tag{7.234}$$

$$\mathrm{d}\varepsilon = \left(\frac{\partial \varepsilon}{\partial T}\right)_{\sigma,\,\zeta} \mathrm{d}T + \left(\frac{\partial \varepsilon}{\partial \sigma}\right)_{T,\,\zeta} \mathrm{d}\sigma + \sum_i \left(\frac{\partial \varepsilon}{\partial \zeta_i}\right)_{T,\,\sigma,\,\zeta_{k\neq i}} \mathrm{d}\zeta_i\,, \tag{7.235}$$

$$\mathrm{d}a_l = \left(\frac{\partial a_\ell}{\partial T}\right)_{\sigma,\,\zeta} \mathrm{d}T + \left(\frac{\partial a_\ell}{\partial \sigma}\right)_{T,\,\zeta} \mathrm{d}\sigma + \sum_i \left(\frac{\partial a_\ell}{\partial \zeta_i}\right)_{T,\,\sigma,\,\zeta_{k\neq i}} \mathrm{d}\zeta_i \tag{7.236}$$

in place of (7.71–7.74) the response functions

$$J_T \equiv \left(\frac{\mathrm{d}\varepsilon}{\mathrm{d}\sigma}\right)_T = \left(\frac{\partial \varepsilon}{\partial \sigma}\right)_{T,\,\zeta} + \sum_i \left(\frac{\partial \varepsilon}{\partial \zeta_i}\right)_{T,\,\sigma,\,\zeta_{k\neq i}} \left(\frac{\mathrm{d}\zeta_i}{\mathrm{d}\sigma}\right)_T\,, \tag{7.237}$$

$$\alpha_\sigma \equiv \left(\frac{\mathrm{d}\varepsilon}{\mathrm{d}T}\right)_\sigma = \left(\frac{\partial \varepsilon}{\partial T}\right)_{\sigma,\,\zeta} + \sum_i \left(\frac{\partial \varepsilon}{\partial \zeta_i}\right)_{T,\,\sigma,\,\zeta_{k\neq i}} \left(\frac{\mathrm{d}\zeta_i}{\mathrm{d}T}\right)_\sigma\,, \tag{7.238}$$

$$\alpha_T \equiv \frac{1}{v}\left(\frac{\mathrm{d}s}{\mathrm{d}\sigma}\right)_T = \frac{1}{v}\left(\frac{\partial s}{\partial \sigma}\right)_{T,\,\zeta} + \frac{1}{v}\sum_i \left(\frac{\partial s}{\partial \zeta_i}\right)_{T,\,\sigma,\,\zeta_{k\neq i}} \left(\frac{\mathrm{d}\zeta_i}{\mathrm{d}\sigma}\right)_T\,, \tag{7.239}$$

$$c_\sigma \equiv T\left(\frac{\mathrm{d}s}{\mathrm{d}T}\right)_\sigma = T\left(\frac{\partial s}{\partial T}\right)_{\sigma,\,\zeta} + T\sum_i \left(\frac{\partial s}{\partial \zeta_i}\right)_{T,\,\sigma,\,\zeta_{k\neq i}} \left(\frac{\mathrm{d}\zeta_i}{\mathrm{d}T}\right)_\sigma \tag{7.240}$$

(J_T: compliance), whereby

$$\left(\frac{\mathrm{d}\zeta_i}{\mathrm{d}\sigma}\right)_T = \dot{\zeta}^i_T/\dot{\sigma}_T\,; \quad \left(\frac{\mathrm{d}\zeta_i}{\mathrm{d}T}\right)_\sigma = \dot{\zeta}^i_\sigma/\dot{T}_\sigma$$

are valid in non-equilibrium (see Sect. 7.3).

In an arrested equilibrium, one measures the compliance

$$J_{T,\,\zeta} = \left(\frac{\partial \varepsilon}{\partial \sigma}\right)_{T,\,\zeta}\,,$$

and in an internal equilibrium the compliance

$$J_T^e = J_{T,\zeta_e} + \sum_i \left(\frac{\partial \varepsilon}{\partial \zeta_i}\right)^e_{T,\sigma,\zeta_{k \neq i}} \left(\frac{d\zeta_i^e}{d\sigma}\right)_T .$$

With the "vectors"

$$\boldsymbol{\zeta}_e \equiv \{\zeta_1^e \dots \zeta_n^e\}$$

and

$$\nabla^e_{T\sigma} x \equiv \left\{ \left(\frac{\partial x}{\partial \zeta_1}\right)^e_{T,\sigma,\zeta_{k \neq 1}} \dots \left(\frac{\partial x}{\partial \zeta_n}\right)^e_{T,\sigma,\zeta_{k \neq n}} \right\}, \tag{7.241}$$

this equation also reads

$$J_T^e = J_{T,\zeta_e} + \left(\nabla^e_{T\sigma}\varepsilon, \left(\frac{d\boldsymbol{\zeta}_e}{d\sigma}\right)_T\right). \tag{7.242}$$

$(\dots,\dots)$ designates the scalar product of the two vectors. (7.236) leads to

$$\left(\frac{\partial a_\ell}{\partial \sigma}\right)_{T,\zeta_e} + \sum_i \left(\frac{\partial a_\ell}{\partial \zeta_i}\right)^e_{T,\sigma,\zeta_{k \neq i}} \left(\frac{d\zeta_i^e}{d\sigma}\right)_T = 0$$

for the internal equilibrium with $a_\ell = 0$, $\boldsymbol{\zeta} = \boldsymbol{\zeta}_e(T,\sigma)$ when $T = \text{const}$. Here, according to (7.17), we have

$$\left(\frac{\partial a_\ell}{\partial \sigma}\right)_{T,\zeta_e} = -\left(\frac{\partial^2 g}{\partial \sigma \partial \zeta_\ell}\right)_e = -\left(\frac{\partial^2 g}{\partial \zeta_\ell \partial \sigma}\right)_e = \upsilon \left(\frac{\partial \varepsilon}{\partial \zeta_\ell}\right)^e_{T,\sigma,\zeta_{k \neq \ell}},$$

$$\left(\frac{\partial a_\ell}{\partial \zeta_i}\right)^e_{T,\sigma,\zeta_{k \neq i}} = -\left(\frac{\partial^2 g}{\partial \zeta_i \partial \zeta_\ell}\right)_e = -G_{i\ell} = -G_{\ell i},$$

i.e.,

$$\upsilon \left(\frac{\partial \varepsilon}{\partial \zeta_\ell}\right)^e_{T,\sigma,\zeta_{k \neq \ell}} - \sum_i G_{\ell i} \left(\frac{d\zeta_i^e}{d\sigma}\right)_T = 0$$

or in the matrix notation

$$\upsilon \nabla^e_{T\sigma}\varepsilon = \boldsymbol{G} \left(\frac{d\boldsymbol{\zeta}_e}{d\sigma}\right)_T,$$

i.e.,

$$\left(\frac{d\boldsymbol{\zeta}_e}{d\sigma}\right)_T = \upsilon \boldsymbol{G}^{-1} \nabla^e_{T\sigma}\varepsilon .$$

If one inserts this expression into (7.242), one obtains the positive definite quadratic form

$$\Delta J_T \equiv J_T^e - J_{T,\zeta_e} = \upsilon\,(\nabla^e_{T\sigma}\varepsilon\,,\,G^{-1}\,\nabla^e_{T\sigma}\varepsilon) > 0 \tag{7.243a}$$

as the contribution of the relevant internal degrees of freedom to the compliance of the internal equilibrium. For this, one can also write

$$\Delta J_T = \sum_i \Delta J_T^i \tag{7.243b}$$

with

$$\Delta J_T^i \equiv \upsilon \sum_k (G^{-1})_{ik} \left(\frac{\partial \varepsilon}{\partial \zeta_i}\right)^e_{T,\,\sigma,\,\zeta_{\ell \neq i}} \left(\frac{\partial \varepsilon}{\partial \zeta_k}\right)^e_{T,\,\sigma,\,\zeta_{\ell \neq k}} . \tag{7.243c}$$

Thus, ΔJ_T is additively composed of n parts. But these parts also cannot be directly assigned to the individual degrees of freedom. They rather correspond to those linear combinations of the internal variables (eigensolutions) in which the matrix G^{-1} occurs in a diagonal form and the quadratic form (7.243a) in a normal form. According to (7.229),

$$G^{-1} = R^{-1} D^{-1} R L \tag{7.244}$$

must be inserted into (7.243) in order to obtain an expression comparable to (7.82) and containing the relaxation times.

In the same way, one gets for the contribution of the internal degrees of freedom to the coefficient of thermal expansion

$$\Delta\alpha \equiv \alpha_e - \alpha_{\zeta_e} = (\nabla^e_{T\sigma}\varepsilon\,,\,G^{-1}\,\nabla^e_{T\sigma}s) = (\nabla^e_{T\sigma}s,\,G^{-1}\,\nabla^e_{T\sigma}\varepsilon) \tag{7.245}$$

and for the contribution of the internal degrees of freedom to the specific heat capacity at constant stress

$$\Delta c_\sigma \equiv c_\sigma^e - c_{\sigma,\,\zeta_e} = T(\nabla^e_{T\sigma}s,\,G^{-1}\,\nabla^e_{T\sigma}s) > 0\,. \tag{7.246}$$

The differences (7.243), (7.245), (7.246) now no longer fulfil the Davies relation (7.85). With a positive definite matrix A, the Cauchy-Schwarz inequality is generally valid [*e.g.*, see Margenau, Murphy (1956) or Courant, Hilbert (1965)]

$$\Big(\sum_i\sum_k A_{ik}\,x_i\,y_k\Big)^2 \leq \Big(\sum_i\sum_k A_{ik}\,x_i\,x_k\Big)\Big(\sum_i\sum_k A_{ik}\,y_i\,y_k\Big)\,, \tag{7.247}$$

whereby the equality sign is only valid when the vectors x and y are proportional. In the case of a simultaneous macroscopic relevance of several internal degrees of freedom, the Davies relation (7.85) is replaced by the Davies inequality

$$\Delta c_\sigma\,\Delta J_T \geq T\upsilon\,(\Delta\alpha)^2\,. \tag{7.248}$$

In the F-representation, the contributions of the relevant internal degrees of freedom to the response functions of the internal equilibrium are given by

$$\Delta M_T \equiv M_T^e - M_{T,\zeta_e} = -\,\upsilon\,(\nabla_{T\varepsilon}^e \sigma, F^{-1}\,\nabla_{T\varepsilon}^e \sigma) < 0 \tag{7.249}$$

$$\Delta \beta \equiv \beta_e - \beta_{\zeta_e} = (\nabla_{T\varepsilon}^e \sigma, F^{-1}\,\nabla_{T\varepsilon}^e s) = (\nabla_{T\varepsilon}^e s, F^{-1}\,\nabla_{T\varepsilon}^e \sigma) \tag{7.250}$$

$$\Delta c_\varepsilon \equiv c_\varepsilon^e - c_{\varepsilon,\zeta_e} = T\,(\nabla_{T\varepsilon}^e s, F^{-1}\,\nabla_{T\varepsilon}^e s) > 0\,. \tag{7.251}$$

Here, the coefficient of thermal stress was defined according to (5.18). The symmetric positive definite matrix F is given by the equilibrium coefficients

$$F_{ik} = F_{ki} \equiv \left(\frac{\partial^2 f}{\partial \zeta_k \partial \zeta_i}\right)_e$$

[see (2.83)]. Corresponding to (7.241), we have further

$$\nabla_{T\varepsilon}^e x \equiv \left\{\left(\frac{\partial x}{\partial \zeta_1}\right)_{T,\,\varepsilon,\,\zeta_{k\neq 1}}^e \cdots \left(\frac{\partial x}{\partial \zeta_n}\right)_{T,\,\varepsilon,\,\zeta_{k\neq n}}^e\right\}. \tag{7.252}$$

According to (7.247), the Davies relation (7.67) is replaced by the Davies inequality

$$-\,\Delta c_\varepsilon \Delta M_T \geq T\upsilon\,(\Delta\,\beta)^2\,. \tag{7.253}$$

[The factor p^2 in (7.67) drops out if the thermal stress is defined according to (5.18); the contribution ΔM_T of the internal degrees of freedom to the elastic modulus is, as in (7.67), negative. The relevance of the internal degrees of freedom always contributes to a "softening" of the material in the internal equilibrium.]

Hence, a possible cause of the experimental observation that the Davies relations (7.67) and (7.85) are not fulfilled may be the relevance of several internal degrees of freedom. This conclusion is only compelling, however, if one is certain that the system was in one single well-defined internal equilibrium state during the quasi-static and very rapid measurements.

According to (7.50),

$$\ddot{\zeta}_j + \sum_\ell M_{j\ell}^e\,\dot{\zeta}_\ell = \upsilon \sum_k L_{jk}\left(\frac{\partial \varepsilon}{\partial \zeta_k}\right)_{T,\,\sigma,\,\zeta_{m\neq k}}^e \dot{\sigma} \tag{7.254a}$$

or in the matrix notation

$$\ddot{\boldsymbol{\zeta}} + \boldsymbol{M}\dot{\boldsymbol{\zeta}} = \upsilon(\boldsymbol{L}\nabla_{T\sigma}^e \varepsilon)\,\dot{\sigma} \tag{7.254b}$$

generally holds for isothermal processes with relatively small deviations from a fixed reference state "e". The transformation (7.228–7.230) into normal coordinates leads to

$$\ddot{\bar{\boldsymbol{\zeta}}} + \boldsymbol{D}\,\dot{\bar{\boldsymbol{\zeta}}} = v\,(\boldsymbol{R}\boldsymbol{L}\,\boldsymbol{\nabla}^{e}_{T\sigma}\varepsilon)\,\dot{\sigma}$$

or

$$\ddot{\bar{\zeta}}_j + \frac{1}{\tau_j^G}\,\dot{\bar{\zeta}}_j = v\,(\boldsymbol{R}\boldsymbol{L}\,\boldsymbol{\nabla}^{e}_{T\sigma}\varepsilon)_j\,\dot{\sigma}\,.$$

If $\bar{\zeta}_j(-\infty) = 0$ can be supposed for all j, the first integral of these equations is

$$\dot{\bar{\zeta}}_j(t) = \mathrm{e}^{-t/\tau_j^G}\,v(\boldsymbol{R}\boldsymbol{L}\,\boldsymbol{\nabla}^{e}_{T\sigma}\varepsilon)_j \int\limits_{-\infty}^{t} \mathrm{e}^{t'/\tau_j^G}\,\dot{\sigma}(t')\,\mathrm{d}t'\,.$$

In the case of a periodic perturbation of the system with

$$\sigma(t) = \Delta\sigma\mathrm{e}^{i\omega t}\,,\quad \dot{\sigma}(t) = i\omega\sigma(t),$$

one obtains

$$\dot{\bar{\zeta}}_j = v\,(\boldsymbol{R}\boldsymbol{L}\,\boldsymbol{\nabla}^{e}_{T\sigma}\varepsilon)_j\,\frac{\tau_j^G}{1 + i\omega\tau_j^G}\,\dot{\sigma}$$

or

$$\dot{\bar{\boldsymbol{\zeta}}} = v\,(\boldsymbol{D}^{-1}\,\boldsymbol{R}\boldsymbol{L}\,\boldsymbol{\nabla}^{e}_{T\sigma}\varepsilon)\,\boldsymbol{\Omega}_G\,\dot{\sigma}\,,$$

whereby $\boldsymbol{\Omega}_G$ designates a diagonal matrix with the elements

$$\Omega^G_{jk} \equiv \frac{\delta_{jk}}{1 + i\omega\tau_j^G}\,.$$

According to (7.244), a transformation back to the internal variables yields

$$\dot{\boldsymbol{\zeta}} = \boldsymbol{R}^{-1}\,\dot{\bar{\boldsymbol{\zeta}}} = v\,(\boldsymbol{G}^{-1}\,\boldsymbol{\nabla}^{e}_{T\sigma}\varepsilon)\,\boldsymbol{\Omega}_G\,\dot{\sigma}$$

and

$$\left(\frac{\mathrm{d}\boldsymbol{\zeta}}{\mathrm{d}\sigma}\right) = \dot{\boldsymbol{\zeta}}_T/\dot{\sigma}_T = v\,(\boldsymbol{G}^{-1}\,\boldsymbol{\nabla}^{e}_{T\sigma}\varepsilon)\,\boldsymbol{\Omega}_G\,.$$

One thus obtains *via* (7.237) the frequency-dependent complex compliance

$$J_T\,(\omega) = J_{T,\,\zeta_e} + v\,(\boldsymbol{\nabla}^{e}_{T\sigma}\varepsilon,\,\boldsymbol{G}^{-1}\,\boldsymbol{\nabla}^{e}_{T\sigma}\varepsilon\,\boldsymbol{\Omega}_G) \tag{7.255a}$$

or with (7.243)

$$J_T(\omega) = J_{T,\zeta_e} + \sum_j \frac{\Delta J_T^j}{1 + i\omega\tau_j^G} \quad . \tag{7.255b}$$

A splitting of the complex compliance corresponding to (7.175) into

$$J_T(\omega) = J_T'(\omega) - iJ_T''(\omega) \tag{7.256a}$$

yields

$$J_T'(\omega) = J_{T,\zeta_e} + \sum_j \frac{\Delta J_T^j}{1 + \omega^2(\tau_j^G)^2} \tag{7.256b}$$

as the real dynamic compliance and

$$J_T''(\omega) = \sum_j \frac{\omega\tau_j^G \, \Delta J_T^j}{1 + \omega^2\,(\tau_j^G)^2} \tag{7.256c}$$

as the loss quantity of the compliance. Examples for the case $n = 2$ and different ratios τ_2^G/τ_1^G of the relaxation times are depicted in Figs. 7.14 and 7.15.

In the case of an isothermal periodic deformation

$$\varepsilon(t) = \Delta\varepsilon e^{i\omega t} \ , \quad \dot{\varepsilon}(t) = i\omega\varepsilon(t) \ ,$$

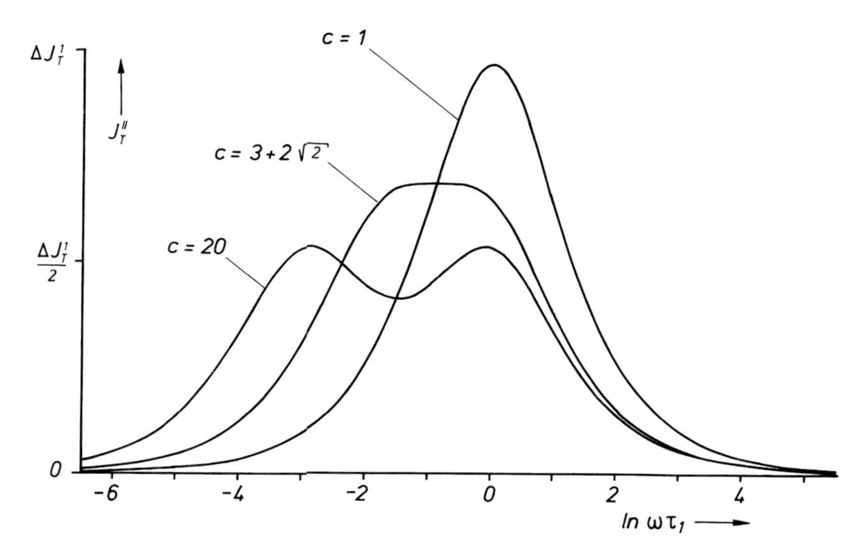

Fig. 7.14. Loss quantity J_T'' of the compliance of a system with $n = 2$ relevant internal degrees of freedom according to (7.256c) for different ratios $c = \tau_2/\tau_1$ of the relaxation times. It was assumed that $\Delta J_T^1 = \Delta J_T^2$

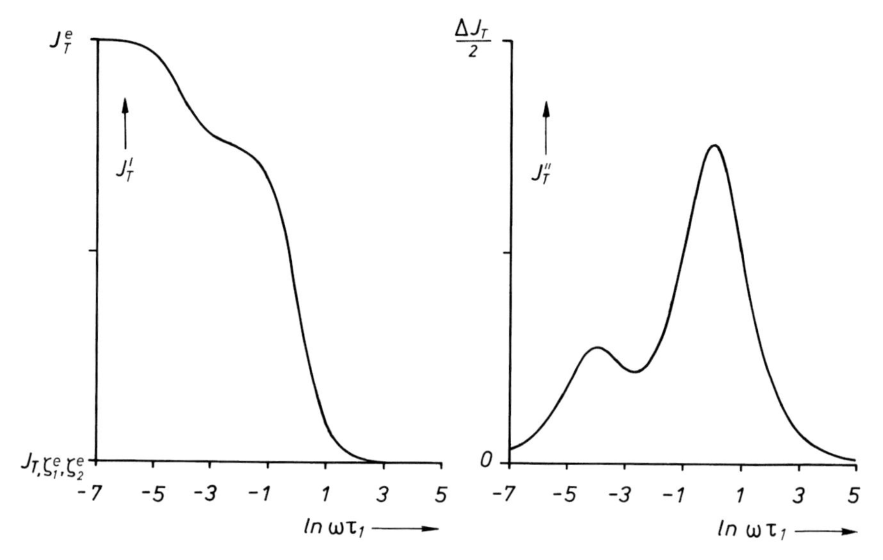

Fig. 7.15. Compliance J_T' and loss quantity J_T'' according to (7.256) with $n = 2$, $\tau_2 = 60\,\tau_1$, and $\Delta J_T^1 = 3\,\Delta J_T^2$

one obtains in the same way from the F-representation the frequency-dependent complex elastic modulus

$$M_T(\omega) = M_{T,\zeta_e} - \upsilon\,(\nabla_{T\varepsilon}^e\sigma, \boldsymbol{F}^{-1}\,\nabla_{T\varepsilon}^e\sigma\,\boldsymbol{\Omega}_F)\,, \tag{7.257}$$

whereby $\boldsymbol{\Omega}_F$ designates a diagonal matrix with the elements

$$\Omega_{jk}^F \equiv \frac{\delta_{jk}}{1 + i\omega\tau_j^F}\,,$$

where the τ_j^F are Debye relaxation times of the normal modes in the F-representation, *i.e.*, the reciprocal eigenvalues of the matrix $\boldsymbol{L}\boldsymbol{F}$. With

$$\Delta M_T^j \equiv -\,\upsilon \sum_k (\boldsymbol{F}^{-1})_{jk} \left(\frac{\partial\sigma}{\partial\zeta_j}\right)^e_{T,\,\varepsilon,\,\zeta_{\ell\neq j}} \left(\frac{\partial\sigma}{\partial\zeta_k}\right)^e_{T,\,\varepsilon,\,\zeta_{\ell\neq k}} < 0 \tag{7.258a}$$

and

$$\Delta M_T = \sum_j \Delta M_T^j \tag{7.258b}$$

according to (7.249), (7.257) can also be written as

$$M_T(\omega) = M_{T,\zeta_e} + \sum_j \frac{\Delta M_T^j}{1 + i\omega\tau_j^F} \tag{7.259a}$$

with the real part (storage modulus)

$$M_T'(\omega) = M_{T,\zeta_e} + \sum_j \frac{\Delta M_T^j}{1+\omega^2(\tau_j^F)^2} \tag{7.259b}$$

and the imaginary part (loss modulus)

$$M_T''(\omega) = -\sum_j \frac{\omega\tau_j^F \Delta M_T^j}{1+\omega^2(\tau_j^F)^2} \,. \tag{7.259c}$$

In the case of a relaxation test of the second kind (Sect. 7.4), one obtains in place of (7.155b) or (7.211b), respectively, the time-dependent relaxation modulus

$$M_T(t) = M_{T,\zeta_e} + \sum_j \Delta M_T^j (1 - \mathrm{e}^{-t/\tau_j^F}) \,, \quad t \geq 0 \tag{7.260a}$$

or

$$M_T(t) = M_T^e - \sum_j \Delta M_T^j \mathrm{e}^{-t/\tau_j^F} \,, \quad t \geq 0 \tag{7.260b}$$

and the memory function

$$N_T(t) \equiv \dot{M}_T(t) = \sum_j \frac{\Delta M_T^j}{\tau_j^F} \mathrm{e}^{-t/\tau_j^F} \,. \tag{7.261}$$

This fulfils the general relation (7.188) of the linear theory of aftereffects. In the theory of aftereffects, the homogeneous Hookean body with n relevant internal degrees of freedom corresponds to the generalized Maxwell model Fig. 7.16 A in the F-representation and to the generalized Kelvin–Voigt model Fig. 7.16 B in the G-representation. Identification of the results from thermodynamics with those of the theory of aftereffects leads to the identities

$$M_j \equiv -\Delta M_T^j \,; \quad \eta_j^F \equiv -\tau_j^F \Delta M_T^j \,; \quad J_j \equiv \Delta J_T^j \,; \quad \eta_j^G \equiv \tau_j^G / \Delta J_T^j \,.$$

The individual Maxwell elements $(M_j; \eta_j^F)$ or Kelvin–Voigt elements $(J_j; \eta_j^G)$ of the models, as can be seen in (7.258), (7.243) and (7.229), cannot be directly assigned to the individual internal degrees of freedom, but to the eigensolutions of the matrices $\boldsymbol{F}^{-1}$ and $\boldsymbol{G}^{-1}$, with whose help the quadratic forms of the response functions can be brought into a normal form. Moreover, similar hierarchic conditions, as given by (7.95), exist between the relaxation times τ_j^F, τ_j^G, τ_j^H ... of the different representations [Meixner (1949), (1953)].

For example, the peculiar elastic behaviour of polymer melts containing very long flexible chain molecules can be made plausible using (7.259). A melt of this kind only exhibits a purely viscous flow at a very low rate of perturbation (*e.g.*, if the melt is left to itself on an inclined plane under the force of gravity). At a

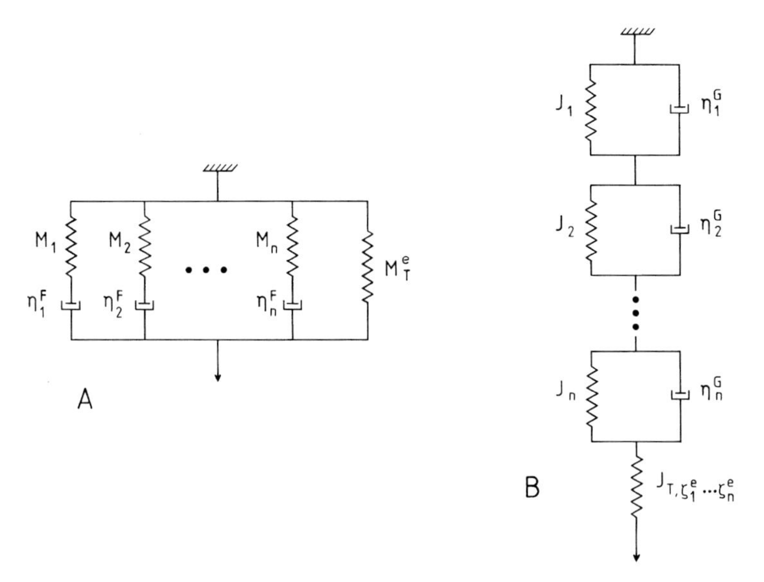

Fig. 7.16. Models of the linear theory of aftereffects. **A**: generalized Maxwell model; **B**: generalized Kelvin–Voigt model

medium rate of perturbation it behaves like a rubber-elastic material (one can form a ball out of a suitable amount of melt and play with it). In the case of an abrupt perturbation, the melt, like a solid, can be shattered. These different modes of behaviour can be measured quantitatively by exposing the melt to a periodic perturbation and, for example, by measuring the elastic modulus relative to the frequency of perturbation [Fig. 7.17; *e.g.*, see Ferry (1980)]. The elastic modulus vanishes, as in a liquid, only at very low frequencies. The melt attains the level of elasticity of a rubber at medium frequencies and the level of elasticity of a glass at high frequencies.

In the thermodynamics of irreversible processes, the polymer melt can be regarded to a first approximation as a liquid with two macroscopically relevant internal degrees of freedom. As liquids do not possess an elastic modulus in internal equilibrium, one must set in (7.259) or Fig. 7.16 A

$$M_T^e = 0, \quad i.e., \quad M_{T,\zeta_e} = -\sum_j \Delta M_T^j \; .$$

[More rigorously, the spring with the modulus M_T^e in Fig. 7.16 A should actually be replaced by a dashpot which describes the real viscosity of the melt (compare Fig. 7.11 A with Fig. 7.10 A). Nevertheless, the viscosity is not of importance in the following argumentation.] The first internal degree of freedom is due to the fact that the long chain molecules in a liquid state are more or less coiled and penetrate each other and then entangle. Due to thermal agitation, however, the points of entanglement are not stable. They come and go. They have a mean lifetime which can be related to a relaxation time τ_1. One can in-

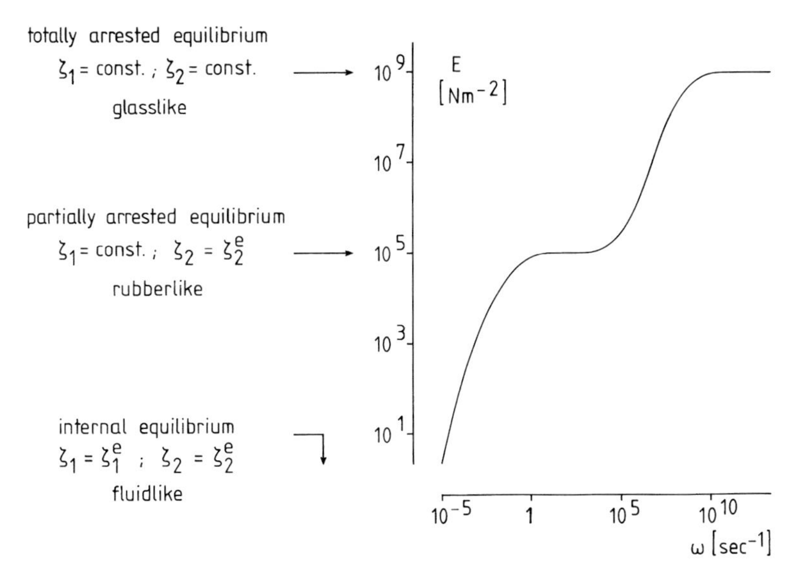

Fig. 7.17. Elastic modulus E as a function of the perturbation frequency ω of a polymer melt composed of very long flexible chain molecules. [Schematically shown; for concrete examples see Ferry (1980)]

troduce the mean concentration ζ_1 of the acute points of entanglement as an internal variable. The second internal degree of freedom originates from the diffusive translational motion of the chain molecules. As a measure of the diffusive translational motion, one can introduce the concentration ζ_2 of the vacancies present in the melt (see Sect. 7.8), which is linked to a relaxation time $\tau_2 \ll \tau_1$. If $\omega\tau_2 \ll \omega\tau_1 \ll 1$ holds, the liquid is practically in internal equilibrium at any time ($\zeta_1 = \zeta_1^e$; $\zeta_2 = \zeta_2^e$). The points of entanglement can disentangle during a half-period of the perturbation. The molecules can slip past each other. The melt behaves like a normal liquid with a practically vanishing elastic modulus. The first internal degree of freedom is arrested if $\omega\tau_2 \ll 1 \ll \omega\tau_1$ holds (ζ_1 = const.; $\zeta_2 = \zeta_2^e$). The entanglements are no longer capable of disentangling during a half-period of the perturbation. The melt behaves like a cross-linked polymer, *i.e.*, like a rubber (Sect. 5.3). If $1 \ll \omega\tau_2 \ll \omega\tau_1$ is finally valid, the second internal degree of freedom, the diffusive translational motion of the molecules, is also arrested (ζ_1 = const.; ζ_2 = const.). The polymer melt behaves like a frozen liquid, *e.g.*, like a glass, with respect to the rapid perturbation (Sect. 7.8).

The real situation is, of course, more complex [*e.g.*, see Lodge (1964); De Gennes (1979); Ferry (1980); Doi, Edwards (1986)]. If one assumes that the diffusive translational motion corresponds to a multitude of internal degrees of freedom dependent on the molecular masses, a whole family of internal variables $\{\zeta_{21} \dots \zeta_{2n}\}$ appears in place of the internal variable ζ_2. In addition, the entanglements are certainly not homogeneous, so that a family of internal variables $\{\zeta_{11} \dots \zeta_{1m}\}$ occurs in place of ζ_1. These degrees of freedom are generally coupled. Within the framework of the linear theory, the system is then characterized by a family of normal modes $\{\bar{\zeta}_1 \dots \bar{\zeta}_{n+m}\}$ whose composition out of primitive molecular mechanisms depends on the thermostatic state of the system and whose spectrum of relaxation times $\{\tau_1 \dots \tau_{n+m}\}$ represents a collective property of the primitive mechanisms.

A problem is that one can describe a quite large class of phenomena with additive expressions of the form (7.233), (7.256), (7.259), and (7.260) if one merely assumes a sufficiently large number of degrees of freedom and a suitable distribution of relaxation times. With a dense sequence of normal modes (7.232), for example, an exact simulation of the scaling invariant hyperbolic relaxation of a single degree of freedom (Sect. 7.4) is also possible [see Mandelbrot (1983), p. 417/18]. It is, therefore, only physically meaningful to apply the linear theory if one is able to precisely indicate the molecular mechanisms and calculate the spectrum of relaxation times and if one is certain that it is a linear problem. If the second derivatives of the Gibbs functions

$$\left(\frac{\partial^2 f}{\partial \zeta_i \partial \zeta_k}\right), \quad \left(\frac{\partial^2 g}{\partial \zeta_i \partial \zeta_k}\right), \dots$$

or the coefficients, for example, of Eq. (7.50) are dependent on the internal variables $\zeta_1 \dots \zeta_n$, the problem becomes non-linear and the concept of normal modes is no longer applicable.

7.8 The Glass Transition

When a liquid is cooled, it usually changes into the crystalline state at a temperature T_M typical for the substance. Thermodynamically, this is a phase transition of the first order (Sect. 8.1). A chemically pure substance, however, only spontaneously crystallizes at T_M if it is cooled slowly enough and a sufficient number of crystal nuclei is present. If a liquid is cooled more quickly, it is possible to maintain the liquid state even below T_M. The liquid is then in a so-called undercooled state which, considered thermodynamically, is a metastable state (see Fig. 2.1 B; the potential barrier between the liquid and the crystalline state corresponds to the work necessary to form the crystal nuclei; Sect. 8.2). In many liquids, the viscosity increases considerably with decreasing temperature and reaches extremely high values in an undercooled state. A high viscosity η, however, means, as can be estimated on the basis of the Einstein–Stokes relation, low translational mobility of the molecules and thus a low rate of crystallization. The Einstein–Stokes relation is:

$$D = \frac{RT}{3\pi\eta d} \tag{7.262}$$

[*D*: coefficient of self-diffusion; *d*: mean diameter of the supposedly spherical molecules; *R*: Boltzmann – or gas constant; *e.g.*, see Landau, Lifshitz (1987)]. If one rapidly cools such liquids below between *ca.* $2T_M/3$ and $T_M/2$, they freeze into an amorphous solid, a glass, without forming an ordered crystal lattice [on the physics of glasses, see, for example, Stevels (1962); MacKenzie (1960–64); Haward (1973); Jäckle (1986)].

The substances with a relatively high tendency towards vitrification include the elements oxygen, sulphur, selenium, tellurium, phosphorus, compounds with these elements such as, for example, quartz (SiO_2), some halides, many organic compounds such as glycerol and sugar, and above all (organic) polymers. Due to the asymmetric irregular arrangement of their chain structure, some polymers are fundamentally incapable of crystallizing. For entropic and kinetic reasons (Sect. 8.2.), other long-chain polymers crystallize only partially *i.e.*, they always contain a glassy part at lower temperatures. In principle, vitrification is possibly a phenomenon common to all substances. Nevertheless, this phenomenon cannot be realized in some substances, as the heat dissipation, which proceeds at a finite rate, acts against the required cooling rate. In polymers which are not capable of crystallizing due to their irregular molecular structure, vitrification is, of course, independent of the cooling rate. Good glass formers (such as B_2O_3 or SiO_2) require cooling rates in the range of 10^{-5} to 10^{-2} K s^{-1}. Metals or metallic alloys freeze to form a glass at cooling rates of about 10^6 K s^{-1}. Here, the sufficiently fast heat dissipation is already a problem. According to computer simulations, the vitrification of argon would require a cooling rate of 10^{12} K s^{-1} [see Jäckle (1986)].

Simon (1930) concluded from the fact that a finite entropic difference is retained between the vitreous state and the crystalline state of a pure substance in the limit $T \to 0$ that one is not dealing with an internal equilibrium state in the case of a glass but rather with a "frozen" non-equilibrium state. Within the definitions of the thermodynamics of irreversible processes, the vitreous state, therefore, has to be classified as an arrested equilibrium state [see Breuer, Rehage (1967), Rehage (1980)]. Further below, we will explicitly assign an internal variable ζ to the internal degrees of freedom frozen (arrested) in the glass. When cooling a liquid at a constant and not too high rate under constant pressure, one schematically obtains, for example, for the volume the following picture (Fig. 7.18): Above the melting temperature T_M, the liquid is in internal equilibrium with respect to the internal variable ζ. A volume $V\,[T, \zeta_e(T)]$ is measured which is uniquely determined by the temperature T. If the liquid state is maintained below T_M, this also at first applies to the volume of the undercooled liquid. At a temperature T_F, however, the volume curve starts to deviate from the equilibrium curve. The liquid falls into non-equilibrium states in which ζ varies independently of T. The volume then follows a non-equilibrium curve $V(T, \zeta)$ whose course is no longer uniquely determined by the temperature. At $T = T_{gl}$, the internal degree of freedom is finally arrested with $\dot{\zeta} = 0$. Below T_{gl}, the volume curve follows the equation $V = V(T, \zeta = \text{const.})$ of the arrested equilibrium. The glass transition region, *i.e.*, the position of the temperature interval $\langle T_{gl}\,;\, T_F\rangle$, as well as the non-equilibrium value of the variable ζ, which freezes at T_{gl}, depend on the cooling rate $\dot{T} = \gamma < 0$.

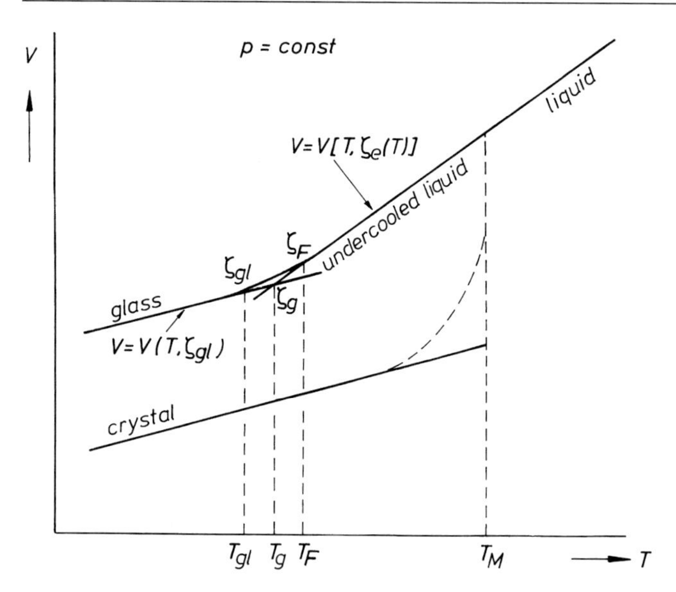

Fig. 7.18. Schematic volume–temperature diagram of a substance when cooled at a finite rate $\dot{T} = \gamma$ [according to Rehage (1980)]. Melting temperature: T_M; glass transition region: $T_F \rightarrow T_{gl}$. See text

If one proceeds from the linearized mechanical equation of state, one obtains the straight lines

$$T \geq T_g: \quad V = V_g\,[1 + \alpha_e\,(T - T_g)]\,,$$

$$T \leq T_g: \quad V = V_g\,[1 + \alpha_{\zeta_e}\,(T - T_g)]\,,$$

as volume curves above and below the glass transition region which intersect at the temperature T_g, the so-called glass temperature (α_e: coefficient of thermal expansion of the internal equilibrium; α_{ζ_e}: coefficient of thermal expansion of the arrested equilibrium; see Sect. 7.3). In this approximation [the so-called "simple freezing-in model"; Rehage (1980)], the glass transition region $\langle T_{gl}; T_F\rangle$ is reduced to a single temperature T_g. The glass transition becomes a discontinuous process. The value ζ_g of the internal variable which freezes is then necessarily the equilibrium value $\zeta_g = \zeta_e(T_g)$. The coefficient of thermal expansion then suffers a discontinuous jump $\Delta\alpha = \alpha_e - \alpha_{\zeta_e}$ at T_g, which is given by (7.83). In the same way, as a result of the linearized thermal equation of state, the heat capacity is subjected to a discontinuous jump Δc_p at T_g according to Eq. (7.84). The glass temperature T_g and $\zeta_g = \zeta_e(T_g)$ are actually purely fictitious quantities. The "simple freezing-in model", however, can be regarded as a limiting case, which attains validity in the limit $\gamma \rightarrow 0$.

What freezes during the transition from the liquid state into the vitreous state is without doubt the molecular disorder (or rather the short-range order) in the liquid and the diffusive translational motion of the molecules. In the following, we will base the description of this process on the simple model of the

Eyring liquid. Eyring regards liquids as a mixture of molecules and vacancies [Hirai, Eyring (1958/59); Eyring, Ree (1961)]. Here, the concentration of the vacancies or the volume fraction of the vacancies (the relative free volume) can serve as a measure of the intensity of the diffusive translational motion. In order to formulate the Gibbs fundamental equation of such a mixture, we proceed from the Flory–Huggins fundamental equation (3.79) of the ideal-athermal polymer solution. This equation would already be useful for our purposes if the vacancies and monomer units were of about the same size. This would be true if the diffusion process could be understood as a hopping of monomer units into holes of the same size. In fact, the diffusion in a relatively compact liquid seems to be more of a cooperative process whose development requires a considerably smaller vacancy size. According to Frenkel (1946), the vacancies in molecular liquids are approximately the size of individual atoms. Nevertheless, the Eq. (3.79) can be used in this case if the degree of polymerization P is replaced by the ratio of volumes

$$\varrho \equiv v_2/v_1 > 0 \tag{7.263}$$

of the two unequal mixing partners [Flory (1941/42); see also Guggenheim (1959, §5.50]. In approximation,

$$G = G_o + RT \left[N_1 \ln \frac{N_1}{N_1 + \varrho N_2} + N_2 \ln \frac{\varrho N_2}{N_1 + \varrho N_2} \right] + G_E \tag{7.264}$$

is the Gibbs fundamental equation for a mixture of partners of different sizes whose volume ratio is given by (7.263). If we refer (7.264) to the Eyring liquid, N_1 designates the mole number of the vacancies and N_2 the mole number of the material particles (molecules of a low-molecular substance or mobile units of a polymer; the mole number N_1 of the massless vacancies can be defined as the fraction $N_1 = \zeta N_2$ of the mole number N_2 of the material particles.

We proceed from Eq. (3.82) for the deduction of an expression for the excess free enthalpy G_E:

$$G_E = z(N_1 + N_2)\, \Delta w_{12} x_1 x_2 = z \Delta w_{12} \frac{N_1 N_2}{N_1 + N_2} \,.$$

Here, we substitute the coordination number z with

$$z N_i \rightarrow o_i N_i$$

by the surfaces capable of contact o_i per mole of the mixing partners [Kanig (1969)]. It is obvious that the interaction between partners of different sizes and shapes can be determined much better by means of the surfaces o_i than by means of the pure number z of the nearest neighbours. Hence, the excess free enthalpy of the mixture should amount to

$$G_E = \Delta w_{12} \frac{o_1 N_1 o_2 N_2}{o_1 N_1 + o_2 N_2} \,, \tag{7.265}$$

where $w_{11} = 0$ naturally holds in a mixture of material particles with vacancies (no mutual interaction of the vacancies), and w_{22} describes the binding energy between adjacent particles, while w_{12} can be interpreted as the binding energy between two material particles separated by vacancies. To simplify matters, we will assume in the following that the excess free enthalpy is independent of temperature and pressure. It will turn out, however, that a variable ratio o_1/o_2 certainly corresponds quantitatively better to the given situations. Temperature independence means that we confine ourselves to a regular mixture with

$$G_E = H_E \tag{7.266}$$

[*cf.* Eq. (3.72)]. As a result of the pressure independence, the molar volumes v_i become equal to the molar volumes v_i^o of the pure substances as well as to the partial molar volumes $\bar{v}_i$ according to (3.53) and (3.17):

$$v_i^o = v_i = \bar{v}_i \,. \tag{7.267}$$

In place of the mole numbers N_i, one can then easily introduce the volume fractions φ_i as variables according to (3.7). From

$$\varphi_1 \equiv \frac{\bar{v}_1 N_1}{\bar{v}_1 N_1 + \bar{v}_2 N_2} = \frac{N_1}{N_1 + \varrho N_2}\,; \quad \varphi_2 = 1 - \varphi_1\,,$$

one obtains

$$N_1 = \frac{\varrho \varphi_1}{1 - \varphi_1} N_2$$

and with this according to (7.264),

$$G = G_o + RTN_2 \left[\frac{\varrho \varphi_1}{1 - \varphi_1} \ln \varphi_1 + \ln (1 - \varphi_1) \right] + G_E \,. \tag{7.268a}$$

(7.265) leads to

$$G_E = N_2 o_1 \Delta w_{12} \frac{\varrho \varphi_1}{1 - \varphi_1 + \lambda \varphi_1} \tag{7.268b}$$

for the excess free enthalpy. Here,

$$\frac{o_1}{o_2} \varrho \equiv \lambda\,, \quad i.e. \ \frac{o_1}{o_2} = \lambda \frac{v_1}{v_2} \tag{7.268c}$$

was used as an abbreviation, and λ is a geometric factor which not only takes into account the different sizes, but also the different shapes of the partners.

The standard term in (7.264) is only determined by the particles, as the vacancies do not possess a chemical potential in the pure state:

$$G_o = \mu_2^o N_2 \, ; \quad \mu_1^o = 0 \tag{7.269}$$

[*cf.* Eq. (3.51 b)].

We will finally introduce the ratio $\zeta \equiv N_1/N_2$ of the mole number of the vacancies to the mole number of the material particles as a dimensionless internal variable of the liquid. The internal variable, which is often rather vaguely referred to as an "ordering parameter" in the literature [*e.g.*, see Kovacs *et al.* (1979), Rehage (1980)], thus obtains a concrete physical meaning. We designate the volume fraction of the vacancies (the relative free volume) with $\varphi \equiv \varphi_1$. The energy (or enthalpy) required to generate one mole of vacancies in the vacancy-free liquid is:

$$h_c \equiv o_1 \Delta w_{12} \, .$$

Furthermore, it is useful to refer all extensive quantities to one mole of particles. We denote these molar quantities G/N_2, V/N_2 ... as before with the corresponding small letters g, v ... We thus obtain as Gibbs's fundamental equation for a melt interspersed with vacancies

$$g = g_o + RT \left[\frac{\varrho \varphi}{1 - \varphi} \ln \varphi + \ln (1 - \varphi) \right] + g_E \tag{7.270 a}$$

with

$$g_o = \mu_2^o (T, p) \tag{7.270 b}$$

and

$$g_E = h_E = h_c \frac{\varrho \varphi}{1 - \varphi + \lambda \varphi} \, . \tag{7.270 c}$$

The interrelation

$$\zeta = \frac{\varrho \varphi}{1 - \varphi} \, ; \quad \mathrm{d}\zeta = \frac{\varrho}{(1 - \varphi)^2} \, \mathrm{d}\varphi \tag{7.271}$$

exists between the internal variable ζ and the relative free volume φ.

According to (7.270), the entropy of the system is given by

$$s \equiv - \left(\frac{\partial g}{\partial T} \right)_{p, \zeta} = - \left(\frac{\partial g_o}{\partial T} \right)_p - R \left[\frac{\varrho \varphi}{1 - \varphi} \ln \varphi + \ln (1 - \varphi) \right] . \tag{7.272}$$

One thus obtains for the enthalpy $h = g + Ts$

$$h = h_o + h_E \quad \text{with} \quad h_o = g_o - T\left(\frac{\partial g_o}{\partial T}\right)_p . \tag{7.273}$$

For the volume v it follows that:

$$v = v_1 \zeta + v_2 = \frac{v_2}{1 - \varphi} . \tag{7.274}$$

The vacancies are not of importance in the heat capacity (7.77) of the arrested equilibrium

$$c_{p,\zeta} \equiv T\left(\frac{\partial s}{\partial T}\right)_{p,\zeta} = -T\left(\frac{\partial^2 g_o}{\partial T^2}\right)_p . \tag{7.275}$$

In the following, we will regard $c_{p,\zeta}$ as a constant, *i.e.*, disregard the dependence of this quantity on temperature and pressure. We will treat the coefficient of thermal expansion α_ζ [Eq. (7.76)] the same way.

The affinity

$$a \equiv -\left(\frac{\partial g}{\partial \zeta}\right)_{T,p} = -\left(\frac{\partial g}{\partial \varphi}\right)_{T,p}\left(\frac{\partial \varphi}{\partial \zeta}\right)_{T,p}$$

[Eq. (7.17)], for which (7.270) and (7.271) result in

$$a = T\sigma_{Tp} - \eta_{Tp} , \tag{7.276}$$

is the determining quantity for the dynamics of the processes in the glass transition region. The partial molar entropy of the internal degree of freedom is:

$$\sigma_{Tp} \equiv \left(\frac{\partial s}{\partial \zeta}\right)_{T,p} = -R\left[\ln \varphi + \left(1 - \frac{1}{\varrho}\right)(1 - \varphi)\right] \tag{7.277}$$

and the partial molar enthalpy of the internal degree of freedom is:

$$\eta_{Tp} \equiv \left(\frac{\partial h}{\partial \zeta}\right)_{T,p} = h_c \left(\frac{1 - \varphi}{1 - \varphi + \lambda\varphi}\right)^2 , \tag{7.278}$$

where η_{Tp} is the energy required for the generation of one mole of vacancies in the vacancy-containing melt. According to (7.274), the partial molar volume is given by

$$\varphi_{Tp} \equiv \left(\frac{\partial v}{\partial \zeta}\right)_{T,p} = v_1 = \frac{1 - \varphi}{\varrho} v . \tag{7.279}$$

The liquid is in equilibrium if $a = 0$, *i.e.*,

$$\eta_{Tp} = T\sigma_{Tp} \tag{7.280}$$

is fulfilled. If h_c, ϱ, λ, T are known, one can determine from (7.280) the relative free volume φ_e in the internal equilibrium, and if h_c, ϱ, λ, φ are known, the equilibrium temperature T_e.

Equations (7.270) and (7.271) yield

$$\frac{1}{\tau_{Tp}} \equiv L\left(\frac{\partial^2 g}{\partial \zeta^2}\right)_{T,p} = L\left[\left(\frac{\partial \eta_{Tp}}{\partial \zeta}\right)_{T,p} - T\left(\frac{\partial \sigma_{Tp}}{\partial \zeta}\right)_{T,p}\right] \tag{7.281 a}$$

as the relaxation time (7.38c) of the system with

$$\left(\frac{\partial \eta_{Tp}}{\partial \zeta}\right)_{T,p} = -\frac{2h_c\lambda}{\varrho}\left(\frac{1-\varphi}{1-\varphi+\lambda\varphi}\right)^3 = -\frac{2h_c\lambda}{\varrho}\left(\frac{\eta_{Tp}}{h_c}\right)^{3/2}, \tag{7.281 b}$$

$$\left(\frac{\partial \sigma_{Tp}}{\partial \zeta}\right)_{T,p} = -\frac{R}{\varrho}\frac{(1-\varphi)^2}{\varphi}\left(1-\varphi+\frac{\varphi}{\varrho}\right). \tag{7.281 c}$$

The relaxation time traverses a singularity (infinity) if

$$\left(\frac{\partial \eta_{Tp}}{\partial \zeta}\right)_{T,p} = T\left(\frac{\partial \sigma_{Tp}}{\partial \zeta}\right)_{T,p}. \tag{7.282}$$

According to (7.276–7.278) and (7.280),

$$h_c\left(\frac{1-\varphi_e}{1-\varphi_e+\lambda\varphi_e}\right)^2 = -RT\left[\ln\varphi_e + \left(1-\frac{1}{\varrho}\right)(1-\varphi_e)\right]$$

holds for the relative free volume φ_e in the internal equilibrium. One can, therefore, also write

$$\frac{1}{\tau_{Tp}^e} = LRT\,\frac{1-\varphi_e}{\varrho}\left[\frac{2\lambda}{1-\varphi_e+\lambda\varphi_e}\left[\ln\varphi_e + \left(1-\frac{1}{\varrho}\right)(1-\varphi_e)\right] + \frac{1-\varphi_e}{\varphi_e}\left(1-\varphi_e+\frac{\varphi_e}{\varrho}\right)\right] \tag{7.283}$$

for the relaxation time with respect to the internal equilibrium (Debye relaxation time). According to (2.88) and (7.29b), the internal equilibrium becomes unstable if $\tau_{Tp}^e < 0$, *i.e.*,

$$\left(\frac{\partial \eta_{Tp}}{\partial \zeta}\right)_{T,p}^e < T\left(\frac{\partial \sigma_{Tp}}{\partial \zeta}\right)_{T,p}^e \tag{7.284}$$

is valid.

According to (7.83), the contribution of the vacancies to the coefficient of thermal expansion in the internal equilibrium is given by

$$\Delta\alpha = \frac{1}{v}\, L\tau^e_{Tp}\,\varphi^e_{Tp}\,\sigma^e_{Tp}\,.$$

With (7.277), (7.279) and (7.283), this leads to

$$\Delta\alpha = -\frac{1}{T}\,\frac{\varphi_e}{\dfrac{2\lambda\varphi_e}{1-\varphi_e+\lambda\varphi_e}+\dfrac{(1-\varphi_e)\left(1-\varphi_e+\dfrac{\varphi_e}{\varrho}\right)}{\ln\varphi_e+\left(1-\dfrac{1}{\varrho}\right)(1-\varphi_e)}}\,. \tag{7.285}$$

One obtains from (7.84)

$$\Delta c_p = LT\tau^e_{Tp}(\sigma^e_{Tp})^2 = vT\Delta\alpha\,\frac{\sigma^e_{Tp}}{\varphi^e_{Tp}}$$

for the contribution of the vacancies to the heat capacity of the internal equilibrium. With (7.277), (7.279), (7.285), this means

$$\Delta c_p = \frac{R\varrho\varphi_e}{1-\varphi_e}\,\frac{\ln\varphi_e+\left(1-\dfrac{1}{\varrho}\right)(1-\varphi_e)}{\dfrac{2\lambda\varphi_e}{1-\varphi_e+\lambda\varphi_e}+\dfrac{(1-\varphi_e)\left(1-\varphi_e+\dfrac{\varphi_e}{\varrho}\right)}{\ln\varphi_e+\left(1-\dfrac{1}{\varrho}\right)(1-\varphi_e)}}\,. \tag{7.286}$$

If $\Delta\alpha$, Δc_p, ϱ and T are given, the relative free volume φ_e and the interaction parameter λ can be determined from (7.285/7.286).

Equations (7.285) and (7.286) hold for every internal equilibrium state, *i.e.*, in the "simple freezing-in model" also at the glass temperature T_g. In this model, the differences $\Delta\alpha$ and Δc_p occur at T_g as differences of the response functions between the liquid and the glassy state. Kanig (1969) [assuming $\varphi_e \ll 1$ and neglecting $1/\varrho$ with respect to 1 in Eq. (7.277)] determined the volume fraction $\varphi^e_g = \varphi_e(T_g)$ and the interaction parameter λ for 11 polymers using $\Delta\alpha, \Delta c_p$ and T_g from experimental data. For these polymers, he found on average

$$\langle\varphi^e_g\rangle = 0.0235 \pm 0.0050\,, \tag{7.287a}$$

$$\langle\lambda\rangle = 3.15 \pm 0.35\,. \tag{7.287b}$$

First of all, it is remarkable that the mean value (7.287a) agrees very well with the value $\varphi^e_g \approx 0.025$ determined from viscosimetric data by Williams, Landel,

Ferry (1955). At least qualitatively, (7.287 a) seems to confirm the thesis of Fox and Flory (1950) in which liquids are in an iso-free volume state at the glass temperature. Equally qualitatively, (7.287 a) seems to confirm the rule of Simha and Boyer (1962) according to which

$$T_g \, \Delta\alpha = \text{const.}$$

should always be fulfilled at the glass temperature. Kanig (1969), however, has found definite systematic differences from the mean value (7.287 a). For example, polymers composed of more flexible chain molecules lead to a smaller φ_g^e-value than polymers composed of stiffer chain molecules. If one regards the mean value (7.287 b) in the same way as a value characteristic for the glass temperature, one should allow a certain variability $\lambda = \lambda(T, \varphi)$ of the interaction parameter above T_g (see further below). Additional terms can then be found in $\Delta\alpha$ and Δc_p which are linked with the derivatives $\partial\lambda/\partial T$ or $\partial\lambda/\partial\varphi$, respectively, and which possibly also somewhat change the result (7.287). In a very extensive table containing 63 polymers, Wrasidlo (1974) showed that a constancy of φ_g^e at the glass temperature does not exist.

The Davies relations (7.85) or (7.248) are often regarded in the literature as relations which are characteristic for the glass transition and then also referred to as Prigogine–Defay relations [*e.g.*, see Elias (1984), Rehage (1980), Jäckle (1986), Donth (1992)]. As shown in Sects. 7.3 and 7.7, the Davies relations involve quantities which refer to an internal equilibrium state. The Davies relations hold for every arbitrary internal equilibrium state. With respect to the glass transition, the Davies relations can, therefore, only be verified up to the experimentally very uncertain temperature T_F (Fig. 7.18). Below T_F, a single internal degree of freedom yields the rate-dependent contributions

$$\Delta' \alpha_p \equiv \alpha_p - \alpha_{p,\zeta} = \frac{1}{v} \, \varphi_{Tp} \, \dot{\zeta}/\dot{T}_p \,, \tag{7.288 a}$$

$$\Delta' c_p \equiv c_p - c_{p,\zeta} = T \sigma_{Tp} \, \dot{\zeta}/\dot{T}_p \tag{7.288 b}$$

to the response functions [Eqs. (7.72–7.77); $\dot{T}_p = \gamma < 0$: cooling rate at constant pressure]. These contributions continuously diminish with decreasing temperature until they finally disappear completely at T_{gl} (because of $\dot{\zeta} = 0$, $\zeta =$ const.) (Fig. 7.23). The Davies relations only become important for the glass transition if an equilibrium value ζ_e freezes out of the internal equilibrium. However, this is only possible in the limit $\gamma \to 0$, *i.e.*, with $\dot{\zeta} = 0$ because of $a = 0$ (see further below).

The fundamental equation (7.270) does not contain anything concerning a possible glass transition. The internal equilibrium of the undercooled melt is stable up to the limit $T_e \to 0$, $\varphi_e \to 0$ (see Fig. 7.19). On the other hand, the liquid is not capable of admitting an arbitrary number of vacancies with increasing temperature. There is a maximum temperature $T_{\max}^e$ and a maximum free volume $\varphi_{\max}^e$ above which $(\partial^2 g/\partial\zeta^2)_{T,p}^e$ or τ_{Tp}^e become negative, *i.e.*, the internal equilibrium of the liquid loses its stability. In the following, we will

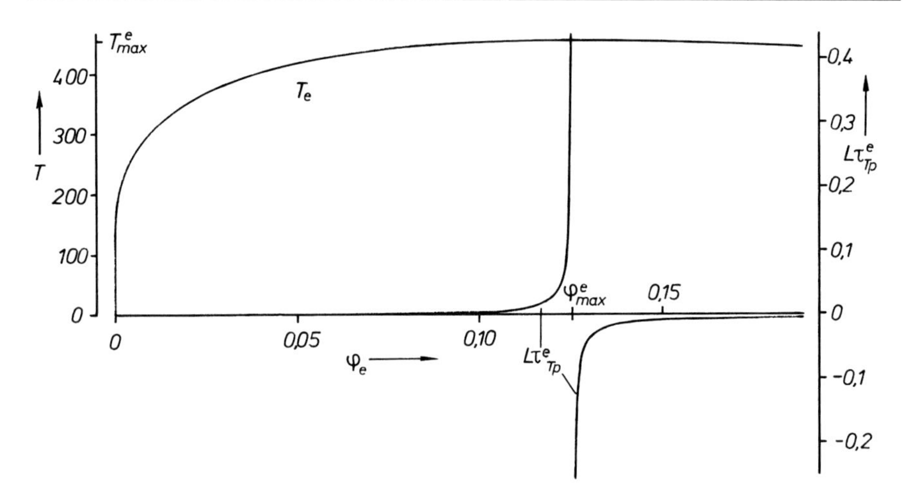

Fig. 7.19. Dependence of the equilibrium temperature T_e[K] and stability function $L\tau^e_{Tp}$[1/J] on the relative free volume φ_e according to (7.280), (7.283), and (7.289)

assume the concrete values

$$h_c = 10 \text{ kJ/mol}\,, \quad \varrho = 10\,, \quad \lambda = 3 \tag{7.289}$$

which are approximately true for polystyrene at T_g [Kanig (1969), Tab. 1]. With (7.289), instability, according to (7.284), already occurs at $T^e_{\max} = 456.303\ \ldots$ K and $\varphi^e_{\max} = 0.125\ldots$, *i.e.*, at values that are certainly too low. This shows clearly that we should actually treat the parameters h_c and λ as the variables $h_c(T)$ and $\lambda(T, \varphi)$. At the critical point $(T^e_{\max};\ \varphi^e_{\max})$, the Debye relaxation time passes through a singularity $\tau^e_{Tp} = \pm\ \infty$ (Fig. 7.19). According to the Eqs. (7.80) and (7.81), the heat capacity c^e_p and the coefficient of thermal expansion α_e also pass through a corresponding singularity at this point. Hence, if we retain the values (7.289) in the following, we overestimate the heat capacity and the coefficient of thermal expansion of the system at higher temperatures. Nevertheless, this should not have an influence on the qualitative development of the processes in the range of the glass transition.

For a description of the freezing process at a constant cooling rate $\dot{T} = \gamma < 0$ and constant pressure p within the framework of the thermodynamics of irreversible processes, one has to proceed from the differential equations

$$\dot{T} = \gamma\,; \quad \dot{\zeta} = L\,a\,; \quad L > 0\,. \tag{7.290a}$$

With (7.271) and (7.276–7.278), the second equation leads to

$$\dot{\varphi} = -\frac{L(1-\varphi)^2}{\varrho}\left[h_c\left(\frac{1-\varphi}{1-\varphi+\lambda\,\varphi}\right)^2 + RT\left[\ln\varphi + \left(1-\frac{1}{\varrho}\right)(1-\varphi)\right]\right] \tag{7.290b}$$

as the differential equation for the relative free volume. Freezing processes with $\dot{\zeta} \to 0$ or $\dot{\varphi} \to 0$ are not characterized by $a \to 0$ (internal equilibrium) but by $L \to 0$ (arrested equilibrium). In (7.290), the coupling parameter must thus be regarded as a variable $L = L(T, \varphi)$ which vanishes at T_{gl} or at least assumes very small values.

We cannot expect an explanation for the functional dependence of the phenomenological coefficient $L(T, \varphi)$ in the macroscopic phenomenological theory. We rather depend on *ad-hoc* formulations. As L has the dimension of a fluidity per volume, a formulation corresponding to the Doolittle proportionality

$$\eta \sim e^{\text{const.}/\varphi}$$

suggests itself to start with [η: viscosity; $1/\eta$: fluidity; Doolittle (1951)]. In a normalized form, we can thus assume

$$L(t) = L(0) \exp\left[c_D \left(\frac{1}{\varphi_0} - \frac{1}{\varphi(t)}\right)\right]. \tag{7.291}$$

(Here and in the following, we indicate the initial state with "0"). No total freezing process can, of course, be expected with (7.291), as L only vanishes with $\varphi \to 0$. Solutions of the equations (7.290) and (7.291), however, already show some characteristics of the glass transition. When cooling at a constant rate $\gamma < 0$, the system reaches a finite value $\varphi \neq 0$ of the free volume at absolute zero and thus a finite entropic value (Fig. 7.20). The final value φ at $T = 0$ is larger, the faster the cooling rate[9]. According to the equations (7.290/7.291), the relaxation in the freezing range (with $T =$ const., $p =$ const.) takes a typically non-linear course. When compared with the comparable linear relaxation

$$\dot{\varphi} = -\frac{1}{\tau^e_{Tp}} [\varphi - \varphi_e(T)], \tag{7.292}$$

one can first observe, as in the hyperbolic relaxation (Fig. 7.4), an acceleration and then a retardation of the process (Fig. 7.21). This result is essentially determined by the Doolittle dependence (7.291). With the formulation

$$L(t) = L(0) \exp\left[c_W \left(\frac{1}{T_0 - T_{gl}} - \frac{1}{T - T_{gl}}\right)\right], \quad T \geq T_{gl}, \tag{7.293}$$

9 The final value of φ also depends on the initial value $L(0)$ of the coupling coefficient. In the following figures, the initial values $L(0)$ or L_o, respectively, were chosen in such a way that a result with the required accuracy can be obtained by means of a PC in an acceptable amount of time. The true time unit remains completely open. In the following, we will only noncommitally denote the time unit with "tu".

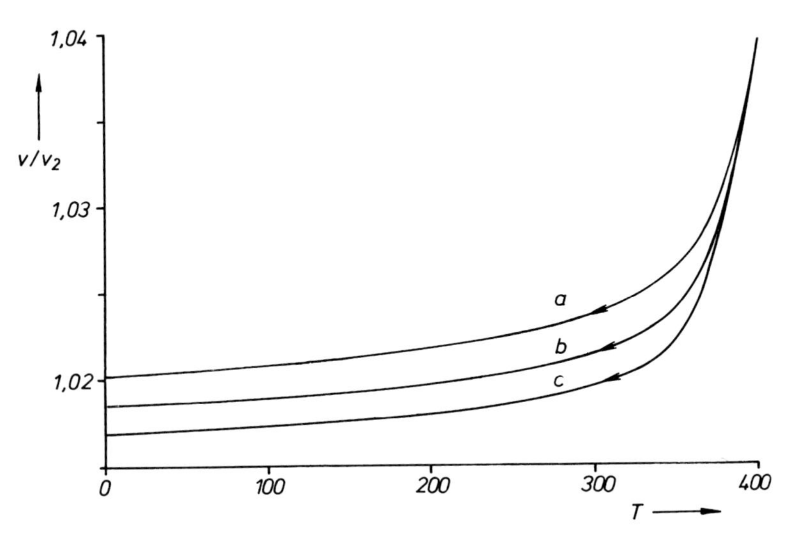

Fig. 7.20. Relative volume v/v_2 according to (7.274), (7.290), (7.291), and (7.289) with $L(0) = 10^{-6}$ mol/J · tu, $c_D = 0.5$ at different cooling rates $\gamma_a = -0.001$ K/tu, $\gamma_b = -0.0001$ K/tu, $\gamma_c = -0.00001$ K/tu (tu: open time unit). The initial state is the equilibrium state at $T = 400$ K

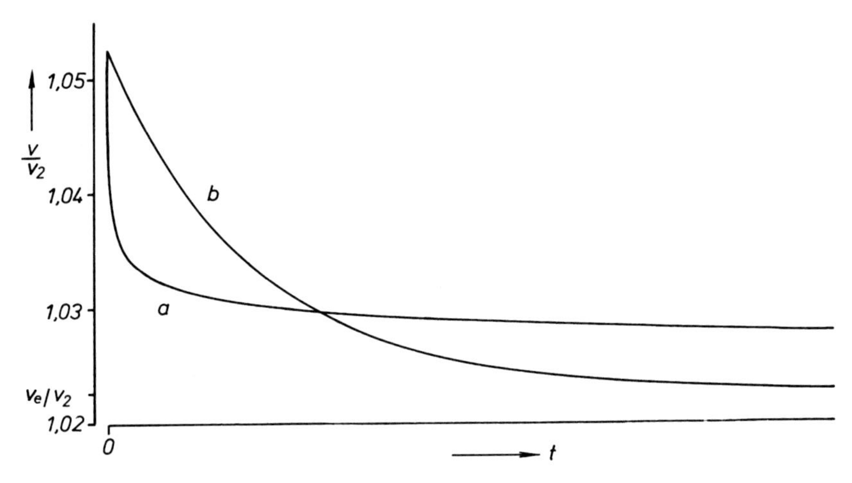

Fig. 7.21. Relaxation of the relative volume v/v_2 [Eq. (7.274)] at $T = 360$ K from the initial value $v = 1.0526\ v_2(\varphi = 0{,}05)$ to the equilibrium value $v_e = 1.0226\ v_2$. **a**: non-linear relaxation according to Eqs. (7.290) and (7.291); **b**: linear relaxation according to Eqs. (7.292) and (7.283). Both cases with (7.289) and $L(0) = 10^{-6}$ mol/J · tu, $c_D = 0.5$

which was redeveloped on the basis of the phenomenological theory of Williams, Landel, Ferry (1955), one can naturally achieve a complete freezing of the system at $T = T_{gl}$. It is, however, not possible to establish a connection between (7.293) and (7.291) in non-equilibrium thermodynamics, as T and φ are mutually independent variables.

It is remarkable that although the relaxation time τ_{Tp} is able to pass through a singularity $\tau_{Tp} = \pm\infty$ at low temperatures while cooling with $\gamma < 0$, this singularity does not gain any influence on the process (Fig. 7.22). If the coupling coefficient L is variable with T and ζ, the second equation (7.290 a) yields by differentiation with respect to the time

$$\ddot{\zeta} = \dot{L}a + L\dot{a}$$

with

$$\dot{L} = \left(\frac{\partial L}{\partial T}\right)_{p,\zeta} \dot{T} + \left(\frac{\partial L}{\partial \zeta}\right)_{T,p} \dot{\zeta}\,.$$

According to (7.37) and (7.276–7.278),

$$\dot{a} = \sigma_{Tp}\,\dot{T} - \frac{1}{L\,\tau_{Tp}}\,\dot{\zeta}$$

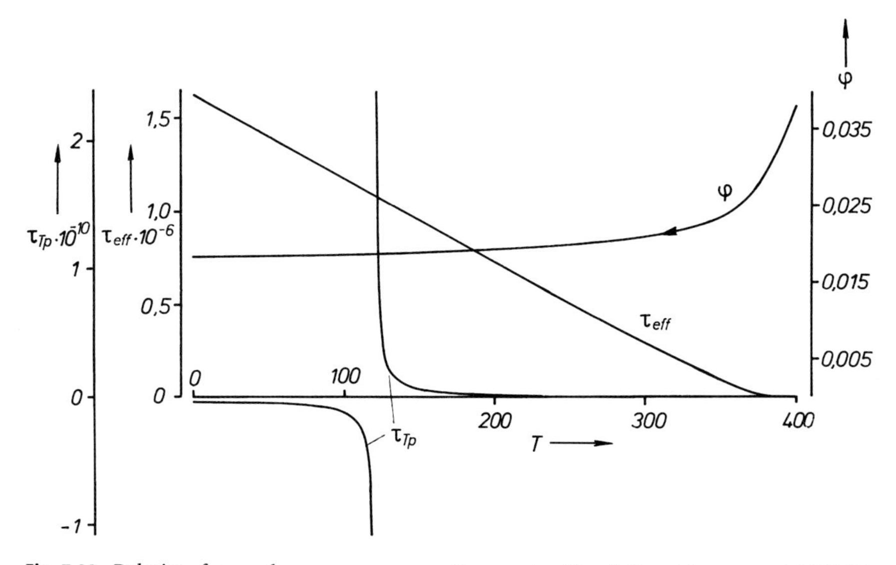

Fig. 7.22. Relative free volume φ upon cooling as in Fig. 7.20 with $\gamma = -0.0001$ K/tu. Relaxation time according to (7.381): τ_{Tp} [tu]; effective relaxation time according to (7.295): τ_{eff} [tu]. Open time unit: tu

is valid. In place of Eq. (7.38), one thus obtains with variable $L(T, \zeta)$ the differential equation

$$\ddot{\zeta} + \left[\frac{1}{\tau_{Tp}} - \left(\frac{\partial \ln L}{\partial T} \right)_{p,\zeta} \dot{T} - \left(\frac{\partial \ln L}{\partial \zeta} \right)_{T,p} \dot{\zeta} \right] \dot{\zeta} = L \sigma_{Tp} \dot{T} .$$

This equation can be brought into the form

$$\ddot{\zeta} + \frac{1}{\tau_{\text{eff}}} \dot{\zeta} = L \sigma_{Tp} \dot{T} \tag{7.294a}$$

of Eq. (7.38) if one introduces with

$$\frac{1}{\tau_{\text{eff}}} \equiv \frac{1}{\tau_{Tp}} - \left(\frac{\partial \ln L}{\partial T} \right)_{p,\zeta} \dot{T} - \left(\frac{\partial \ln L}{\partial \zeta} \right)_{T,p} \dot{\zeta} \tag{7.294b}$$

the effective relaxation time τ_{eff} for the freezing process. Hence, the effective relaxation time of a freezing process under constant pressure depends not only on the independent variables T, ζ but also on their rates $\dot{T}, \dot{\zeta}$:

$$\tau_{\text{eff}} = \tau_{\text{eff}} (T, \dot{T}, \zeta, \dot{\zeta}) .$$

With the "Doolittle formulation" (7.291), one obtains

$$\left(\frac{\partial \ln L}{\partial T} \right)_{p,\zeta} = 0 , \quad \left(\frac{\partial \ln L}{\partial \zeta} \right)_{T,p} = \frac{c_D}{\varrho} \left(\frac{1-\varphi}{\varphi} \right)^2$$

i.e.,

$$\frac{1}{\tau_{\text{eff}}} = \frac{1}{\tau_{Tp}} - \frac{L c_D}{\varrho} \left(\frac{1-\varphi}{\varphi} \right)^2 a . \tag{7.295}$$

At low temperatures, the first term $1/\tau_{Tp}$ on the right-hand side of (7.295) is completely dominated by the second term, so that the singularity of τ_{Tp} does not achieve any importance (Fig. 7.22). In the case of the "WLF-formulation" (7.293), one obtains

$$\left(\frac{\partial \ln L}{\partial T} \right)_{p,\zeta} = \frac{c_W}{(T - T_{gl})^2} ; \quad \left(\frac{\partial \ln L}{\partial \zeta} \right)_{T,p} = 0$$

and thus

$$\frac{1}{\tau_{\text{eff}}} = \frac{1}{\tau_{Tp}} - \frac{c_W \gamma}{(T - T_{gl})^2} . \tag{7.296}$$

Although $\tau_{Tp} = +\infty$ results because of $L \to 0$, $\tau_{\text{eff}} = 0$ holds at $T = T_{gl}$.

However, the actual problem to begin with when solving the differential equations (7.290) is the boundary conditions. If we, as in Fig. 7.20, assume an equilibrium state (φ_e^o, T_e^o), the initial values of the response functions are determined by the differences $(\Delta c_p)_o$, $(\Delta\alpha)_o$ given at (φ_e^o, T_e^o) independent of the cooling rate $\gamma < 0$. Because of

$$\lim_{\substack{T \to T_e^o \\ \varphi \to \varphi_e}} \frac{\dot{\zeta}}{\dot{T}} = 0 \,,$$

the initial values of the differences (7.288) to be compared can only be adjusted to these values if one assumes that φ or T at $t = t_o$ undergo a jump which is dependent on the cooling rate. Hence, the boundary conditions to be fulfilled

$$(\Delta' c_p)_o = (\Delta c_p)_o \,; \quad (\Delta' \alpha_p)_o = (\Delta\alpha)_o \,; \quad \dot{a}_o = 0 \tag{7.297}$$

cannot be formulated in the case of a smooth development of φ and T. For example, the first of the conditions (7.297) is explicitly

$$\frac{1}{\gamma} L(0)\, T_o \sigma_{Tp}^o a_o = L(0)\, T_e^o\, \tau_{Tp}^e(\varphi_e^o)\, [\sigma_{Tp}^e(\varphi_e^o)]^2$$

or, if one assumes a smooth development of the free volume with $\varphi_o = \varphi_e^o$, according to (7.276)

$$\frac{1}{\gamma}\, T_o (T_o \sigma_{Tp}^e - \eta_{Tp}^e) = T_e^o \tau_{Tp}^e \sigma_{Tp}^e \,.$$

This boundary condition would only be fulfilled if $T_o \neq T_e^o$ could be supposed.

One sees here that one possibly has to differentiate between the thermostatic temperature T and a thermodynamic temperature T^* in the thermodynamics of irreversible processes [Meixner (1969)]. This differentiation can be seen in complete analogy to hydromechanics in which it is normal to distinguish between the hydrostatic and the hydrodynamic pressures. The thermostatic temperature T is a characteristic quantity of the arrested equilibrium states which serve as a basis for the description of the mode of action of a relevant internal degree of freedom (Sect. 7.1). On the other hand, T^* is the actual temperature of the system which corresponds to the temperature of the surroundings of the system (exchange temperature). When differentiating between the static and the dynamic temperatures, one must write in place of (7.28):

$$\frac{\mathrm{d}_a s}{\mathrm{d}t} = \frac{1}{T^*}\,\dot{q} \,, \tag{7.298}$$

and $T = T^*$ is, of course, true in the internal as well as in the arrested equilibrium.

According to (6.10), T is determined by the phase volume of all internal degrees of freedom with the exception of the degrees of freedom declared as macroscopically relevant. In a liquid, this temperature bath is specifically given by the degrees of freedom of diffusive translational motion of the molecules; in a solid, by the degrees of freedom of the phonon gas. If the relevant internal degree of freedom, for example, represents the rotation of a side group of a polymer molecule, one can assume that this degree of freedom hardly disturbs the temperature bath given by T. In this case, $T \approx T^*$ will hold. If, on the other hand, the relevant internal degree of freedom, as in the case of the glass transition, refers to the diffusive translational motion itself, one can expect a considerable difference between T and T^*. The static and dynamic temperatures must be distinguished in any case when formulating boundary conditions of the kind (7.297).

If one keeps the external pressure constant and neglects the possible difference between the hydrostatic and hydrodynamic pressures, one obtains $\dot{q} = \dot{h}$. With (7.273), Eq. (7.298) is then given fully by

$$\frac{\mathrm{d}_a s}{\mathrm{d}t} = \frac{1}{T^*}\,\dot{h} = \frac{1}{T^*}\,(c_{p,\zeta}\dot{T} + \eta_{Tp}\dot{\zeta})\,. \tag{7.299}$$

According to (7.272), (7.275), and (7.277), the temporal change in the total entropy in the G-representation is given by

$$\dot{s} = \frac{1}{T}\,c_{p,\zeta}\dot{T} + \sigma_{Tp}\dot{\zeta}\,. \tag{7.300}$$

Hence, the entropy produced in the interior of the system

$$\frac{\mathrm{d}_i s}{\mathrm{d}t} = \dot{s} - \frac{\mathrm{d}_a s}{\mathrm{d}t}$$

[Eq. (7.27)] must amount to

$$\frac{\mathrm{d}_i s}{\mathrm{d}t} = c_{p,\zeta}\left(\frac{1}{T} - \frac{1}{T^*}\right)\dot{T} + \left(\sigma_{Tp} - \frac{1}{T^*}\,\eta_{Tp}\right)\dot{\zeta}\,. \tag{7.301}$$

With the form of the dynamic laws chosen in Sect. 7.2, the product $T(\mathrm{d}_i s/\mathrm{d}t)$ appears as a bilinear form of the effective fluxes and forces. Accordingly, we can conclude inversely from (7.301) that upon distinction between the static and dynamic temperatures the dynamic laws exist:

$$\dot{T} = L_{11}\,c_{p,\zeta}\left(1 - \frac{T}{T^*}\right) + L_{12}\,T\left(\sigma_{Tp} - \frac{1}{T^*}\,\eta_{Tp}\right),$$

$$\dot{\zeta} = L_{21}\,c_{p,\zeta}\left(1 - \frac{T}{T^*}\right) + L_{22}\,T\left(\sigma_{Tp} - \frac{1}{T^*}\,\eta_{Tp}\right).$$

If we disregard the possible interference phenomena with $L_{12} = L_{21} = 0$ and set $L_{22} \equiv L$, $L_{11}\, c_{p,\zeta} \equiv L_T$, we get the simple dynamic laws

$$\dot{T} = L_T \left(1 - \frac{T}{T^*}\right), \tag{7.302a}$$

$$\dot{\zeta} = LT \left(\sigma_{Tp} - \frac{1}{T^*}\, \eta_{Tp}\right). \tag{7.302b}$$

With a constant cooling or heating rate, we have in addition

$$\dot{T}^* = \gamma = \text{const.} \tag{7.302c}$$

For the internal variable ζ, there is as before a simple dynamic law of the form (7.29):

$$\dot{\zeta} = L a^* . \tag{7.303}$$

The "static affinity" a [Eq. (7.276)], however, must be replaced by the "dynamic affinity"

$$a^* = \frac{T}{T^*}\,(T^* \sigma_{Tp} - \eta_{Tp}) . \tag{7.304a}$$

The interrelation

$$a^* = a + \eta_{Tp} \left(1 - \frac{T}{T^*}\right) \tag{7.304b}$$

results between the affinity a and the affinity a^*. According to Tool (1946), [also compare Davies, Jones (1953)], one can further introduce a fictive temperature

$$T_f \equiv \eta_{Tp} / \sigma_{Tp} . \tag{7.305}$$

This is the temperature which the system in a non-equilibrium state (T, φ) would have if it were in internal equilibrium [see (7.280)]. With (7.305), one can also write

$$a^* = T \eta_{Tp} \left(\frac{1}{T_f} - \frac{1}{T^*}\right) \tag{7.304c}$$

in place of (7.304a).

With regard to the boundary conditions (7.297), one thus obtains if, with $\varphi_o = \varphi_e^o$, $T_o = T_f^o = T_e^o$ one demands a smooth transition from the equilibrium to the non-equilibrium:

$$\lim_{T^* \to T_e^o} \frac{\dot{\zeta}}{\dot{T}} = \frac{L(0)}{L_T(0)}\, T_e^o \sigma_{Tp}^e (\varphi_e^o) .$$

The boundary conditions are thus fulfilled if one sets

$$L_T(0) = T_e^o / \tau_{Tp}^e (\varphi_e^o) \,. \tag{7.306a}$$

Now $L_T(t)$, like $L(t)$, has to be regarded as a function of the mutually independent variables, and $T \to T^*$, $\dot{T} \to \gamma$, *i.e.*, $L_T \to +\infty$ should hold in the limit $T^* \to T_{gl}$. In the case of low cooling or heating rates $|\gamma|$, as we will exclusively regard them in the following, however, $T \approx T^*$ is always valid. It is not very inaccurate, at least qualitatively, if we fix L_T at

$$L_T(t) = L_T(0) = \text{const.} \tag{7.306b}$$

With constant $L_T > 0$, however, the value T for the static temperature must be replaced by the value $2T^* - T$ at the instant of a sign reversal for γ[10]. In place of (7.293), we must now set (in a non-normalized form)

$$L(t) = L_o \exp\left(-\frac{c_W}{T^* - T_{gl}}\right), \quad T^* \geq T_{gl} \,. \tag{7.307}$$

Solutions of the differential equations (7.302) with the values (7.289), the coupling coefficients (7.306) and (7.307), and the boundary conditions (7.297) are depicted in the Figs. 7.23 to 7.27. The value 350 K was chosen for T_{gl} in order to keep a greater distance from the critical temperature $T_{\max}^e$ at 456 K.

Figure 7.23 shows the relative volume

$$\frac{v}{v_2} = \frac{1}{1 - \varphi} \tag{7.308}$$

according to (7.274) and the contribution of the vacancies to the heat capacity

$$\Delta' c_p = \frac{LT}{\dot{T}} \sigma_{Tp} a^* = \frac{LTT^*}{L_T (T^* - T)} \sigma_{Tp} a^* \tag{7.309}$$

according to (7.288b) for different cooling rates γ. With decreasing temperature, the curves first approximately follow the equilibrium curves. The faster the cooling, the faster the separation of the curves from the equilibrium curves and the earlier the freezing of the liquid. The width of the freezing range increases with the cooling rate.

If the liquid is cooled down to the freezing range at a specified rate $\gamma < 0$ and then heated up again with the same rate $\gamma > 0$, the volume goes through a hysteresis cycle (Fig. 7.24). This can be explained by the fact that, according to

10 With respect to a correct consideration of the boundary conditions for $T^* \to T_{gl}$ see Baur (1998).

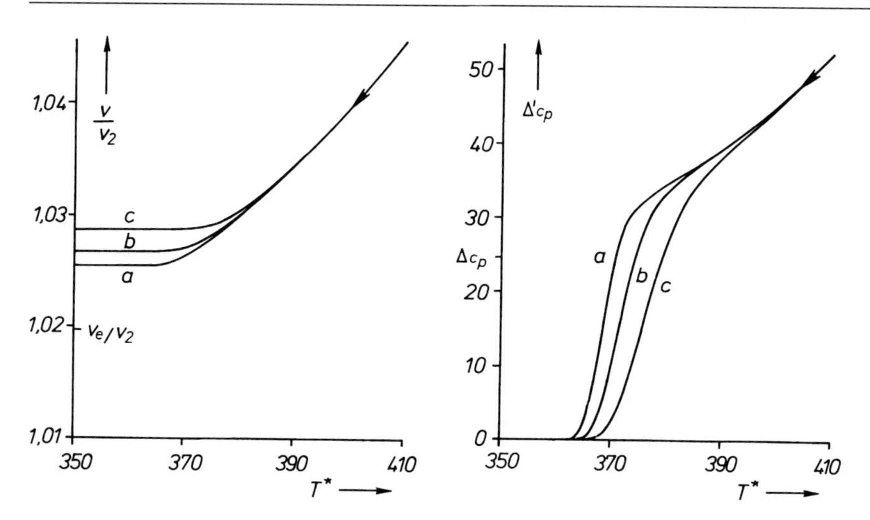

Fig. 7.23. Relative volume v/v_2 and contribution $\Delta' c_p$ [J/mol · K] of the vacancies to the heat capacity according to (7.308) and (7.309) with (7.302), (7.306), (7.307), and (7.289) for different cooling rates: $\gamma_c = -0.001$ K/tu, $\gamma_b = -0.0001$ K/tu, $\gamma_a = -0.00001$ K/tu. $T_{gl} = 350$ K, $L_o = 5 \cdot 10^{-4}$ mol/J · tu, $c_W = 250$ K. The initial state is the internal equilibrium state at $T^* = 410$ K. v_e and Δc_p denote the equilibrium values which would be achieved with $\dot{T} = \gamma \to 0$ at $T^* = 350$ K

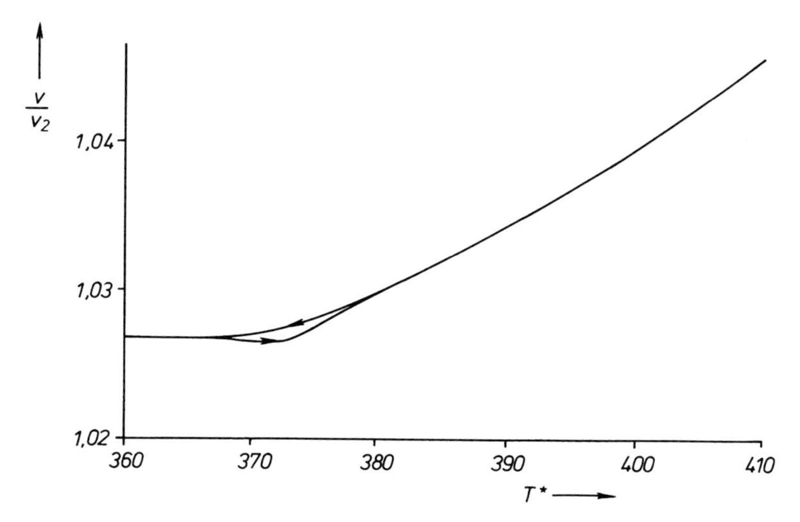

Fig. 7.24. Relative volume v/v_2 upon cooling and heating at the rate $|\gamma| = 0.0001$ K/tu. The system was cooled down to $T^* = 360$ K. The freezing temperature was at $T^* = 350$ K. Equations and numerical values as in Fig. 7.23

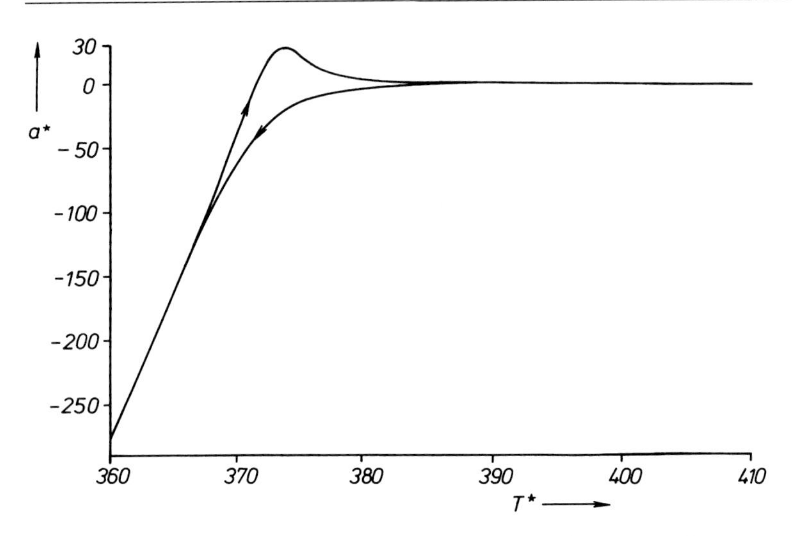

Fig. 7.25. Affinity a^* [J/mol] according to (7.304) during the process Fig. 7.24

(7.303), upon cooling the affinity a^* necessarily assumes negative values. When heating from the non-equilibrium value $\varphi(T^*, \gamma)$ reached at T^*, a^* first remains negative. The relative free volume φ continues to decrease when heating until a^* has reached positive values. As the affinity necessarily has to approach the equilibrium value from the positive side $a^* > 0$ when heating, a^* first intersects the equilibrium axis $a^* = 0$ and passes through a maximum before it disappears again (Fig. 7.25).

Both effects become noticeable in the heat capacity when cooling and heating at the same rate $|\gamma_a|$. As long as a^* is negative, $\Delta' c_p$ decreases upon heating. Subsequently, $\Delta' c_p$, like a^*, traverses a maximum before it flows into the equilibrium curve Δc_p (Fig. 26a). If one heats at a lower rate $\gamma_b < \gamma_a|$, the glass thaws in a lower, narrower temperature range (Fig. 7.26b). If one heats at a higher rate $\gamma_c > |\gamma_a|$, the thawing process takes place in a higher, broader temperature range (Fig. 26c).

If one cools the melt down to the freezing range at a rate $\gamma_a < 0$, it reaches a free volume $\varphi_a(T^*, \gamma_a)$ at the temperature T^* which is greater than the equilibrium volume $\varphi_e(T^*)$ of the liquid. The related affinity $a^*(T^*, \gamma_a)$ is negative. If one anneals the melt at T^*, the free volume strives for the equilibrium value φ_e (T^*) and the affinity for the equilibrium value $a^* = 0$. If one heats the melt in an intermediate stage $\varphi_e(T^*) < \varphi(t) < \varphi_a(T^*, \gamma_a)$ at the rate $\gamma = |\gamma_a|$, the hysteresis becomes ever weaker and the maximum of the heat capacity ever larger with increasing annealing time t. When the melt has reached equilibrium with $\varphi_e(T^*)$, $a^* = 0$, the hysteresis disappears completely when heating (Fig. 7.27b). As the liquid is in a metastable state below T_M (Fig. 7.18), it is possible that the free volume decreases even beyond the equilibrium value $\varphi_e(T^*)$ of the liquid state when annealing and approaches the equilibrium value $\varphi_e^c(T^*)$

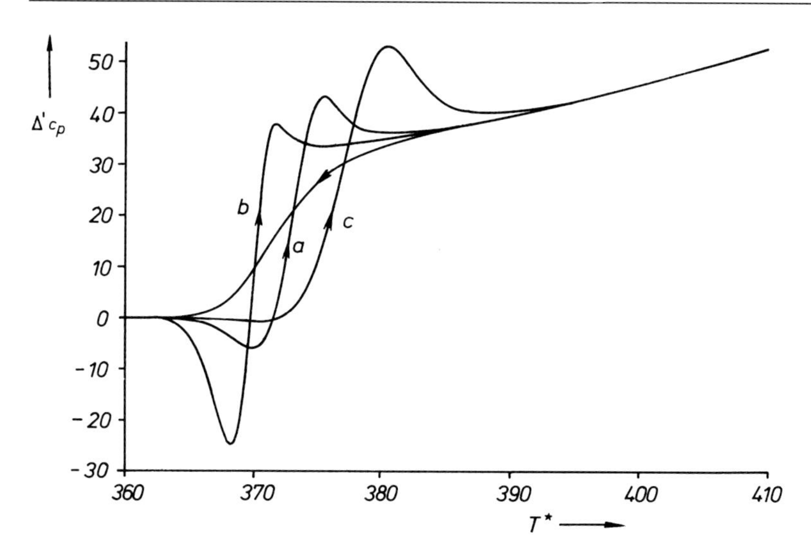

Fig. 7.26. Heat capacity $\Delta' c_p$ [J/mol · K] according to (7.309) upon cooling and subsequent heating. The system was cooled with $\gamma = -0.0001$ K/tu down to 360 K and heated with $\gamma_a = |\gamma|$, $\gamma_b = 0.1\,|\gamma|$, $\gamma_c = 10\,|\gamma|$. Otherwise as in Fig. 7.23

$< \varphi_e(T^*)$ of the crystalline state. When judged with respect to the potential g of the liquid, the states $\varphi(t) < \varphi_e(T^*)$ are non-equilibrium states with a positive affinity. If one heats the melt from such a state, the maximum of the heat capacity increases further (Fig. 7.27c). $a^* \to +\infty$ and $\Delta' c_p \to +\infty$ result with $\varphi \to 0$. When judged with respect to the potential $g_c < g$ of the crystalline state, however, these states are non-equilibrium states with a negative affinity, as they strive for the smaller equilibrium value $\varphi_e^c(T^*)$. The increase of the maximum of the heat capacity with the annealing time, which is measured when heating a glassy frozen melt, has repeatedly led to the conclusion [see Wrasidlo (1974)] that the glass transition must be a masked phase transition in the sense of the equilibrium thermodynamics. But this is definitely not true. The maximum of the heat capacity is clearly connected with the maximum of the affinity $a^* \neq 0$ which is characteristic for the non-equilibrium.

The thermodynamics of irreversible processes allows an understanding, at least qualitative, of the dynamics of the processes in the glass transition region even on the basis of simple dynamic laws and the simple, certainly only approximately valid Gibbs fundamental equation (7.270). Here, the coupling coefficient L and the affinity a^* are determining quantities. It becomes clear that these are typical non-linear phenomena. Therefore, there is no point in explaining these processes by the superposition of a spectrum of linear relaxation mechanisms. Moreover, the various relaxation times, which can also be defined in the non-linear case for each of the thermodynamic quantities [11], are found to

11 The relaxation times τ_{Tp}^e, τ_{Tp}, τ_{eff} which we use specifically refer to the affinity a.

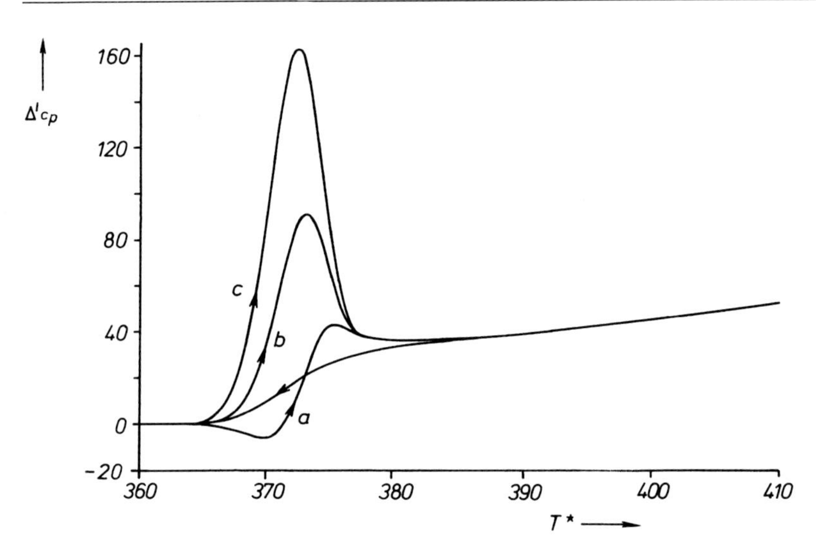

Fig. 7.27. Heat capacity $\Delta' c_p$ [J/mol · K] according to (7.309) upon cooling at the rate $\gamma_a = -0.0001$ K/tu down to 360 K, "annealing" at $T^* = 360$ K and subsequent heating at the rate $\gamma = |\gamma_a|$. The initial values when heating were **a:** $\varphi = 0.0260697$ (the value which was attained at $T^* = 360$ K upon cooling); **b:** $\varphi = \varphi_e = 0.0221387$ (the equilibrium value of the liquid state at $T^* = 360$ K); **c:** $\varphi = 0.018 < \varphi_e$. Otherwise as in Fig. 7.23

be auxiliary quantities whose practical usefulness disappears when the differential equations describing the system are explicitly known. However, we do not learn anything about the physical causes of glassy freezing in this manner. With the formulations (7.293) or (7.307), the existence of a glass transition temperature was postulated *a priori*.

The theoretical limiting case $\gamma \to 0$ is without doubt of importance for an explanation of the nature of the glass transition. This limiting case is a problem of equilibrium thermodynamics in which the quantities L and a^* are irrelevant and the internal variable or the free volume become the dependent variables $\zeta_e(T, p)$ or $\varphi_e(T, p)$, respectively. In equilibrium thermodynamics, one is confronted with the question of which class of transition phenomena the glass transition belongs to. There are voices in the literature which classify the glass transition as a transition of the second order in the sense of Ehrenfest's scheme [*e.g.*, see Gibbs, DiMarzio (1958); Adam, Gibbs (1965)]. In the following, the differences between a transition of the second order and a freezing process will be described.

We assume that the liquid is characterized at constant pressure by the Gibbs fundamental equation

$$g = g(T, \zeta), \quad p = \text{const.} \tag{7.310}$$

If the liquid is in internal equilibrium, the internal variable $\zeta_e(T)$ can be eliminated *via* the equilibrium conditions. One thus obtains the equilibrium poten-

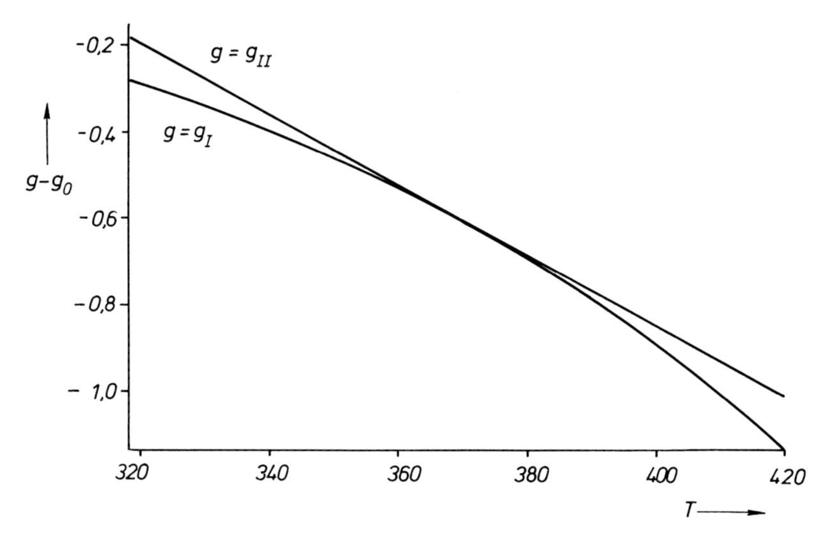

Fig. 7.28. Free enthalpy $g - g_o$ [kJ/mol] according to (7.270), (7.280), and (7.289). $g_I = g[T, \varphi_e(T)]$: free enthalpy of the internal equilibrium. $g_{II} = g[T, \varphi_e(T_{gl})]$: free enthalpy of the arrested equilibrium with $T_{gl} = 370$ K

tial $g = g_I(T)$ which is valid for the liquid state. If the liquid freezes at the temperature T_{gl}, (7.310) leads to the potential of the arrested equilibrium $g = g_{II}(T, \zeta^e_{gl})$, ζ^e_{gl} is the value of the internal variable frozen under the condition $\gamma = 0$ at T_{gl}. As ζ^e_{gl} is a constant, one can also write $g = g_{II}(T)$. The curve $g_I(T)$ is concavely curved *versus* the T-axis if the internal equilibrium states are stable [see Eqs. (2.49) and (4.21)]. The curve $g_{II}(T)$ then always lies above the curve $g_I(T)$: $g_I(T) > g_I(T)$. However, at the temperature $T = T_{gl}$, the two curves contact with $g_I = g_{II}$(see Fig. 7.28). If the system follows the curve g_I above the temperature T_{gl} and the curve g_{II} below the temperature T_{gl}, a break occurs in the functions of state v and s at T_{gl}. Jumps Δc_p, $\Delta\alpha$ occur in the response functions which satisfy the Davies relation (7.85)

$$\Delta c_p \Delta \kappa_T = T v (\Delta\alpha)^2 \tag{7.311}$$

These characteristics of the equilibrium glass transition are formally also characteristics of the second order phase transition according to Ehrenfest (Sect. 8.1). So far, however, the relations have only been considered assuming that $p =$ const. Consideration of the behaviour with respect to pressure changes is also essential for the classification of a transition process.

The position of the critical temperature T_{gl} of the glass transition, like the position of the critical temperature T_c of a second order transition, depends on the pressure p. If one considers the glass transition as a freezing process, there is the additional criterion that the value ζ^e_{gl} of the internal variable or the value of the relative free volume φ^e_{gl} frozen at T_{gl}, also depends on the pressure. If one compares the Gibbs potentials of the glass at different pressures, the parameter ζ_{gl}^{e} can no longer be ignored. Hence, the complete Gibbs fundamental equation for the glass is $g = g_{II}(T, p, \zeta_{gl}^{e})$, whereas the corresponding branch in the second order transition is only a function of the temperature and the pressure: $g = g_{II}(T, p)$. This expresses that the branch g_{II} refers to arrested equilibrium states in the case of the freezing process and to internal equilibrium states in the case of a second order transition.

The derivation of the so-called Ehrenfest equations (Sect. 8.1) is based on the changes $(\mathrm{d}v)_c$ and $(\mathrm{d}s)_c$ of the volume and the entropy along the transition curve $T_c(p)$. In the same way, we can examine the changes $(\mathrm{d}v)_g$ and $(\mathrm{d}s)_g$ along the pressure-dependent curve $T_{gl}(p)$ of the glass transition. The changes have to be independent of an assessment with respect to the internal equilibrium or with respect to the arrested equilibrium. In the internal equilibrium, we obtain

$$(\mathrm{d}v)_g = \left(\frac{\partial v}{\partial T}\right)_p (\mathrm{d}T)_g + \left(\frac{\partial v}{\partial p}\right)_T (\mathrm{d}p)_g$$

$$= v\alpha_e\,(\mathrm{d}T)_g - v\kappa_T^e\,(\mathrm{d}p)_g$$

for the volume, and in the arrested equilibrium

$$(\mathrm{d}v)_g = \left(\frac{\partial v}{\partial T}\right)_{p,\,\zeta_e} (\mathrm{d}T)_g + \left(\frac{\partial v}{\partial p}\right)_{T,\,\zeta_e} (\mathrm{d}p)_g + \left(\frac{\partial v}{\partial \zeta}\right)_{T,\,p}^{e} (\mathrm{d}\zeta_e)_g$$

$$= v\alpha_{\zeta_e}\,(\mathrm{d}T)_g - v\kappa_{T,\,\zeta_e}(\mathrm{d}p)_g + \varphi_{Tp}^{e}(\mathrm{d}\zeta_e)_g$$

(see Sects. 2.3 and 7.3). Setting these equations equal yields with (7.82) and (7.83)

$$v\Delta\kappa_T(\mathrm{d}p)_g = v\Delta\alpha\,(\mathrm{d}T)_g - \varphi_{Tp}^{e}(\mathrm{d}\zeta_e)_g$$

or

$$\left(\frac{\mathrm{d}T}{\mathrm{d}p}\right)_g = \frac{\Delta\kappa_T}{\Delta\alpha} + \frac{\varphi_{Tp}^{e}}{v\Delta\alpha}\left(\frac{\mathrm{d}\zeta_e}{\mathrm{d}p}\right)_g . \tag{7.312}$$

Correspondingly, we get

$$(\mathrm{d}s)_g = \frac{1}{T}\,c_p^e\,(\mathrm{d}T)_g - v\alpha_e(\mathrm{d}p)_g$$

for the change $(\mathrm{d}s)_g$ of the entropy along the transition curve in the internal equilibrium and

$$(\mathrm{d}s)_g = \frac{1}{T}\,c_{p,\,\zeta_e}\,(\mathrm{d}T)_g - v\alpha_{\zeta_e}\,(\mathrm{d}p)_g + \sigma_{Tp}^{e}\,(\mathrm{d}\zeta_e)_g$$

in the arrested equilibrium. With (7.83) and (7.84), this leads to

$$\left(\frac{\mathrm{d}T}{\mathrm{d}p}\right)_g = \frac{Tv\Delta\alpha}{\Delta c_p} + \frac{T\sigma_{Tp}^e}{\Delta c_p}\left(\frac{\mathrm{d}\zeta_e}{\mathrm{d}p}\right)_g . \tag{7.313}$$

According to (7.83) and (7.84), we have further

$$\varphi_{Tp}^e/\sigma_{Tp}^e = Tv\Delta\alpha/\Delta c_p . \tag{7.314}$$

Not only the left-hand sides but also the last terms on the right-hand side are equal in Eqs. (7.312) and (7.313). (7.312) and (7.313) thus lead to the Davies relation (7.311) or (7.85).

Equations (7.312) and (7.313) replace the Ehrenfest equations (8.6) and (8.7) if one understands the glass transition as a freezing process with one relevant internal degree of freedom. The freezing process can only be classified as a transition of the second order if $(\mathrm{d}\zeta_e/\mathrm{d}p)_g = 0$ is valid, *i.e.*, if the frozen values ζ_{gl}^e or φ_{gl}^e are always the same regardless of the pressure. Experimental results obtained with polystyrene indicate clearly that the terms burdened with $(\mathrm{d}\zeta_e/\mathrm{d}p)_g$ in Eqs. (7.312) and (7.313) cannot be neglected [Breuer, Rehage (1967)]. In addition, Eqs. (7.312) and (7.313) show that the macroscopic relevance of a single internal degree of freedom can already violate the Ehrenfest equations. In order to explain the fact that the Ehrenfest equations are not valid in the case of the glass transition, it is by no means necessary to assume that several internal degrees of freedom, for which

$$\frac{\Delta\kappa_T}{\Delta\alpha} > \frac{Tv\Delta\alpha}{\Delta c_p}$$

then holds, are effective [Eq. (7.248)]. Nevertheless, the corrections in the Ehrenfest equations which are caused by an individual internal degree of freedom are, because of (7.314), equal in both equations. The observation of different discrepancies [O'Reilly (1962); Goldstein (1963); Gupta, Moynihan (1976); Moynihan, Lesikar (1981)] speaks for the relevance of several internal degrees of freedom if we are dealing with true equilibrium quantities.

It should be noted that the contributions of the internal degree of freedom to the response functions

$$\Delta' k_T \equiv k_T - k_{T,\zeta} , \qquad \Delta' c_v \equiv c_v - c_{v,\zeta} ,$$

$$\Delta' \beta_v \equiv \beta_v - \beta_\zeta , \qquad \Delta' \beta_T \equiv \beta_T - \beta_\zeta ,$$

$$\Delta' \kappa_T \equiv \kappa_T - \kappa_{T,\zeta} , \qquad \Delta' c_p \equiv c_p - c_{p,\zeta} ,$$

$$\Delta' \alpha_p \equiv \alpha_p - \alpha_\zeta , \qquad \Delta' \alpha_T \equiv \alpha_T - \alpha_\zeta ,$$

which are measured at finite rates $\dot{T} \neq 0, \dot{v} \neq 0$ or $\dot{p} \neq 0$, do not fulfil the Davies or Prigogine–Defay relations. According to (7.52–7.56) and (7.71–7.74) we have in place of (7.67) and (7.85)

$$-\Delta' c_v \Delta' k_T = Tvp^2 \Delta' \beta_v \Delta' \beta_T ,$$

$$\Delta' c_p \Delta' \kappa_T = Tv \Delta' \alpha_p \Delta' \alpha_T .$$

However, in non-equilibrium, there is

$$\Delta' \beta_v \neq \Delta' \beta_T , \quad \Delta' \alpha_p \neq \Delta' \alpha_T .$$

CHAPTER 8

Phase Transitions

In the Gibbs formalism of equilibrium thermodynamics, one usually supposes that the Gibbs functions and the state and response functions which can be derived from them are smooth and continuous. The derivatives of these functions with respect to the mutually independent variables exist and are also smooth and continuous. One observes experimentally, however, that a thermodynamic system, for example, in the case of variable temperature or variable pressure, can also pass through singularities in which some state or response functions undergo discontinuous jumps. Such singularities, for example, occur during the condensation of gases, the crystallization or vaporization of liquids, the melting of crystals, the demixing of homogeneous solutions and mixtures, the transition to ferromagnetism or superconduction, or the transition from the coil to the helix structure of chain molecules in solution. The latter occurs specifically in the case of polymers. One speaks of phase transitions as some properties of the phases present in these processes undergo abrupt changes.

Singularities of this kind are in a strictly mathematical sense only possible in the limit $N \to \infty$ or, because of N/V = finite, $V \to \infty$ [Yang, Lee (1952); see also Huang (1963) or Pathria (1972)] as a consequence of the representation of the free energy (6.16/6.17), the free enthalpy (6.34/6.35) and especially the corresponding representation of the Gibbs function assigned to the independent variables T, V, μ

$$J(T, V, \mu) \equiv pV = kT \ln Q''(T, V, \mu)\,, \tag{8.1 a}$$

$$Q''(T, V, \mu) = \sum_{N=0}^{\infty} e^{\mu N/kT}\, Q(T, V, N) \tag{8.1 b}$$

[Q: partition function (6.17)]. Hence, phase transitions measured in finitely extended systems are never infinitely sharp and never singularities in the mathematical sense. There is no doubt, however, that the experimentally determined phase transitions are directly related to the singularities which can be theoretically ascertained in the limit $N \to \infty$.

Another condition for the occurrence of a discontinuous singularity in the state and response functions is revealed in the representations (6.16/6.17), (6.34/6.35) and (1): the potential energy Φ in the Hamiltonian function of the total system of the molecules is not allowed to vanish. Hence, the fundamental prerequisite of a phase transition is the existence of intermolecular interactions. These are cooperative phenomena. The only exception which appears to contradict this thesis is the so-called Einstein condensation of an ideal Bose gas [Huang (1963); Pathria (1972)]. In this case, the partition function solely depends on the kinetic energy of the particles. The symmetry of the wave function of the total system, which has to be presupposed in the case of a Bose gas, however, already includes *a priori* a certain particle cooperation. The individuality of the particles is completely lost during condensation.

8.1 Phenomenological Classification

As we have just observed, the occurrence of phase transitions is decisively dependent on the intermolecular interactions (in larger molecules, as in polymers, possibly also on the intramolecular interactions). The determining quantities are certain internal molecular degrees of freedom. The diversity of possible interaction types corresponds to a diversity of transition types. The variables assigned to the internal degrees of freedom are eliminated in pure equilibrium thermodynamics. Hence, the actual distinctive features of the different transition types remain concealed. Within the framework of equilibrium thermodynamics, a classification of the phase transitions can thus only schematically orientate itself on the external facts. Such classifications originate particularly from Ehrenfest (1933) and Tisza (1951).

In the following, we will consider a closed, homogeneous, one-component system ($N = \text{const.}$). To classify the transition phenomena, Ehrenfest proceeds from the G-representation, *i.e.*, from the specific or molar Gibbs free enthalpy $g(T, p)$. A transition is designated as a transition of the nth order if g and all derivatives

$$\frac{\partial^m g}{\partial T^m}, \frac{\partial^m g}{\partial p \, \partial T^{m-1}}, \dots, \frac{\partial^m g}{\partial p^{m-1} \partial T}, \frac{\partial^m g}{\partial p^m}$$

up to the $(n-1)$th order ($m = 1, \dots, n-1$) develop continuously at the transformation point (T_c, p_c) and the nth derivatives ($m = n$) exhibit a discontinuous jump.

The function $g(T, p)$ is thus continuous during a first-order transition. The first derivatives

$$\frac{\partial g}{\partial T} = -s \quad \text{and} \quad \frac{\partial g}{\partial p} = v \tag{8.2}$$

[see Eqs. (2.13) and (2.14)], *i.e.*, the entropy s and the volume v, suffer a discontinuous jump at the transformation point. The internal energy u and the enthalpy h also pass with s and v through a discontinuity. The second derivatives

$$\frac{\partial^2 g}{\partial T^2} = -\frac{1}{T} c_p, \quad \frac{\partial^2 g}{\partial T \partial p} = v\alpha, \quad \frac{\partial^2 g}{\partial p^2} = -v\kappa_T \tag{8.3}$$

[see Eqs. (2.47) to (2.49)], *i.e.*, the response functions, then necessarily assume unlimited values at the critical point. The discontinuity of the first derivatives (8.2) leads to a sharp bend in the Gibbs function g at the critical point (T_c, p_c). This sharp bend can be explained by the fact that the development of $g(T, p)$ is not uniquely determined. Two branches g_1 and g_2 of the free enthalpy, which exist at least in close proximity to (T_c, p_c), intersect at this point (Fig. 8.1, left; see also Fig. 6.7). The system in internal equilibrium has to follow the energetically most favourable branch as a result of the minimum principle (2.87) valid for the free enthalpy. If $g_2 > g_1$ holds for $T < T_c$, the system follows the branch g_1 below T_c and the branch g_2 above T_c. A characteristic feature of a first-order transition is that the phases assigned to the branches g_1 and g_2 coexist in internal equilibrium at the transformation point (T_c, p_c).

According to Gibbs's phase rule (Sect. 4.2), two phases in a closed, homogeneous one-component system at a given pressure p_c can only be in internal equilibrium at a single temperature T_c. A change in the pressure p_c necessarily leads to a change in the transition temperature T_c. The transition temperature is a function of the transition pressure: $T_c = T_c(p_c)$. On the curve $T_c(p_c)$, the change in the potential g_1 is given by

$$\mathrm{d}g_1 = \frac{\partial g_1}{\partial T}(\mathrm{d}T)_c + \frac{\partial g_1}{\partial p}(\mathrm{d}p)_c = -s_1(\mathrm{d}T)_c + v_1(\mathrm{d}p)_c$$

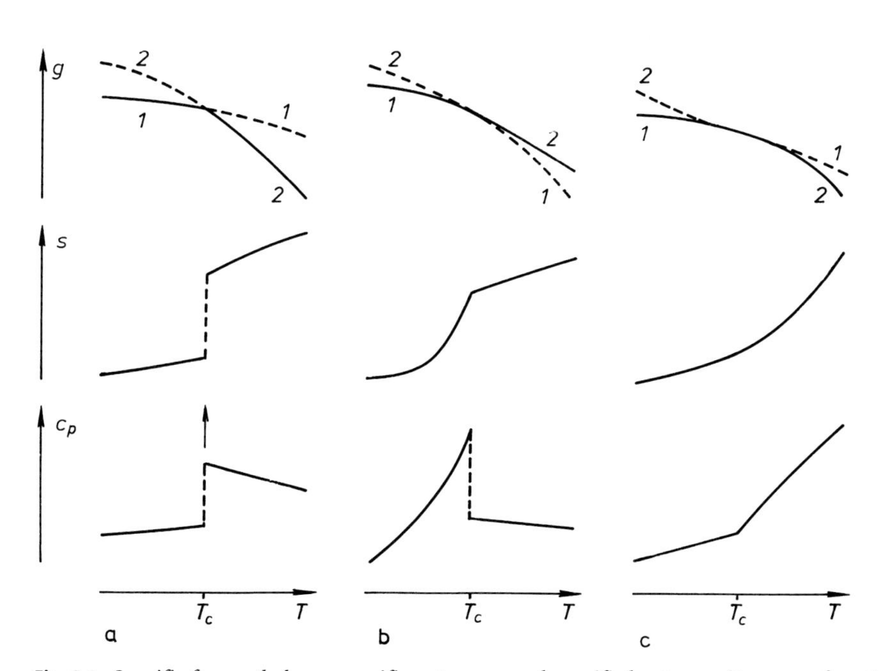

Fig. 8.1. Specific free enthalpy g, specific entropy s, and specific heat capacity c_p as a function of the temperature in the vicinity of a critical transition temperature T_c (schematically shown). From left to right: normal transitions of the first, second, and third order according to Ehrenfest

and the change in the potential g_2 by

$$dg_2 = -s_2\,(dT)_c + v_2(dp)_c\,.$$

On the other hand, $g_1 = g_2$, *i.e.*, $dg_1 = dg_2$ is always true on this curve. The two equations above must be equal. This leads to

$$(s_2 - s_1)\,(dT)_c = (v_2 - v_1)\,(dp)_c\,.$$

The dependence of the transition temperature on the transition pressure is given by

$$\left(\frac{dT}{dp}\right)_c = \frac{v_2 - v_1}{s_2 - s_1}\,. \tag{8.4a}$$

The discontinuous volume jump is $\Delta v \equiv v_2 - v_1$, the discontinuous entropy jump $\Delta s \equiv s_2 - s_1$. Equation $\Delta g \equiv g_2 - g_1 = 0$ holds for the free enthalpy and thus

$$T_c\,\Delta s = \Delta h \equiv h_2 - h_1\,.$$

(8.4a) is identical with the Clausius–Clapeyron equation

$$\left(\frac{dT}{dp}\right)_c = \frac{T_c \Delta v}{\Delta h}\,. \tag{8.4b}$$

Another characteristic feature of a first-order transition is that it is associated with a latent heat Δh which is released at the transition temperature

$$T_c = \Delta h / \Delta s\,. \tag{8.5}$$

In the second-order transition, $g(T, p)$ and the first derivatives (8.2) are continuous functions of T and p, the second derivatives (8.3), on the other hand, are subject to a discontinuous jump at (T_c, p_c). If one attempts once again to explain the second-order transition using the existence of two branches g_1 and g_2, the following can be observed: If the curves represent mechanically and thermally stable internal equilibrium states, they are, according to (2.49) and (4.21), necessarily concavely curved *versus* the T-axis. Because of the continuity of the first derivatives, however, the curves are then never able to intersect at the critical point (T_c, p_c), but are only able to touch. In close proximity to the critical point, the one curve must always lie above the other curve (Fig. 8.1, middle). There are no energetic reasons for a transition. A transition from one branch to the other is possible, however, if the system loses its stability with respect to an internal variable on the one branch at (T_c, p_c) (see Sect. 2.4) and then changes over to the other stable branch. In contrast to the first-order transition, only one of the two phases (the stable phase) exists in the internal equilibrium at (T_c, p_c) in a second-order transition. Because of the continuity of the enthalpy h, the process is not connected with the release of latent heat.

The transition temperature T_c is also found to be a function of the transition pressure p_c in the second-order transition. One obtains for the change in volumes on the curve $T_c(p_c)$

$$\mathrm{d}v_1 = \frac{\partial v_1}{\partial T}(\mathrm{d}T)_c + \frac{\partial v_1}{\partial p}(\mathrm{d}p)_c = v_1\alpha_1(\mathrm{d}T)_c - v_1\kappa_T^{(1)}(\mathrm{d}p)_c\,,$$

$$\mathrm{d}v_2 = v_2\alpha_2(\mathrm{d}T)_c - v_2\kappa_T^{(2)}(\mathrm{d}p)_c$$

[see Eqs. (2.47) and (2.48)]. Because of the continuity of the volume, $v_1 = v_2 = v_c$, *i.e.*, $\mathrm{d}v_1 = \mathrm{d}v_2$ must hold on the curve $T_c(p_c)$. Thus,

$$v_2\alpha_2(\mathrm{d}T)_c - v_2\kappa_T^{(2)}(\mathrm{d}p)_c = v_1\alpha_1(\mathrm{d}T)_c - v_1\kappa_T^{(1)}(\mathrm{d}p)_c\,.$$

If one designates the differences in the response functions between the two phases with $\Delta\alpha \equiv \alpha_2 - \alpha_1$ and $\Delta\kappa_T \equiv \kappa_T^{(2)} - \kappa_T^{(1)}$, one obtains the first Ehrenfest equation

$$\left(\frac{\mathrm{d}T}{\mathrm{d}p}\right)_c = \frac{\Delta\kappa_T}{\Delta\alpha}\,. \tag{8.6}$$

In the same way,

$$\mathrm{d}s_1 = \frac{\partial s_1}{\partial T}(\mathrm{d}T)_c + \frac{\partial s_1}{\partial p}(\mathrm{d}p)_c = \frac{1}{T_c}c_p^{(1)}(\mathrm{d}T)_c - v_1\alpha_1(\mathrm{d}p)_c\,,$$

$$\mathrm{d}s_2 = \frac{1}{T_c}c_p^{(2)}(\mathrm{d}T)_c - v_2\alpha_2(\mathrm{d}p)_c$$

hold for the change in entropies [Eqs. (2.48) and (2.49)]. With $v_1 = v_2 = v_c$, $s_1 = s_2 = s_c$ and $\Delta c_p \equiv c_p^{(2)} - c_p^{(1)}$, one obtains the second Ehrenfest equation

$$\left(\frac{\mathrm{d}T}{\mathrm{d}p}\right)_c = \frac{v_c T_c \Delta\alpha}{\Delta c_p}\,. \tag{8.7}$$

A comparison of the Eqs. (8.6) and (8.7) leads to the Ehrenfest relation

$$\Delta c_p \Delta\kappa_T = v_c T_c (\Delta\alpha)^2\,. \tag{8.8}$$

We should emphasize with regard to the formally identical Davies relation (7.85) that the relation (8.8) is only valid for the curve $T_c(p_c)$ of the phase transition. The symbols Δ designate the difference in the response functions of the internal equilibrium between a stable and an unstable phase. The Davies relations, on the other hand, are valid for every arbitrary internal equilibrium state. There, the symbols Δ refer to the difference of the response functions between the free and the, with respect to an internal variable, arrested internal equilibrium state.

According to Ehrenfest's classification, a transition of the third order (Fig. 8.1, right) and transitions of an even higher order are in principle also conceivable. Nevertheless, to date it is still contentious whether a third-order transition as defined by Ehrenfest has ever been observed. Transitions of a higher order are obviously not of importance. It is certain, however, that there are types of transition which are not included in the Ehrenfest classification. Originally, it was assumed that the limiting values of the response functions of the Ehrenfest transition types remain finite when they reach the critical point (Fig. 8.1). Transitions where this is not the case were designated as anomalous transitions (Fig. 8.2, left). We know in the meantime that anomalous transitions, in particular the anomalous second-order transitions [Tisza (1951)], are realized much more often in nature than normal transitions. In addition, transition phenomena are known experimentally as well as molecular-theoretically which are completely beyond the scope of the Ehrenfest scheme (Fig. 8.2, right). These include, for example, the so-called Onsager transition, the diffuse transition (see also Fig. 8.17) as well as the glass transition, as long as it represents a freezing process in internal equilibrium. There are no discontinuities in diffuse transitions. In the glass transition, the Ehrenfest equations (8.6) and (8.7) are replaced by the equations (7.312) and (7.313) or by even more complicated equations, if several internal degrees of freedom freeze simultaneously at the critical point (T_{gl}, p_{gl}).

The most prominent difference between second-order transitions and first-order transitions according to Ehrenfest is that only one phase exists at the

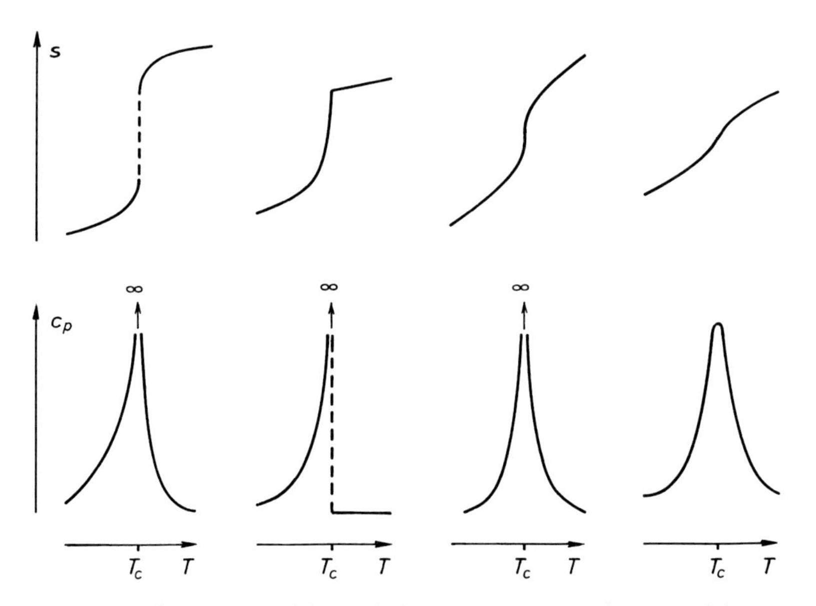

Fig. 8.2. Specific entropy s and specific heat capacity c_p as a function of the temperature in the vicinity of a critical transition temperature T_c. From left to right: anomalous first-order transition, anomalous second-order transition, Onsager transition, diffuse transition

transition point in the internal equilibrium and that there is no heat of transition. These characteristics alone are often used to classify a transition as a second-order transition.

8.2 Melting and Crystallization of Polymers

Melting and crystallization are prototypes of a first-order transition according to Ehrenfest. The sharpness and character of an ideal melting and crystallization process, however, are often blurred by numerous influences [*e.g.*, see Ubbelohde (1965)]. In addition to interferences that occur in the case of low-molecular-mass substances, polymers composed of long flexible chain molecules are further characterized by a quite decisive feature: the fact that the crystallizing units, the chain links or units of a comparable size, are considerably smaller than the molecules themselves. This already leads to the development of non-autonomous boundary phases at the stage of nucleus formation and of self-induced restraints during crystalline growth. The occurrence of non-autonomous phases and restraints means, however, that the processes of melting and crystallization of polymers from flexible chain molecules cannot be interpreted on the basis of classical classifications of equilibrium thermodynamics (Sect. 8.1).

The ideal polymer crystal in internal equilibrium consists of fully stretched parallel chains. As in every crystal, it will have some entropy-induced defects at higher temperatures. The mole number n of the defects can be estimated, for example, using

$$n \sim N\,\mathrm{e}^{-\varepsilon/RT}$$

[ε: energy or enthalpy required to generate one mole of defects, *e.g.*, compare Ubbelohde (1952) and Sect. 8.3]. Such crystals, however, cannot be realized, at least in macroscopic dimensions. Under normal conditions, polymers composed of long flexible chain molecules never crystallize completely. A partially crystalline texture, in which the individual chain molecules pass through the crystalline as well as the amorphous phase, forms out of the melt. Folded chain lamellae with amorphous surface layers precipitate out of solution [*e.g.*, see Mandelkern (1964); Wunderlich (1973), (1976)].

Figure 8.3 shows the electron micrograph of an ultrathin section of a stained partially crystalline polyethylene sample obtained from a melt. The bright strips represent the areas in the section occupied by extended crystal lamellae perpendicular to the image plane which are not penetrated by the staining medium. The dark strips represent stained amorphous layers located between the crystalline lamellae. The lamellar thickness fluctuates about 0.04 μm. According to X-ray measurements, the crystallized chain segments are nearly at right angles to the surface layers of the lamellae. This means that individual long-chain molecules necessarily pass through the crystalline as well as the amorphous regions. The dimensions of the chain coils interlocked in the melt

Fig. 8.3. Electron micrograph of an ultrathin section of a polyethylene sample (LPE). The weight average (3.12) of the sample was $\langle m \rangle_w = 100000$. The sample was annealed at 125 °C for four weeks, subsequently stained with chlorosulphonic acid at 60 °C for five hours, and finally treated with uranyl acetate at room temperature [Kanig (1991), Fig. 1]. The enlargement was 1:50000

are maintained during crystallization [Schelten et al. (1976), (1977)]. Within the chain coils, aggregates develop from trans-bonds, which pass through the melt in a smectic arrangement. For energetic reasons, the gauche-bonds are then forced towards the periphery of the trans-aggregates (see Sect. 6.3). This leads to the development of amorphous loops which prevents further growth of the trans-aggregates in the chain direction. The smectic structures finally join up to form compact crystal lamellae [Kanig (1991); additional literature can be found there]. In general, the melting of this partially crystalline texture occurs continuously over a more or less broad temperature range. At first glance, the heat capacity as a function of the temperature rather resembles a diffuse transition or a normal second-order transition. Three facts which are of major importance in these processes will be emphasized in the following: 1. the influence of the mixing effect particularly on the crystallization from the solution, 2. the crystal nucleus formation and the influence of the low lamellar thickness on melting and crystallization, and 3. the formation of a non-autonomous boundary phase and its influence on the habit of the partially crystalline texture. For a comparison, however, we will first once again consider the ideal system.

For the description of a closed, partially crystalline texture in thermal and mechanical equilibrium, one can introduce a parameter of order α, *e.g.*, the crystallinity, as an internal variable and proceed from Gibbs fundamental equation

$$g = g(T, p, \alpha)\,. \tag{8.9}$$

We have a stable or metastable internal equilibrium with respect to the parameter of order [corresponding to Eqs. (2.77) and (2.78)] if

$$\left(\frac{\partial g}{\partial \alpha}\right)_e = 0\,; \quad \left(\frac{\partial^2 g}{\partial \alpha^2}\right)_e \geq 0 \tag{8.10a}$$

or, because of $g = h - Ts$,

$$\left(\frac{\partial h}{\partial \alpha}\right)_e = T\left(\frac{\partial s}{\partial \alpha}\right)_e\,; \quad \left(\frac{\partial^2 h}{\partial \alpha^2}\right)_e \geq T\left(\frac{\partial^2 s}{\partial \alpha^2}\right)_e \tag{8.10b}$$

are valid. The first of the conditions (8.10) allows a determination of the equilibrium value

$$\alpha_e = \alpha_e(T, p)$$

of the parameter of order as a function of the temperature and the pressure or, as in the Ehrenfest scheme, an elimination of this parameter altogether.

The response functions of the system are given by expressions of the form (7.71 –7.74), the heat capacity according to (7.74), for example, by

$$c_p = T\left[\left(\frac{\partial s}{\partial T}\right)_{p,\alpha} + \left(\frac{\partial s}{\partial \alpha}\right)_{T,p}\left(\frac{\mathrm{d}\alpha}{\mathrm{d}T}\right)_p\right], \tag{8.11a}$$

whereby

$$\left(\frac{\mathrm{d}\alpha}{\mathrm{d}T}\right)_p = \dot{\alpha}/\dot{T}, \quad p = \text{const.} \tag{8.11b}$$

generally depends on the ratio of the transition rate to the heating or cooling rate. A classification of the transition corresponding to the Ehrenfest scheme assumes a quasi-static perturbation in the non-arrested internal equilibrium. In this case, one obtains according to Eq. (7.78b) with (7.42)

$$\left(\frac{\mathrm{d}\alpha_e}{\mathrm{d}T}\right)_p = \left(\frac{\partial s}{\partial \alpha}\right)^e_{T,p} \Big/ \left(\frac{\partial^2 g}{\partial \alpha^2}\right)^e_{T,p} \tag{8.12}$$

and

$$c_p^e = T\left[\left(\frac{\partial s}{\partial T}\right)_{p,\alpha_e} + \left[\left(\frac{\partial s}{\partial \alpha}\right)^e_{T,p} \Big/ \left(\frac{\partial^2 g}{\partial \alpha^2}\right)^e_{T,p}\right]\right]. \tag{8.13}$$

If one interprets the partially crystalline structure as a two-phase system consisting of an autonomous amorphous phase "a" and an autonomous crystalline phase "c", Gibbs fundamental equation of the system must, according to (4.28), be representable in the form

$$G = \mu_a N_a + \mu_c N_c\,. \tag{8.14}$$

(N_a, N_c: mole numbers of the crystallizable or crystallized chain units). One can then introduce the crystallinity

$$\alpha \equiv \frac{N_c}{N_a + N_c}; \quad N_a + N_c = N = \text{const.} \tag{8.15}$$

With $\mu_a \equiv g_a, \mu_c \equiv g_c$, one obtains

$$g(T, p, \alpha) = (1-\alpha)\, g_a(T, p) + \alpha g_c(T, p)\,. \tag{8.16a}$$

Correspondingly, one obtains for the enthalpy and the entropy of the system

$$h(T, p, \alpha) = (1-\alpha)\, h_a(T, p) + \alpha h_c(T, p)\,, \tag{8.16b}$$

$$s(T, p, \alpha) = (1-\alpha)\, s_a(T, p) + \alpha s_c(T, p)\,. \tag{8.16c}$$

Because of the autonomy of the phases, Eqs. (8.16) represent linear relations with respect to α. If one abbreviates with $\Delta h \equiv h_a - h_c$ and $\Delta s \equiv s_a - s_c$, the equilibrium and stability conditions (8.10) result in

$$\Delta h = T \Delta s; \quad \frac{\partial^2 g}{\partial \alpha^2} = \frac{\partial^2 h}{\partial \alpha^2} = \frac{\partial^2 s}{\partial \alpha^2} = 0\,. \tag{8.17}$$

At a given pressure, the two phases coexist in internal equilibrium only at a single temperature

$$T_M = \Delta h / \Delta s \,. \tag{8.18}$$

The equilibrium is neutral (see Sect. 2.4), so that the parameter of order α is able to assume every arbitrary value between 0 and 1 at T_M. With $g = h - Ts$ and (8.16b, c), one obtains in place of (8.16a)

$$g = g_a - \alpha(\Delta h - T\Delta s)$$

or with (8.18)

$$g = g_a - \left(1 - \frac{T}{T_M}\right) \alpha \Delta h \,. \tag{8.19}$$

If g_a and Δh are approximately constant in the vicinity of T_M and if $\Delta h > 0$, one obtains the representation shown in Fig. 8.4 for $g(T, \alpha)$, p = const. One can see that the minimum value of the free enthalpy is always at $\alpha = 1$ in the range $T < T_M$ and always at $\alpha = 0$ in the range $T > T_M$. When passing through T_M, α_e suffers a jump and $d\alpha_e/dT$ or c_p^e, respectively, reach a singular infinity point according to (8.12), (8.13), and (8.17). The system of autonomous phases is only in internal equilibrium at $T \neq T_M$ if it is either completely amorphous or completely crystalline. This corresponds to the mono-variance required by the Gibbs phase rule for two-phase, one-component systems (Sect. 4.2).

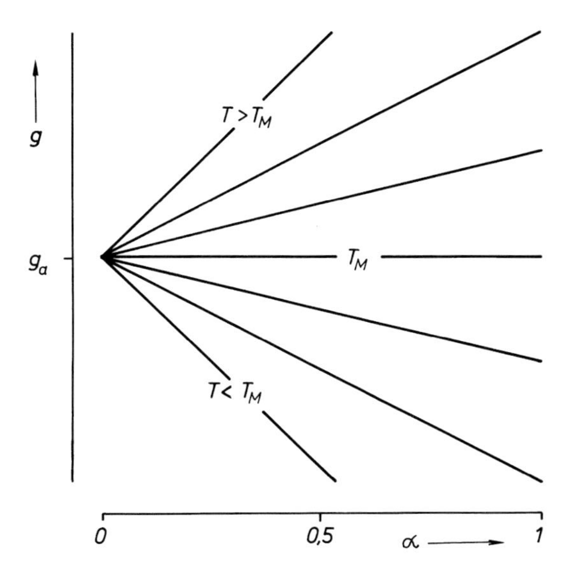

Fig. 8.4. Specific free enthalpy g of the system (crystal; melt) as a function of the crystallinity α according to (8.19) for different temperatures T. $g_a(T)$ = const., $\Delta h(T)$ = const. > 0 were assumed

In a mixture of components $i = 1 \ldots r$, Eq. (8.14) is replaced by Gibbs fundamental equation

$$G = \sum_i (\mu_i^a N_i^a + \mu_i^c N_i^c) . \tag{8.20}$$

This mixture is in internal equilibrium if

$$\Delta\mu_i \equiv \mu_i^a - \mu_i^c = 0 \tag{8.21}$$

holds for every component [Eq. (3.41) or (4.5), respectively]. If both phases are autonomous, the chemical potentials μ_i^a and μ_i^c are not dependent on the composition of the other phase:

$$\mu_i^a = (T, p, N_1^a \ldots N_r^a) \quad \text{or} \quad \mu_i^c = \mu_i^c(T, p, N_1^c \ldots N_r^c) \text{ , respectively.}$$

According to (3.50), one can then set for the chemical potentials of the amorphous phase:

$$\mu_i^a = \mu_i^{oa} (T, p) + RT \ln \gamma_i^a x_i^a. \tag{8.22}$$

If, for example, as in a solution of a polymer in a low-molecular-mass substance, there is no formation of mixed crystals, one can further replace the chemical potential of the ith component in the crystalline phase by the chemical potential of the pure substance:

$$\mu_i^c = \mu_i^{oc} (T, p). \tag{8.23}$$

(8.21–8.23) result in

$$\Delta\mu_i = \Delta\mu_i^o + RT \ln \gamma_i^a x_i^a = 0 \tag{8.24}$$

for the internal equilibrium of the system.

The differences $\Delta\mu_i^o$ of the chemical potentials of the pure components can be split into an enthalpy term and an entropy term:

$$\Delta\mu_i^o = \Delta h_i^o - T \Delta s_i^o .$$

In addition, one obtains for the melting temperature T_{Mi}^o of the pure substances

$$T_{Mi}^o \Delta s_i^o = \Delta h_i^o . \tag{8.25}$$

In the case of an approximately constant Δh_i^o, the temperature dependence of the differences $\Delta\mu_i^o$ is given by the Schröder–Van Laar equation

$$\Delta\mu_i^o = \left(1 - \frac{T}{T_{Mi}^o}\right) \Delta h_i^o . \tag{8.26}$$

Equations (8.24–8.26) yield the expression

$$T_{Mi} = \frac{\Delta h_i^o}{\Delta s_i^o - R \ln \gamma_i^a x_i^a} \tag{8.27a}$$

or

$$\frac{1}{T_{Mi}} = \frac{1}{T^o_{Mi}} - \frac{R}{\Delta h^o_i} \ln \gamma^a_i x^a_i \tag{8.27b}$$

for the temperatures $T = T_{Mi}$ in which the crystals of the component No. i are in internal equilibrium with the melt. It is always valid that $T_{Mi} \leq T^o_{Mi}$ when $0 \leq \gamma^a_i x^a_i \leq 1$ (so-called melting-point depression). The temperatures

$$T_{Mi} = T_{Mi}\,(x^a_1 \ldots x^a_r)$$

depend *via* the activity coefficients γ^a_i on the composition of the melt. Hence, the internal equilibrium of the two-phase system at constant pressure is no longer fixed at a single temperature. This corresponds to Gibbs phase rule (Sect. 4.2), according to which a two-phase mixture with constant mass possesses r macroscopic degrees of freedom. As the composition of the melt changes during melting or crystallization of one or more components, the equilibrium temperatures T_{Mi} also change during these processes. Melting or crystallization processes are thus capable of extending continuously over an extensive temperature range (see Figs. 8.5 and 8.6).

Let us specifically consider the solution of a polymer in a low-molecular-mass substance which fulfils the Flory–Huggins equations (3.94). If we assume $x^*_i = \varphi_i$, (see Sect. 3.1), we obtain using (8.27b) and (3.94) for the melting temperature of the solvent

$$\frac{1}{T_{M1}} = \frac{1}{T^o_{M1}} - \frac{R}{\Delta h^o_1} \left[\ln \varphi_1 + \left(1 - \frac{1}{P}\right)(1 - \varphi_1) + \chi(1 - \varphi_1)^2\right] \tag{8.28a}$$

and for the melting temperature of the polymer

$$\frac{1}{T_{M2}} = \frac{1}{T^o_{M2}} - \frac{R}{\Delta h^o_2} \left[\ln (1 - \varphi_1) - (P - 1)\,\varphi_1 + P\chi\varphi_1^2\right], \tag{8.28b}$$

where Δh^o_2 is the melting enthalpy of the pure polymer per mole of chain molecules. If one designates the melting enthalpy per mole of chain links of the polymer as $\Delta h^o_u = \Delta h^o_2/P$, one can also write in place of (8.28b)

$$\frac{1}{T_{M2}} = \frac{1}{T^o_{M2}} - \frac{R}{\Delta h^o_u} \left[\frac{1}{P} \ln (1 - \varphi_1) - \left(1 - \frac{1}{P}\right)\varphi_1 + \chi\varphi_1^2\right]. \tag{8.28c}$$

One deals with the simplest case if there is miscibility over the whole range of $\varphi_1 = 1 - \varphi_2$, *i.e.*, if

$$\chi < \frac{1}{2}\left(1 + \frac{1}{\sqrt{P}}\right)^2$$

holds for Huggins' interaction parameter [Eq. (3.105)]. In this case, the polymer solution forms a so-called eutectic system.

Figures 8.5 and 8.6 show two examples. If the melting temperature T^o_{M1} of the pure solvent is of the same order of magnitude as the melting temperature T^o_{M2} of the pure polymer (Fig. 8.5), the two curves $T_{M1}(\varphi_1)$ and $T_{M2}(\varphi_1)$ intersect at the eutectic point (T_E, φ_{E1}). $T_E < T^o_{M1}, T^o_{M2}$ is valid. If one starts with a solution in the mixing ratio $\varphi^o_1 = 1 - \varphi^o_2 < \varphi_{E1}$, and if this solution is cooled quasi-statically, the first polymer crystals precipitate after exceeding the temperature $T_{M2}(\varphi^o_1)$. This leads to a slight decrease in the concentration of the polymer in the solution and an increase in the equilibrium value φ_1 by an infinitesimal amount. If one cools further, the solution follows the equilibrium curve $T_{M2}(\varphi_1)$ towards higher φ_1-values. This causes a continuous precipitation of additional polymer crystals. When one finally reaches the eutectic temperature T_E, the rest of the solution crystallizes spontaneously and discontinuously in the so-called eutectic ratio $\varphi_{E1}/(1 - \varphi_{E1})$. If one starts with a solution in the mixing ratio $\varphi^o_1 = 1 - \varphi^o_2 > \varphi_{E1}$, only the solvent crystallizes initially after exceeding the temperature $T_{M1}(\varphi^o_1)$. Polymer crystals are precipitated only when the eutectic temperature is reached, but then completely. If one starts with a solution in the eutectic mixing ratio $\varphi^o_1 = \varphi_{E1}, \varphi^o_2 = 1 - \varphi_{E1}$, the solution crystallizes discontinuously and completely at T_E. In principle, if the melting temperature T^o_{M1} of the pure solvent lies considerably below the melting temperature T^o_{M2} of the pure polymer (Fig. 8.6), the same conditions exist. However, when $P \gg 1$, the eutectic point is shifted towards the value $\varphi_{E1} \approx 1$. When cooling along the curve

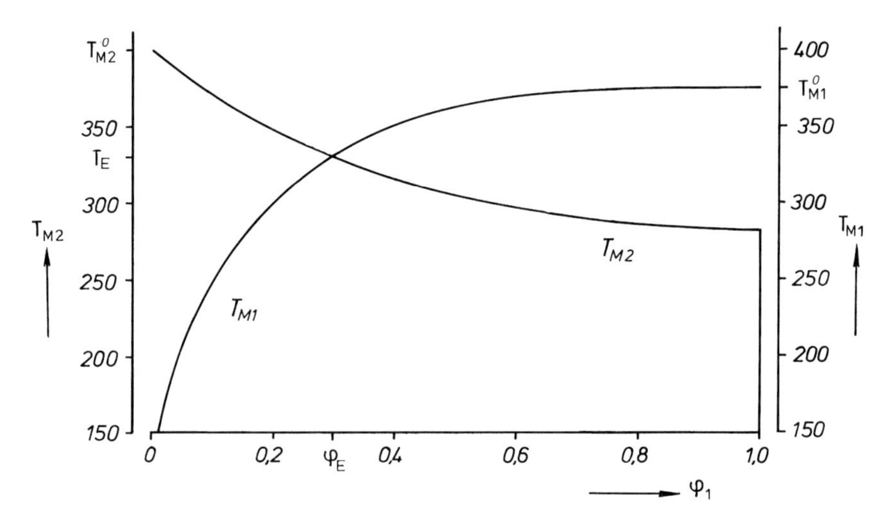

Fig. 8.5. Melting temperature T_{M2} of the polymer and melting temperature T_{M1} of the solvent in a polymer solution as a function of the volume fraction φ_1 of the solvent according to (8.28). Melting temperature of the pure polymer: T^o_{M2}; melting temperature of the pure solvent: T^o_{M1}. Volume fraction and melting temperature of the eutectic composition: φ_E, T_E. Assumed parameters: $\Delta h^o_1 = 6$ kJ/mol, $T^o_{M1} = 375$ K, $\Delta h^o_u = 4$ kJ/mol, $T_{M2} = 400$ K, $\chi = 0.5$, and $P = 1000$

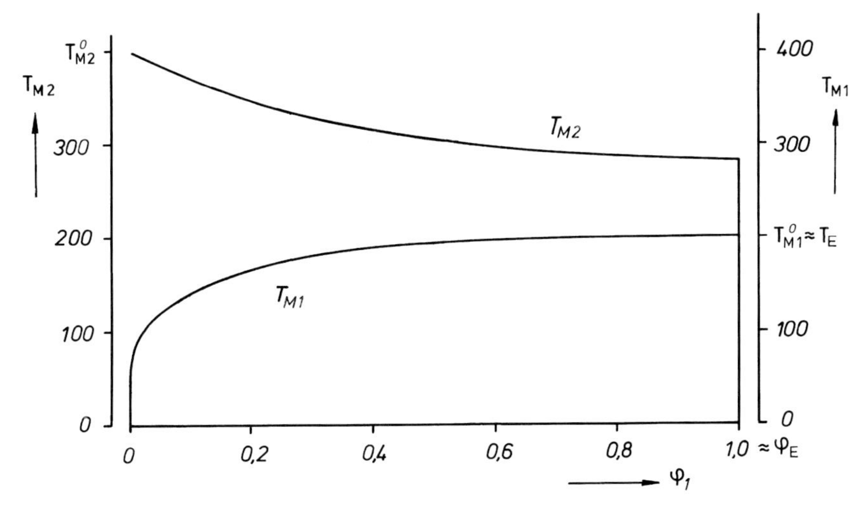

Fig. 8.6. As in Fig. 8.5, with $\Delta h_1^o = \Delta h_u^o = 4$ kJ/mol, $T_{M1}^o = 200$ K, $T_{M2}^o = 400$ K, $\chi = 0.5$, and $P = 1000$. With these values, the eutectic point is at $\varphi_1 \approx 1$, $T_E \approx T_{M1}^o = 200$ K

$T_{M2}(\varphi_1)$, only polymer crystals are precipitated continuously. Molecules of the solvent first crystallize when almost all of the polymer molecules in the solution have disappeared, after reaching the eutectic temperature $T_E \approx T_{M1}^o$.

In very small crystals with linear dimensions in the range of less than one micrometer, the surfaces, in addition to the volume, also play a role which cannot be disregarded [*e.g.*, see Prigogine, Defay (1951); Strickland-Constable (1968)]. In this case, the surfaces O_i of the crystals must be taken into account as an additional variable. The Gibbs fundamental equation of the two-phase, one-component system (crystal; melt) is then in the G-representation

$$G = G(T, p, N_a, N_c, \dots O_i \dots) .$$

As the surfaces are extensive quantities, Eq. (8.14) is replaced by

$$G = \mu_a N_a + \mu_c N_c + \sum_i \sigma_i O_i . \tag{8.29a}$$

The equations of state

$$\frac{\partial G}{\partial O_i} \equiv \sigma_i (T, p, N_a, N_c, \dots O_i \dots) \tag{8.29b}$$

define σ_i as a partial free enthalpy of the part O_i of the surface. σ_i is practically identical with the surface tension present at the surface O_i. The σ_i are usually supposed to be independent of the variables O_i as well as of the mass variables N_a, N_c:

$$\sigma_i = \sigma_i (T, p) .$$

If the chemical potentials μ_a, μ_c are also independent of the surfaces O_i, one can write in place of (8.29)

$$G = G_\infty + \sum_i \sigma_i O_i \,, \tag{8.30}$$

where G_∞ is the free enthalpy of the system of large crystals, as given by (8.14).

Internal equilibrium exists in the system with small crystals if

$$\frac{\partial G}{\partial N_c} = \frac{\partial G_\infty}{\partial N_c} + \sum_i \sigma_i \frac{\partial O_i}{\partial N_c} = 0 \tag{8.31}$$

is valid. Because $N_a + N_c = \text{const.}$, (8.14) results in

$$\frac{\partial G_\infty}{\partial N_c} = \mu_c^\infty - \mu_a^\infty = -\Delta\mu_\infty = -\Delta h_\infty + T\Delta s_\infty \tag{8.32}$$

for the system with large crystals (if the phases are autonomous), whereby

$$\Delta s_\infty = \Delta h_\infty / T_M^\infty$$

holds at the melting point T_M^∞ of the large crystals. Corresponding to the Schröder–Van Laar equation (8.26), one obtains

$$\frac{\partial G_\infty}{\partial N_c} = \Delta h_\infty \left(\frac{T}{T_M^\infty} - 1 \right) . \tag{8.33}$$

If one inserts (8.33) into the equilibrium condition (8.31), one obtains

$$T_M = T_M^\infty \left(1 - \frac{1}{\Delta h_\infty} \sum_i \sigma_i \frac{\partial O_i}{\partial N_c} \right) \tag{8.34}$$

for the melting temperature T_M of the small crystals. In general, σ_i, $\partial O_i / \partial N_c$, Δh_∞ are positive quantities. Hence, the melting point of small crystals lies at lower temperatures than that of the large crystals. Like the mixing effect, the surface effect leads to a melting point depression. In addition, the entropy of transition Δs is often much less sensitive to the surface effects than the heat of transition Δh. In this case, one can assume to a first approximation

$$\Delta s \approx \Delta s_\infty = \Delta h_\infty / T_M^\infty \,,$$

i.e.,

$$\Delta h = T_M \, \Delta s \approx \frac{T_M}{T_M^\infty} \, \Delta h_\infty \,.$$

If one inserts (8.34), one obtains for the heat of fusion of the small crystals

$$\Delta h = \Delta h_\infty - \sum_i \sigma_i \frac{\partial O_i}{\partial N_c} . \tag{8.35}$$

According to the Curie–Gibbs theorem [Curie (1885), (1887); Gibbs (1928)], the equilibrium shape of a small crystal is determined by the fact that the surface energy at constant volume V_c assumes a minimum value. Thus,

$$(\mathrm{d} \sum_i \sigma_i O_i)_{V_c} = \sum_i \sigma_i (\mathrm{d}O_i)_{V_c} = 0 \tag{8.36}$$

must hold in equilibrium. From this minimum postulate, one can derive the existence of a centre C in the interior of the crystals whose distances h_i from the crystal surfaces O_i satisfy the relation

$$\pi h_i = \sigma_i \tag{8.37}$$

[Liebmann (1914), Stranski (1943), v. Laue (1943); see Fig. 8.7]. Here, π is an isotropic equilibrium pressure which only depends on the absolute size of the small crystal. Equation (8.37) is equivalent to the Wulff theorem (or the Gibbs–Wulff theorem), according to which the ratio of the surface tension σ_i to the so-called central distance h_i is constant for the equilibrium shape of small crystals:

$$\frac{\sigma_1}{h_1} = \frac{\sigma_2}{h_2} = \frac{\sigma_3}{h_3} = \ldots \tag{8.38}$$

[Wulff (1901); Liebmann (1914); v. Laue (1943)]. As long as surface effects are of importance, an equilibrium crystal thus always maintains its geometric form during growth.

The volume V_c of the crystallites is composed of the volumes of the pyramids with base O_i and height h_i (see Fig. 8.7). Hence, we have

$$V_c = \frac{1}{3} \sum_i h_i O_i , \tag{8.39a}$$

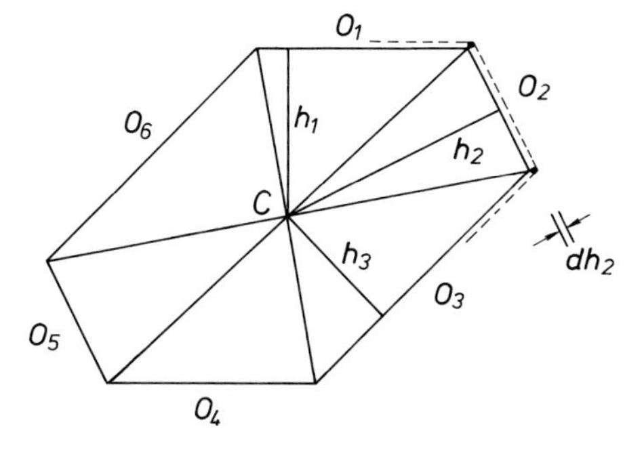

Fig. 8.7. Two-dimensional model of the equilibrium form of a small crystal. Surfaces: O_i; so-called central distances: h_i. In the three-dimensional case, the volume of the crystal is composed of pyramids with the volumes $h_i O_i/3$

$$\mathrm{d}V_c = \frac{1}{3} \sum_i h_i \,\mathrm{d}O_i + \frac{1}{3} \sum_i O_i \,\mathrm{d}h_i \,. \tag{8.39b}$$

When evaluating an infinitesimal increase in the crystal volume, on the other hand, the contribution at the corners of the pyramid base (in Fig. 8.7 marked in black at two points) can be neglected to a good approximation. One then obtains

$$\mathrm{d}V_c = \sum_i O_i \,\mathrm{d}h_i \,. \tag{8.40}$$

Equation (8.39b) then also leads to

$$\mathrm{d}V_c = \frac{1}{2} \sum_i h_i \,\mathrm{d}O_i \tag{8.41}$$

or

$$\frac{\partial V_c}{\partial N_c} \equiv v_c = \frac{1}{2} \sum_i h_i \frac{\partial O_i}{\partial N_c} \,. \tag{8.42}$$

Insertion of the expression for h_i given by (8.37) results in

$$\sum_j \sigma_i \frac{\partial O_i}{\partial N_c} = 2\pi v_c = \frac{2 v_c \sigma_i}{h_i} \tag{8.43}$$

with arbitrary i on the right-hand side. By insertion of (8.43) into (8.34), one finally obtains the Thomson–Gibbs equation

$$T_M = T_M^\infty \left(1 - \frac{2 v_c \sigma_i}{h_i \,\Delta h_\infty}\right) \tag{8.44}$$

for the determination of the melting point T_M of small crystals and from (8.35), the heat of fusion of small crystals

$$\Delta h = \Delta h_\infty - \frac{2 v_c \sigma_i}{h_i} \,. \tag{8.45}$$

Equation (8.44) is an analogue to the Thomson equation for the vaporization of a liquid droplet. The melting point depression of small crystals is larger the larger the surface tension σ_i is and the smaller the geometric dimensions h_i of the crystals are. In the case of a parallelepiped with the edge lengths a_1, a_2, a_3 and the central distances $h_i = a_i/2$, one can also specifically write in place of (8.43):

$$\sum_i \sigma_i \frac{\partial O_i}{\partial N_c} = \frac{4 v_c}{3} \left(\frac{\sigma_1}{a_1} + \frac{\sigma_2}{a_2} + \frac{\sigma_3}{a_3}\right) . \tag{8.46}$$

The free enthalpy of the pure melt is $G_a^\infty \equiv \mu_a^\infty N$. The change in the free enthalpy of the system caused by the formation of the small crystals, is thus, according to (8.29) or (8.30), respectively, given by

$$G - G_a^\infty = -\Delta\mu_\infty N_c + \sum_i \sigma_i O_i \tag{8.47a}$$

and with (8.32), (8.33) by

$$G - G_a^\infty = \left(\frac{T}{T_M^\infty} - 1\right) \Delta h_\infty N_c + \sum_i \sigma_i O_i \,. \tag{8.47b}$$

Here, the first term in the sum on the right side is always negative for $T < T_M^\infty$, the second term in the sum, on the other hand, always positive. The proportionality $O \sim V_c^m$ with $0 < m < 1$ usually exists between the surface O and the volume V_c of a crystal. For example, $m = 2/3$ holds for a sphere or a parallelepiped. Because of $V_c \sim N_c$, one can then also assume $O \sim N_c^m$. The different exponents with which N_c enters the first and the second terms on the right side of (8.47) have the consequence, together with the usual numerical values for the coefficients Δh_∞ and σ_i, that with a very small N_c the positive term first prevails in (8.47), but with a larger N_c, the negative term predominates. The free enthalpy G first increases with N_c above G_a^∞, traverses a maximum value, and finally drops below G_a^∞ (Fig. 8.8). This shows that the creation of new surfaces in the melt requires an expenditure of energy (work necessary to form the crystal nuclei). However, the formation of nuclei does not yet occur at $T = T_M^\infty$, as the

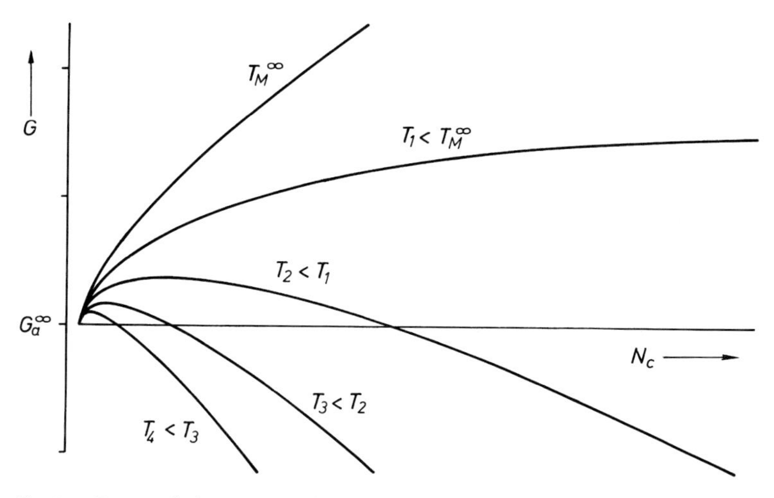

Fig. 8.8. Free enthalpy G according to (8.47b) as a function of the mole number N_c of the crystallized units for different temperatures T. Free enthalpy of the pure melt: G_a^∞; melting temperature of the fully developed crystal: T_M^∞. It was assumed that: Δh_∞ = const. and $\sum \sigma_i O_i \sim N_c^{2/3}$

positive term in (8.47) predominates here. Undercooling is required in order to initiate crystallization by homogeneous nucleation (*i.e.*, nucleation without the help of foreign substances). In general, however, the crystal nuclei are not at all stable. The equilibrium condition (8.31) is only fulfilled at the maxima of the curves in Fig. 8.8, and the equations (8.36–8.46) only hold for these states. Thermal fluctuations induce the formation of crystal-like molecular aggregates in the melt at temperatures $T < T_M^\infty$. These non-equilibrium aggregates usually disappear as fast as they appear. Such an aggregate achieves internal equilibrium (with $\partial G/\partial N_c = 0$) only if it accidentally reaches the critical number N_c, which corresponds to the maximum of the free enthalpy. This equilibrium, however, is labile ($\partial^2 G/\partial N_c^2 < 0$). This means that the equilibrium aggregate has only two possibilities: it either disappears again or it spontaneously grows into a large crystal. If the amorphous and the crystalline phases are autonomous, stable or metastable small crystals can only originate from a restraint of spontaneous growth, *e.g.*, by a growth failure or by a lack of mass transport.

In the melt of a polymer composed of long flexible chain molecules, restraints of growth develop rather rapidly by themselves if the trans-aggregates find each other considerably faster than the entangled mutually penetrating chain coils are capable of disentangling. The chain segments containing gauche-bonds are forced in the chain direction towards the periphery of the trans-aggregates, where they form loops which prevent further growth of the crystal aggregates in the chain direction. In the case of a parallelepiped with the edge lengths a_1, a_2, a_3 and a restraint of growth in the 3-direction, we have to take into account $V_c = a_1 a_2 a_3 =$ const. and, in addition, the subsidiary condition $a_3 =$ const. In the minimum requirement (8.36)

$$\sum_i \sigma_i (\mathrm{d}O_i)_{V_c} = 2\,[\sigma_1 d\,(a_2 a_3) + \sigma_2 d\,(a_1 a_3) + \sigma_3 d\,(a_1 a_2)]_{V_c} = 0\,,$$

$$\mathrm{d}a_3 = 0 \qquad \text{and} \qquad a_1\,\mathrm{d}a_2 = -\,a_2\,\mathrm{d}a_1$$

are then valid. The minimum condition for the surface energy becomes

$$\sum_i \sigma_i (\mathrm{d}O_i)_{V_c, a_3} = 2\left(-\,\sigma_1\,\frac{a_2\,a_3}{a_1} + \sigma_2\,a_3\right)\mathrm{d}a_1 = 0\,. \tag{8.48}$$

The Wulff theorem now only holds in the restricted form

$$\frac{\sigma_1}{a_1} = \frac{\sigma_2}{a_2}\,. \tag{8.49}$$

If we assume $\sigma_i =$ const., we obtain

$$\frac{\partial O_1}{\partial V_c} = \frac{\partial\,(a_2 a_3)}{\partial\,(a_1 a_2 a_3)} = \frac{\partial a_2}{\partial\,(a_1 a_2)} = \frac{\partial a_1}{\partial a_1^2} = \frac{1}{2a_1}\,,$$

$$\frac{\partial O_2}{\partial V_c} = \frac{\partial (a_1 a_3)}{\partial (a_1 a_2 a_3)} = \frac{1}{2a_2},$$

$$\frac{\partial O_3}{\partial V_c} = \frac{\partial (a_1 a_2)}{\partial (a_1 a_2 a_3)} = \frac{1}{a_3}.$$

In the case of the parallelepiped with restrained growth in the 3-direction, (8.46) is replaced by the equation

$$\sum_i \sigma_i \frac{\partial O_i}{\partial N_c} = v_c \sum_i \sigma_i \frac{\partial O_i}{\partial V_c} = v_c \left(\frac{\sigma_1}{a_1} + \frac{\sigma_2}{a_2} + \frac{2\sigma_3}{a_3} \right). \tag{8.50}$$

This equation holds for the crystal nucleus which is in internal equilibrium at the maximum of G (Fig. 8.8). As the equilibrium is labile, this nucleus disappears again or grows spontaneously in the 1- and 2-directions forming a lamella with a large lateral extension. In case of a large lateral extension a_1 and a_2, the first two terms of the sum on the right-hand side of Eq. (8.50) can be neglected. If one then designates the lamellar thickness with $\ell \equiv a_3$ and the surface tension assigned to the top surfaces of the lamella by $\sigma \equiv \sigma_3$, one obtains for the laterally grown lamella

$$\sum_i \sigma_i \frac{\partial O_i}{\partial N_c} = \frac{2 v_c \sigma}{\ell}. \tag{8.51}$$

If one inserts (8.51) into (8.34), one obtains

$$T_M = T_M^\infty \left(1 - \frac{2 v_c \sigma}{\ell \, \Delta h_\infty} \right) \tag{8.52}$$

as the melting point of the lamella with restrained growth in the chain direction. This is the relation given by Lauritzen, Hoffman (1960) for the determination of the melting point of a crystalline polymer lamella. One should take into account that in contrast to the Thomson–Gibbs equation (8.44), the denominator of the second term on the right in (8.52) contains twice the central distance $\ell = 2h_3$ instead of the central distance. Equations (8.51) and (8.35) further lead to

$$\Delta h = \Delta h_\infty - \frac{2 v_c \sigma}{\ell} \tag{8.53}$$

as the heat of fusion of a polymer lamella.

Cooling of a polymer with a finite rate usually leads to the formation of lamellae with a certain distribution of thicknesses. If one reheats such a sample, the lamellae melt at different temperatures according to (8.52), the thin ones first, the thick ones last. The melting process occurs over a larger temperature range. However, annealing such a sample at a temperature $T < T_M^\infty$ for a while sometimes induces a slight increase in the lamellar thickness [*e.g.*, see Wunderlich (1973)]. This indicates that the restraints in the chain direction are not absolute in character.

There are chain segments in the amorphous top surface of the lamellae and in the amorphous boundary phase around the nuclei which are bound to the crystalline phase. These are more or less loose loops as well as loose chain ends or chain segments which extend from one lamella to the neighbouring lamella or from one nucleus to the next (so-called tie molecules). The number of conformational isomers capable of forming these segments and thus the entropy of the boundary phase (see Sect. 6.2) depend on the length of the amorphous segments, whether only one end or both ends of the segments are fixed and, if both ends of the segments are fixed, on how large the end-to-end distance of the segments is. The length of the chain segments necessarily decreases in the amorphous phase if a lamella, for example, extends during annealing, or a nucleus grows in the chain direction. This leads to a change in the number of possible conformational isomers of the amorphous segments and in the entropy of the amorphous boundary phase. Hence, the entropy of the boundary phase depends on the lamellar thickness or on the longitudinal dimension of the nucleus, or, a little less specifically, on the mole number N_c of the crystallized chain units or their crystallinity α. The amorphous boundary phase, as long as it contains chain molecules which at the same time also traverse the crystalline phase, is non-autonomous.

One can expect that the enthalpy h_a of the non-autonomous amorphous phase also depends on the extent of the crystalline phase. This dependence, however, may be neglected as a first approximation [Schrader, Zachmann (1970)]. One can continue to assume a linear relation of the form (8.16b) for the enthalpy h of the system (crystal; amorphous boundary phase). Because of the dependence

$$s_a = s_a(\alpha)\,, \tag{8.54}$$

however, the equations (8.16 a) and (8.16 c) are replaced by the non-linear equations

$$g(T, p, \alpha) = (1 - \alpha)\, g_a(T, p, \alpha) + \alpha g_c(T, p)\,, \tag{8.55a}$$

with

$$g_a(T, p, \alpha) = h_a(T, p) - Ts_a(T, p, \alpha)\,, \tag{8.55b}$$

and

$$s(T, p, \alpha) = (1 - \alpha)\, s_a(T, p, \alpha) + \alpha s_c(T, p)\,. \tag{8.55c}$$

As the dependence (8.54) only applies to changes in the crystallinity α in the chain direction, $\partial\alpha$, $d\alpha$ have to be interpreted correspondingly in the following.

The equilibrium and stability conditions (8.10) now lead to

$$\Delta h = -T\left(\frac{\partial s}{\partial \alpha}\right)_e ; \quad \left(\frac{\partial^2 g}{\partial \alpha^2}\right)_e = -T\left(\frac{\partial^2 s}{\partial \alpha^2}\right)_e \geq 0 . \tag{8.56}$$

As $(\partial s/\partial \alpha)_e$ is no longer a constant, the equilibrium of the partially crystalline state is at constant pressure no longer necessarily confined to a single temperature T_M. The partially crystalline equilibrium state is stable or metastable if $-\partial s/\partial \alpha = \partial s/\partial(1-\alpha)$ is a positive function, increasing with α. In addition, the equilibrium is no longer neutral, so that $\alpha_e(T, p)$ becomes a continuous function of temperature and pressure. According to (8.12), one obtains with (8.56)

$$\left(\frac{d\alpha_e}{dT}\right)_p = \frac{\Delta h}{T^2} \Big/ \left(\frac{\partial^2 s}{\partial \alpha^2}\right)^e_{T,p} . \tag{8.57}$$

In a stable internal equilibrium, the crystallinity decreases with increasing temperature. Equations (8.13) and (8.56) lead to

$$c_p^e = T\left(\frac{\partial s}{\partial T}\right)^e_{p,\alpha} - \left(\frac{\Delta h}{T}\right)^2 \Big/ \left(\frac{\partial^2 s}{\partial \alpha^2}\right)^e_{T,p} \tag{8.58}$$

for the heat capacity.

Alfrey, Mark (1942), Frith, Tuckett (1944), and Münster (1954) were the first to point out that the amorphous phase is not independent of the crystalline phase in the case of polymers in a partially crystalline state. Tung, Buckser (1958); Roe, Smith, Krigbaum (1961); Elyashevich, Baranov, Frenkel (1977) then followed. The most extensive computations were performed by Zachmann [(1964–1969); Schrader, Zachmann (1970)]. The statistical determination of the entropy of the non-autonomous boundary phase is an extremely complicated problem in which one has to take into account that the conformational isomerism of the chain segments is reduced in many ways, *e.g.*, by the adjacent crystal phase [compare Juilfs, Künne (1971)]. Only those cases in which the amorphous phase consists of identical segments have been calculated up to now.

In the left column of Fig. 8.9, $-\partial s/\partial \alpha$ is schematically represented as a function of α as determined by Zachmann for the cases of a crystalline segment bundle with loosely dangling chain ends (Fig. 8.9, top), a crystalline lamella with sharp folds (Fig. 8.9, middle) and a crystalline lamella with loose loops or tie molecules (Fig. 8.9, bottom). The resulting development of the free enthalpy (8.55) as a function of α is given on the right for different temperatures [see Baur (1979), (1986)].

A fundamental result is that the crystalline segment bundle with loosely dangling chain ends is unstable. In this case, $-\partial s/\partial \alpha$ is a function which decreases with increasing $\alpha\,(-\partial^2 s/\partial \alpha^2 < 0)$. A stable equilibrium is only possible with $\alpha = 0$ or $\alpha = 1$. In addition, the amorphous and crystalline states are sepa-

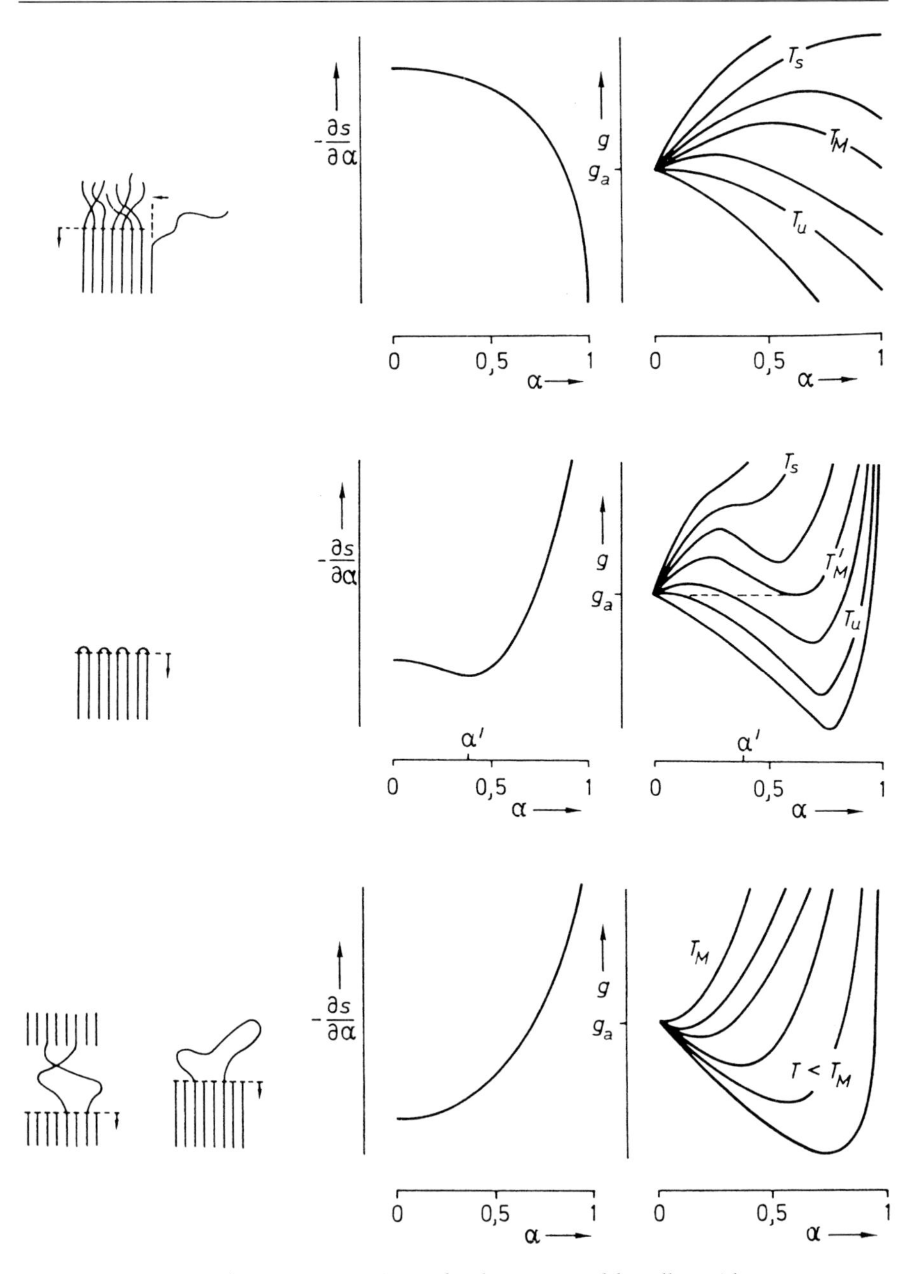

Fig. 8.9. Energetic and entropic situations of polymer crystal lamellae with a non-autonomous boundary phase during a change in the crystallinity α in the chain direction. Specific free enthalpy: g; specific entropy: s. Top: segment bundle with loosely dangling chain ends; middle: so-called adjacent reentry model; bottom: so-called switchboard model. [According to Baur (1980)]

rated by a potential barrier in a temperature range $\langle T_u; T_s \rangle$. Only an extremely improbable, large fluctuation would, therefore, be able to cause the formation of an ideal polymer crystal composed of fully extended chains at T_M. Only considerable undercooling can induce the spontaneous growth of an ideal polymer crystal starting from $T_u < T_M$ by growth in the chain direction. The melting of an ideal crystal starting from the top surface, on the other hand, requires overheating up to $T_s > T_M$. [With respect to the phenomena of overheating in polymer crystals see Wunderlich (1980).]

In the case of crystalline lamellae with sharp folds [adjacent re-entry model; see Keller (1957)], $-\partial s/\partial \alpha$ is also a function decreasing with increasing α up to a value $0 < \alpha' < 1$. Here, stable lamellae are thus only possible above α'. Spontaneous growth of the lamellae in the chain direction also only occurs here after an undercooling of the system down to T_u. Melting requires overheating up to T_s. At a temperature T_M', which does not necessarily correspond to T_M, the pure melt and a partially crystalline lamella can be in internal equilibrium ($g_e = g_a$).

In the case of a lamella with loose loops or tie molecules [so-called switchboard model; see Flory (1962); Fischer, Schmidt (1962); Fischer (1978); Stamm, Fischer, Dettenmaier, Convert (1979)], $-\partial s/\partial \alpha$ is a function increasing with α. The lamella is always stable. Its formation does not require undercooling. The minimum of the free enthalpy (the stable equilibrium position of the lamella) continuously shifts towards higher α-values with decreasing temperature. In reality, the scale of possible α-values is, of course, limited downwards by the instability of the nuclei and upwards by the restraints induced by loose loops. One sees that the formation of lamellae with a switchboard structure is preferred not only for kinetic (see above) but also for entropic reasons when cooling a polymer melt.

It should be pointed out again that the situations represented in Fig. 8.9 only refer to changes in the crystallinity α in the chain direction. The lateral growth or the lateral melting of the lamellae are only indirectly affected.

8.3 Condis Crystals

At constant temperature and pressure, the internal equilibrium of a crystal is characterized by a minimum of the free enthalpy $G = H - TS$ [see Eq. (2.88)]. At $T = 0$, the energy $H = U + pV$ is then also at a minimum. With increasing temperature, however, the entropy of the crystal increasingly contributes to the minimization of the free enthalpy. As a result, an ideal crystal incorporates more and more defects with increasing temperature [*e.g.*, see Ubbelohde (1952); Fowler, Guggenheim (1960)].

In particular, the incorporation of chain ends as well as the incorporation of chain defects which developed during the polymerization process results in the formation of fixed defects in polymer crystals. The conformational isomerism of the chain molecules leads to thermally excitable defects whose number varies with the temperature. The development of individual gauche-bonds in a

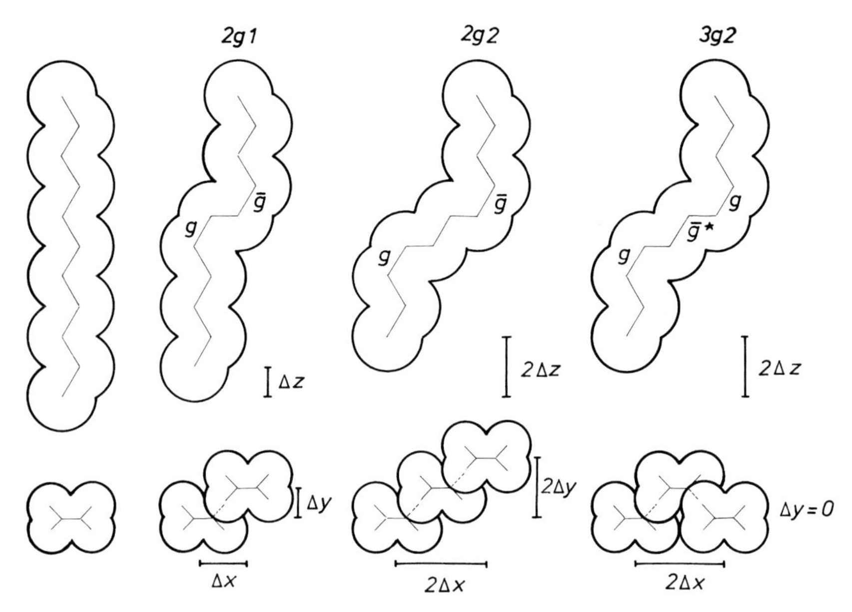

Fig. 8.10. Kink isomers of an *n*-alkyl chain. From left to right: all-trans chain, 2g1-kink, 2g2-kink, 3g2-kink. Top: side view; bottom: top view. Δz indicates the contraction as compared to the all-trans chain, Δx, Δy the lateral displacement of one chain end with respect to the other chain end. [According to Pechhold (1968)]

crystal, however, is not possible for stereometric reasons (see Sect. 6.2). Although the development of gauche-bands by a cooperative process is conceivable, it is very unlikely at lower temperatures because of the minimal entropy gain. The same applies to many other stereometrically "bulky" conformational isomers. However, there are combinations of gauche-bonds which hardly disturb the geometry of the bundle- or lamella-like polymer crystals. These are the so-called kinks, as first described by Blasenbrey, Pechhold (1967) and by McMahon, McCullough, Schlegel (1967). Some of these kinks are represented in Fig. 8.10 using the Pechhold nomenclature (1968). These are the trans-gauche combinations

$$\ldots \text{tttgt}\bar{\text{g}}\text{ttt} \ldots : \text{2g1 - kink} ,$$

$$\ldots \text{ttgttt}\bar{\text{g}}\text{tt} \ldots : \text{2g2 - kink} ,$$

$$\ldots \text{ttgt}\bar{\text{g}}\text{tgtt} \ldots : \text{3g2 - kink} .$$

The displacements which one end of an *n*-alkyl chain undergoes with respect to the other by incorporation of a 2g1-kink are given by

$$\Delta x = \pm\, 3.088 \cdot \left(\frac{1}{3}\right)^{1/2} \text{Å} \approx \pm\, 1.76\ \text{Å} ,$$

$$\Delta y = \pm\, 1.544 \cdot \left(\frac{2}{3}\right)^{1/2} \text{Å} \approx \pm\, 1.26 \text{ Å} ,$$

$$\Delta z = -\, 1.544 \cdot \left(\frac{2}{3}\right)^{1/2} \text{Å} \approx -\, 1.26 \text{ Å}$$

in a Cartesian coordinate system (x, y, z) (z: chain direction). The incorporation of such defects into the crystal becomes increasingly easier with increasing thermal expansion. The kinks are mainly generated *via* the torsional modes of the chains (Sect. 6.4.4).

The 2g1-kink is the stereometrically most favourable conformational defect. Among the thermally generated defects, it therefore predominates at low temperatures and still prevails at higher temperatures. In general, the number of conformational defects continuously increase with increasing temperature according to the equilibrium conditions. This is particularly true if the concentration of the defects is so low that a mutual interaction of the defects can be excluded. In some crystals composed of linear flexible chain molecules, a discontinuous creation of defects can, however, already be observed before the melting point is reached. The crystals discontinuously change over into a "*con*formational *dis*ordered" phase while maintaining the positional and orientational order of the chain molecules. Wunderlich [Wunderlich, Grebowicz (1984); Wunderlich *et al.* (1988)] designated such crystal modifications as condis crystals. Using a simple model, we will examine in the following the conditions which make a discontinuous defect transition possible. We will confine ourselves to the specific situations which approximately occur in the crystals of n-paraffins and polyethylene.

Let us examine a lamella composed of N parallelly arranged, equally long n-alkyl chains. As long as the chains are in an extended form, the top surfaces of this lamella are assumed to be planar. The number of next neighbours of a chain (the coordination number) is z. As a possible chain defect, we will exclusively consider the 2g1-kink. Several 2g1-kinks in a single chain can only occur in those combinations which induce the smallest possible deviation of the chain sections in the lateral x- and y-directions (see Fig. 8.11, right). Likewise, only those chain pairs will develop, for energetic and stereometric reasons, which induce the smallest possible lateral disturbance (Fig. 8.12, left). Hence, the kinks have a tendency towards a sort of block formation, *i.e.*, towards aggregation. Because a 2g1-kink causes a disturbance along the whole chain, one has to regard the complete chains as elementary statistical units (see Fig. 8.10). A self-regulation of this disturbance can only be expected in the case of long chains and then only in a lateral direction.

We will designate the number of chains containing j kinks as N_j. The maximum number of kinks a chain is able to incorporate is L. We will designate the number of pairs of adjacent chains, where one contains j and the other k kinks, with N_{jk}. The concentration of j-chains then amounts to

$$x_j \equiv N_j / N ,$$

Fig. 8.11. n-Alkyl chains with the same number of 2g1-kinks but different lateral displacements Δx

Fig. 8.12. Pairs of n-alkyl chains containing one 2g1-kink each but having different binding energies. The left pair has the strongest binding energy and is, therefore, energetically preferred. The 2g1-kinks tend to aggregate

the concentration of (j, k)-pairs to

$$x_{jk} \equiv 2N_{jk}/zN\,.$$

The total number of chain pairs is given by $zN/2$. The $x_j, x_{jk}\,(j, k = 0 \ldots L)$ can be regarded as internal variables of the lamella. They fulfil the subsidiary conditions

$$\sum_j x_j = 1\,, \tag{8.59 a}$$

$$x_{jk} = x_{kj}\,, \tag{8.59 b}$$

$$x_{jj} + \frac{1}{2} \sum_{k \neq j} x_{jk} = x_j\,. \tag{8.59 c}$$

We abbreviate (as is also done in the following)

$$\sum_j \equiv \sum_{j=0}^{L}$$

The subsidiary condition (8.59c) is due to the fact that the probability of finding a j-chain is equal to the probability of finding a (j, j)-pair or a (j, k)-pair. The factor 1/2 in front of the sum must be included, as $x_{jk}\,(j \neq k)$ gives the probability of finding a (j, k)-pair or a (k, j)-pair. As expected, (8.59a) and (8.59c) together lead to

$$\sum_{j \geq k}\sum x_{jk} = 1\,. \tag{8.59 d}$$

Due to the subsidiary conditions (8.59), only $L\,(L+3)/2$ of the altogether $(L+1)$ $(L+2)$ internal variables are independent of each other.

We assume the degrees of freedom of conformation to be separable from the other degrees of freedom of the molecules. The partition function of the lamella then decomposes into a product $Q \cdot Q'$, in which the factor Q only refers to the conformational isomerism (see Sect. 6.1). In the following, we will only consider Q, and in Q, we will only consider the maximum term using $E = U$ and $g = \Omega$, so that according to (6.16) and (6.20), we obtain

$$F(T, V, N) = U - kT \ln g(U, V, N)\,. \tag{8.60}$$

Let us first disregard the interaction of the chains. The pair concentrations x_{jk} are then unimportant. We can then assume the j-chains to be statistically distributed. The number of possible arrangements of the chains amounts to $N!$. If γ_j is the number of distinguishable conformational isomers of a j-chain, then N_j/γ_j of the j-chains are equal. The number of distinguishable possible arrangements of the chains is then

$$g = \frac{N!}{\prod_{j=0}^{L} \left(\frac{N_j}{\gamma_j}\right)!^{\gamma_j}}\,. \tag{8.61}$$

By means of Stirling's formula (p. 109; assumption: a large lateral extension of the lamella), we obtain

$$\ln g = -N \sum_j x_j \ln\left(\frac{x_j}{\gamma_j}\right). \tag{8.62}$$

When one neglects the intermolecular interaction, the internal energy of the system amounts to

$$U = N \sum_j u_j x_j = N\left(u_0 + \Delta u \sum_j{}' j x_j\right), \tag{8.63}$$

where u_j designates the internal energy of a j-chain, u_0 the internal energy of a defect-free extended chain, and Δu the excess energy of a 2g1-kink. The primed sum is an abbreviation for

$$\sum_j{}' \equiv \sum_{j=1}^{L}.$$

With (8.62) and (8.63), one obtains from (8.60) the mean free energy per chain

$$f \equiv \frac{F}{N} = u_0 + \Delta u \sum_j{}' j x_j + kT \sum_j x_j \ln\left(\frac{x_j}{\gamma_j}\right). \tag{8.64}$$

If we replace the Boltzmann constant k with the gas constant and interpret N as being the mole number of the chains and x_j as the mole fraction, (8.64) yields the free energy of the lamella per mole of chain molecules. One should take into account that because of (8.59 a) one of the x_j in the last sum on the right-hand side of (8.64) is a dependent variable. If we choose x_0 as the dependent variable, we obtain with

$$x_0 = 1 - \sum_j{}' x_j; \quad \gamma_0 = 1 \tag{8.65 a}$$

$$f = u_0 + \Delta u \sum_j{}' j x_j + kT\left(x_0 \ln x_0 + \sum_j{}' x_j \ln\left(\frac{x_j}{\gamma_j}\right)\right). \tag{8.65 b}$$

According to (2.82 b) and (8.65), the lamella is in internal equilibrium if

$$\frac{\partial f}{\partial x_j} = j\Delta u + kT \ln\left(\frac{x_j}{x_0 \gamma_j}\right) = 0 \tag{8.66}$$

holds for all $j \neq 0$. The equilibrium concentrations x_j^e of the j-chains are thus given by

$$x_j^e = x_0^e \gamma_j \, e^{-j\Delta u/kT} \quad (j \neq 0). \tag{8.67}$$

By summation, one obtains

$$\sum_j{}' x_j^e = 1 - x_0^e = x_0^e \sum_j{}' \gamma_j \, e^{-j\Delta u/kT} ,$$

and from this, the equilibrium concentration of the defect-free chains is

$$x_0^e = \frac{1}{1 + \sum_j{}' \gamma_j \, e^{-j\Delta u/kT}} . \tag{8.68}$$

Using this, one can also write in place of (8.67)

$$x_j^e = \frac{\gamma_j \, e^{-j\Delta u/kT}}{1 + \sum_j{}' \gamma_j \, e^{-j\Delta u/kT}} , \tag{8.69}$$

whereby this equation now holds for all $j = 0 \ldots L$. Insertion of (8.67) into (8.65b) leads to

$$f_e = u_0 + kT \ln x_0^e \tag{8.70}$$

as the equilibrium free energy of the lamella per chain. In internal equilibrium, the thermodynamics of the ideal lamella (the lamella with negligible intermolecular interaction) is solely determined by the concentration of the defect-free chains. The lamella behaves similarly to a 2-level system (Sect. 6.3). In particular, the heat capacity of the lamella passes through a Schottky anomaly.

One obtains a complete agreement with the 2-level system using the formulation

$$\gamma_j = \binom{L}{j} . \tag{8.71}$$

This formulation is based on the idea that each chain can be divided into fixed segments, each of which can be occupied by a kink. The number of arrangements of j kinks on L places is then equal to γ_j. In this case, L is proportional to the chain length. Equation (8.71) disregards the kink-induced disturbance of the rotational symmetry of the chains and the possible overlap of the potential defect sites. On the basis of the binomial theorem, one obtains by insertion of (8.71) into (8.68)

$$x_0^e = \frac{1}{(1 + e^{-\Delta u/kT})^L} \tag{8.72}$$

and thus from (8.70)

$$f_e = u_0 - LkT \ln (1 - e^{-\Delta u/kT}) \tag{8.73}$$

[see Eq. (6.99)]. This is the Gibbs fundamental equation for a 2-level system composed of *NL* units with a non-degenerated excited state (*NL* = number of chain segments which can be occupied by kinks). The Schottky maximum of this lamella is located at

$$T_{SA} = \Delta u/k\eta \quad \text{with} \quad \eta \approx 2.400 \tag{8.74}$$

(see Sect. 6.3). With (8.71) and (8.72), the equilibrium concentration of the kink-containing chains is given by

$$x_j^e = \frac{\binom{L}{j} e^{-j\Delta u/kT}}{(1 + e^{-\Delta u/kT})^L} . \tag{8.75}$$

Examples are depicted in Figs. 8.13 and 8.14. Figure 8.13 shows that the formation of kinks is more pronounced at even lower temperatures, the larger L is, *i.e.*, the longer the chain molecules are. This is a pure quantity effect which is solely induced by the increase in segments which can be occupied by kinks. Referring to an individual segment, the same conditions exist on average in all lamellae, independent of L.

The explanation of a discontinuous defect transition requires consideration of the intermolecular interaction. With respect to this, we will confine ourselves once again to the simple Bragg–Williams approximation in the following [see Fowler–Guggenheim (1960); for a treatment of the problem within the frame-

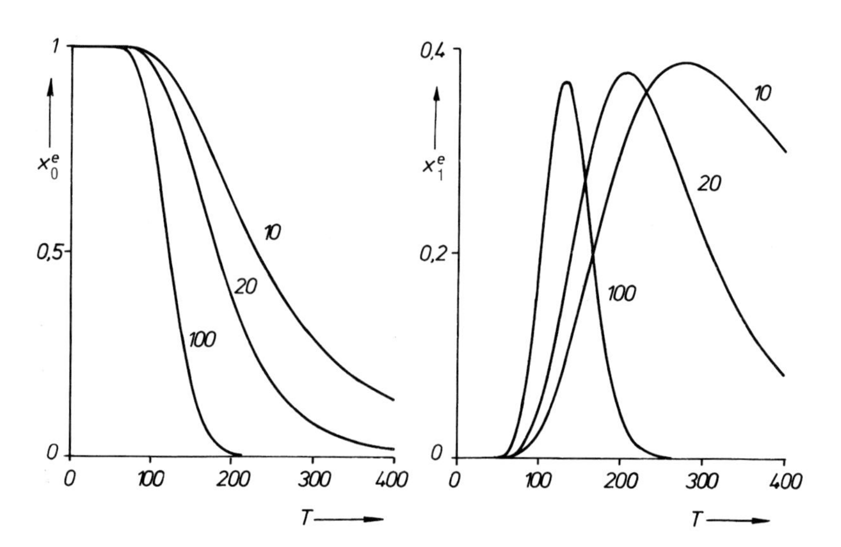

Fig. 8.13. Concentration x_0^e of defect-free chains and concentration x_1^e of chains containing a 2g1-kink as a function of the temperature T [K] according to (8.75). The parameter is L. It was assumed that: $\Delta u = 5$ kJ/mol

work of the quasi-chemical approximation of Fowler and Guggenheim or the Bethe–Peierls approximation see Baur (1975), (1977)]. In the Bragg–Williams approximation, one has to assume that the j-chains are statistically distributed. As this assumption strictly only applies to interaction-free particles, it would not be very meaningful to presuppose a complicated field of interaction. Rather, it seems possible to drastically reduce the number of interaction parameters and to replace them by a few mean quantities. In the following, we will characterize the field of interaction by three general parameters:

w_{00}: for the binding energy between two adjacent defect-free chains,
w_{D0}: for the binding energy between a defective chain and an adjacent defect-free chain,
w_{DD}: for the binding energy between two adjacent defective chains.

Corresponding to these parameters, the pair concentrations

$$\left.\begin{aligned} x_{00} &= x_0^2 \\ x_{D0} &= 2x_0(1-x_0) \\ x_{DD} &= (1-x_0)^2 \end{aligned}\right\} \tag{8.76}$$

are decisive.

In place of (8.63), the internal energy is now given by

$$u = u_0 + \Delta u \sum_j{}' j x_j + \frac{z}{2}\,[w_{00}\,x_0^2 + 2\,w_{D0}\,x_0\,(1-x_0) + w_{DD}\,(1-x_0)^2] \tag{8.77a}$$

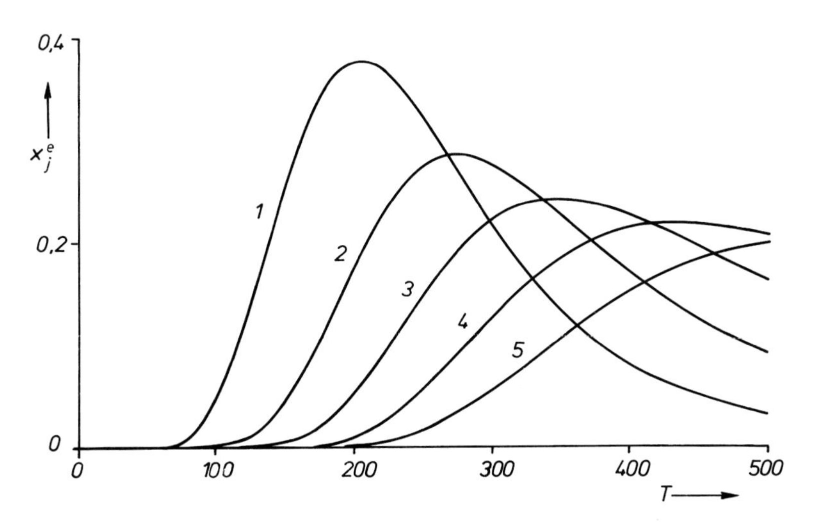

Fig. 8.14. Concentration x_j^e of chains containing j 2g1-kinks as a function of the temperature T[K] according to (8.75); $L = 20$; $\Delta u = 5$ kJ/mol; j is the parameter

with

$$x_0 = 1 - \sum_j{}' x_j \,. \tag{8.77b}$$

As (8.61) or (8.62), respectively, remain valid because of the statistical distribution of the chains, (8.60) leads to the free energy of the lamella

$$f = u_0 + \Delta u \sum_j{}' j x_j + \frac{z}{2}\,[w_{00}\,x_0^2 + 2\,w_{D0}\,x_0(1 - x_0) + w_{DD}(1 - x_0)^2] + kT\left[x_0 \ln x_0 + \sum_j{}' x_j \ln\left(\frac{x_j}{\gamma_j}\right)\right]. \tag{8.78}$$

With the abbreviation

$$\Delta w \equiv 2\,w_{D0} - w_{DD} - w_{00}\,, \tag{8.79}$$

one obtains as the condition for internal equilibrium for all $j \neq 0$

$$\frac{\partial f}{\partial x_j} = j\,\Delta u + z\,\Delta w\,x_0 + z\,(w_{DD} - w_{D0}) + kT \ln\left(\frac{x_j}{x_0\,\gamma_j}\right) = 0 \tag{8.80}$$

i.e.,

$$x_j^e = x_0^e\,\gamma_j \exp\left[-\frac{1}{kT}\,(j\,\Delta u + z\,\Delta w\,x_0^e + z\,(w_{DD} - w_{D0}))\right]. \tag{8.81}$$

If we once again use the formulation (8.71), the summation of the equilibrium concentrations x_j^e $(j \neq 0)$ results in

$$\sum_j{}' x_j^e = 1 - x_0^e = x_0^e\,[(1 + \mathrm{e}^{-\Delta u/kT})^L - 1] \cdot \exp\left[-\frac{z}{kT}\,(\Delta w\,x_0^e + w_{DD} - w_{D0})\right].$$

The equilibrium concentration of the defect-free chains as a function of the temperature is determined by the implicit transcendental equation

$$a_0 \equiv \ln\frac{1 - x_0^e}{x_0^e} + \frac{z}{kT}\,\Delta w\,x_0^e - \ln\,[(1 + \mathrm{e}^{-\Delta u/kT})^L - 1] + \frac{z}{kT}\,(w_{DD} - w_{D0}) = 0\,. \tag{8.82}$$

Equilibrium concentrations x_0^e according to (8.82) using

$$w_{DD} - w_{D0} = \frac{1}{2}\,(w_{DD} - w_{00} - \Delta w)$$

are shown in Fig. 8.15 for different values of the parameters $w_{DD} - w_{00}$ and Δw. Consideration of the term $w_{DD} - w_{D0} = (w_{DD} - w_{00})/2$ with $\Delta w = 0$ produces a shift of the defect formation towards higher temperatures as compared to the

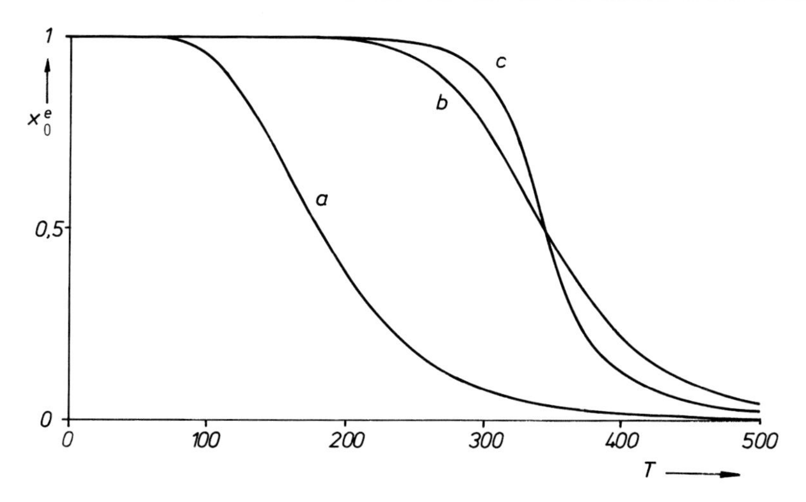

Fig. 8.15. Concentration x_0^e of defect-free chains as a function of the temperature T[K] according to (8.82) with $L = 20, \Delta u = 5$ kJ/mol and $z = 6$. **a:** $w_{DD} - w_{00} = 0, \Delta w = 0$ (ideal lamella, identical with $L = 20$ in Fig. 8.13, left); **b:** $w_{DD} - w_{00} = 3$ kJ/mol, $\Delta w = 0$; **c:** $w_{DD} - w_{00} = 3$ kJ/mol, $\Delta w = 1$ kJ/mol

ideal lamella ($w_{DD} - w_{00} = 0; \Delta w = 0$). The additional consideration of the terms with $\Delta w > 0$ (aggregation tendency of the kinks) leads to the formation of the main amount of defects in a narrower temperature range.

The internal equilibrium states according to (8.80–8.82) are stable if the matrix of the coefficients

$$\left(\frac{\partial^2 f}{\partial x_j \partial x_k}\right)_e = kT\left(\frac{1}{x_0^e} + \frac{\delta_{jk}}{x_j^e}\right) - z\Delta w \tag{8.83}$$

is positive definite [see Sect. 2.4, especially Eq. (2.83)]. The principal minors of the matrix and the determinant of the matrix are positive if it is valid that:

$$\left.\begin{aligned}
&z\Delta w < kT\left(\frac{1}{x_0^e} + \frac{1}{x_j^e}\right)\\
&z\Delta w < kT\left(\frac{1}{x_0^e} + \frac{1}{x_j^e + x_k^e}\right)\\
&z\Delta w < kT\left(\frac{1}{x_0^e} + \frac{1}{x_j^e + x_k^e + x_\ell^e}\right)\\
&\cdots\cdots\cdots\cdots\cdots\cdots\cdots\cdots\\
&z\Delta w < kT\left(\frac{1}{x_0^e} + \frac{1}{1 - x_0^e}\right).
\end{aligned}\right\} \tag{8.84}$$

With (8.76), one can also write

$$kT > \frac{z}{2}\,\Delta w\, x^e_{D0} \tag{8.85}$$

for the last of the inequalities. With a predominant tendency towards aggregation of the kinks ($\Delta w > 0$), the lamella only remains stable if the pair concentration x^e_{D0} stays below the ratio $2kT/z\Delta w$.

By insertion of (8.81) into (8.78), one obtains with (8.65a)

$$f_e = u_0 + \frac{z}{2}\,w_{00} + kT \ln x^e_0 + \frac{z}{2}\,\Delta w\,(1 - x^e_0)^2 \tag{8.86}$$

as the free energy of the lamella in internal equilibrium. This equation is formally identical with the Gibbs fundamental equation (6.113). Nevertheless, the two equations differ in that the concentrations x^e_0 and x^e_t are governed by different equilibrium conditions. Formally, (8.86) also exhibits the same properties as (6.113), *i.e.*, it exhibits, particularly in certain parameter ranges, the many-valued development of f_e, which is typical for discontinuous transition phenomena of the first order. In order to determine the potential transition temperature T_u, however, (6.118) is replaced by the transcendental equation

$$kT_u \ln\left[(1 + \mathrm{e}^{-\Delta u/kT_u})^L - 1\right] = \frac{z}{2}\,(w_{DD} - w_{00})\,. \tag{8.87}$$

One obtains in place of (6.119/6.120)

$$\frac{z}{2}\,\Delta w > 2kT_u = \frac{z\,(w_{DD} - w_{00})}{\ln\left[(1 + \mathrm{e}^{-\Delta u/kT_u})^L - 1\right]} \tag{8.88}$$

as a necessary condition for the occurrence of a discontinuous transition at T_u. This means that a discontinuous defect transition sets in when the aggregation exceeds the critical value

$$\Delta w_{\text{crit}} = 4\,kT_u/z\,. \tag{8.89}$$

According to (2.46) and (8.86), the heat capacity of the lamella in internal equilibrium is given by

$$c^e_V = -\,kT\left[\frac{2}{x^e_0}\,\frac{\mathrm{d}x^e_0}{\mathrm{d}T} + \left(\frac{z}{k}\,\Delta w - \frac{T}{(x^e_0)^2}\right)\left(\frac{\mathrm{d}x^e_0}{\mathrm{d}T}\right)^2 + \left(\frac{T}{x^e_0} - \frac{z}{k}\,\Delta w\,(1 - x^e_0)\right)\frac{\mathrm{d}^2 x^e_0}{\mathrm{d}T^2}\right]. \tag{8.90}$$

x_0^e, as a function of the temperature, is determined by the implicit function $a_0(T, x_0^e) = 0$ [Eq. (8.82)]. We then have

$$\frac{\mathrm{d}x_0^e}{\mathrm{d}T} = -\left(\frac{\partial a_0}{\partial T}\right)_{x_0^e} \Bigg/ \left(\frac{\partial a_0}{\partial x_0^e}\right)_T, \tag{8.91 a}$$

$$\frac{\mathrm{d}^2 x_0^e}{\mathrm{d}T^2} = -\left[\frac{\partial^2 a_0}{\partial T^2}\left(\frac{\partial a_0}{\partial x_0^e}\right)^2 - 2\,\frac{\partial^2 a_0}{\partial x_0^e \partial T}\,\frac{\partial a_0}{\partial x_0^e}\,\frac{\partial a_0}{\partial T} + \frac{\partial^2 a_0}{\partial (x_0^e)^2}\left(\frac{\partial a_0}{\partial T}\right)^2\right] \Bigg/ \left(\frac{\partial a_0}{\partial x_0^e}\right)^3. \tag{8.91 b}$$

The heat capacity of the lamella in internal equilibrium, c_V^e, can be represented in an analytically closed form. We have not given this expression here purely and simply because of its length. The heat capacities according to (8.90/8.91) for the cases represented in Fig. 8.15 are given in Fig. 8.16. The Schottky maximum increases with increasing difference $w_{DD} - w_{00} > 0$ and shifts towards higher temperatures. With an increasing tendency towards aggregation $\Delta w > 0$, the Schottky maximum becomes narrower and narrower until a very sharp diffuse transition phenomenon finally develops (Fig. 8.17). After exceeding the critical value (8.89), the Schottky maximum rips open on its high-temperature flank. The diffuse transition develops into a normal first-order transition.

One can assume that the interaction parameters, which refer to the chains as a whole, depend on the chain length, *i.e.*, on L. It is definitely true for w_{00} that:

$$w_{00} = L\, v_{00} - \Delta v_{00}\,, \tag{8.92}$$

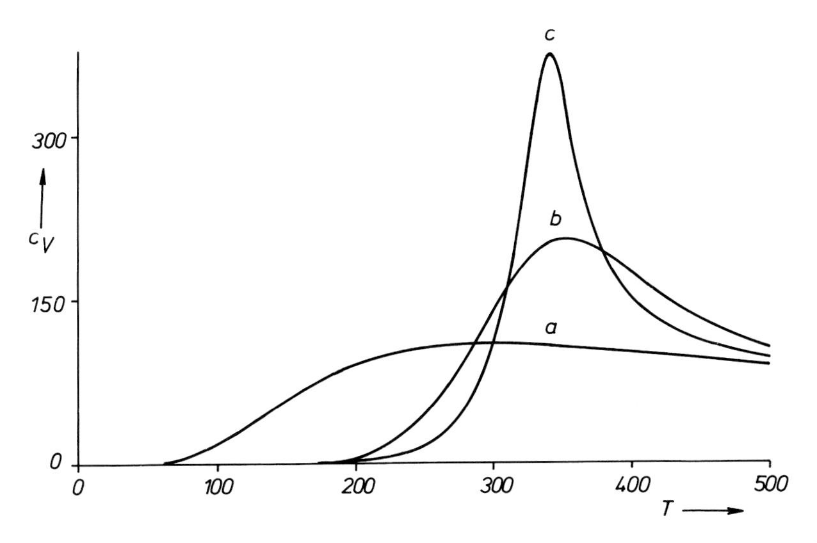

Fig. 8.16. Specific heat capacity c_V^e [J/mol · K] as a function of the temperature T[K] according to (8.90) with $L = 20, \Delta u = 5$ kJ/mol and $z = 6$. As in Fig. 8.15 **a:** $w_{DD} - w_{00} = 0, \Delta w = 0$ (ideal 2-level system; see Fig. 6.5); **b:** $w_{DD} - w_{00} = 3$ kJ/mol, $\Delta w = 0$; **c:** $w_{DD} - w_{00} = 3$ kJ/mol, $\Delta w = 1$ kJ/mol

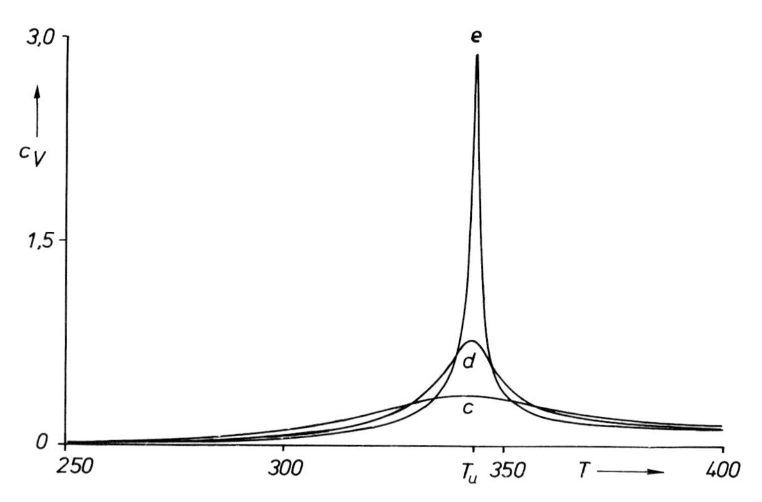

Fig. 8.17. Specific heat capacity c_V^e [kJ/mol · K] as a function of the temperature T [K] according to (8.90) with $L = 20$, $\Delta u = 5$ kJ/mol, $z = 6$ and $w_{DD} - w_{00} = 3$ kJ/mol. **c**: $\Delta w = 1$ kJ/mol (as c in Fig. 8.16); **d**: $\Delta w = 1.5$ kJ/mol; **e**: $\Delta w = 1.8$ kJ/mol. According to (8.87), the potential transition temperature T_u is at 343.2 K, and according to (8.89), the critical interaction parameter Δw_{crit} at 1.9018 kJ/mol

where v_{00} is the binding energy of two adjacent defect-free chain segments, and Δv_{00} is a correction term which takes into account the effects on the surfaces of the lamella (in *n*-paraffins, *e.g.*, the effects of the end-groups, in polyethylene the effects of the amorphous top layers of the lamella). If the defective chains are stereometrically arranged in such a way that the smallest possible lateral disturbance develops (Fig. 8.12, left), one can also assume

$$w_{DD} = L\,v_{DD} - \Delta v_{DD}\,, \tag{8.93}$$

where v_{DD} is the mean binding energy per segment of two adjacent defective chains. In this case, the surface correction Δv_{DD} also contains the effect of the kink-induced reductions in chain length (Fig. 8.10).

Insertion of (8.92) and (8.93) into the equation (8.87) determining the potential transition temperature yields

$$kT_u \ln\left[(1 + \mathrm{e}^{-\Delta u/kT_u})^L - 1\right] = \frac{z}{2}\left[L\,(v_{DD} - v_{00}) - (\Delta v_{DD} - \Delta v_{00})\right]. \tag{8.94}$$

The transition temperature increases with the chain length or rather the lamellar thickness (Fig. 8.18) if

$$v_{00} < v_{DD} < 0 \tag{8.95}$$

is valid, *i. e.*, if the kinks on average lead to a loosening of the segment bonding. In shorter chains, the position of the transition temperature also depends quite sensitively on the conditions at the surfaces. In addition, the transition temperature possesses a finite upper limit

$$T_u(L) \leq T_{u,\max} = \text{finite} .$$

In order to determine the limiting temperature $T_{u,\max}$, one obtains from (8.94) in the limit $L \to \infty$

$$kT_{u,\max} \ln(1 + e^{-\Delta u/kT_{u,\max}}) = \frac{z}{2}(v_{DD} - v_{00}) . \tag{8.96}$$

In analogy to (8.92) and (8.93), one can finally also assume

$$w_{D0} = L v_{D0} - \Delta v_{D0} . \tag{8.97}$$

The existence condition (8.88) for a discontinuous defect transition is then

$$\Theta(L) \equiv \frac{z}{4k}(L\Delta v - \Delta\Delta v) > T_u(L) \tag{8.98a}$$

with

$$\Delta v \equiv 2v_{D0} - v_{DD} - v_{00} \tag{8.98b}$$

and

$$\Delta\Delta v \equiv 2\,\Delta v_{D0} - \Delta v_{DD} - \Delta v_{00} . \tag{8.98c}$$

As a result of (8.98), there is with $\Delta v > 0$ also a lower, L-dependent limit

$$T_{u,\min}(L_{\min}) < T_u(L) \leq T_{u,\max}(L \to \infty) \tag{8.99}$$

for the temperature $T_u(L)$, at which a discontinuous transition can actually set in (see Fig. 8.18). Hence, the lamellae are only capable of a discontinuous defect transition starting from a specific thickness $L_{\min}$ determined by the interaction forces.

The situations in n-paraffins under normal pressure can be described very well using the model which has just been developed [see Baur (1977)]. In this model, the differences between the even-numbered and the odd-numbered n-paraffins are almost entirely due to different surface terms $\Delta v_{DD} - \Delta v_{00}$. The fact that no discontinuous defect transition occurs at normal pressure in the lamellae of polyethylene, which form under normal crystallization conditions (Sect. 8.2), must also be attributed to the surface terms, *i.e.*, to the effect of the amorphous top layers. In n-paraffins, the discontinuous transition of the crystal lamellae disappears upon elevation of the pressure. In polyethylene, on the other hand, such a transition only occurs at elevated pressure [see Wunderlich et al. (1988)]. In order to explain these phenomena, one would have

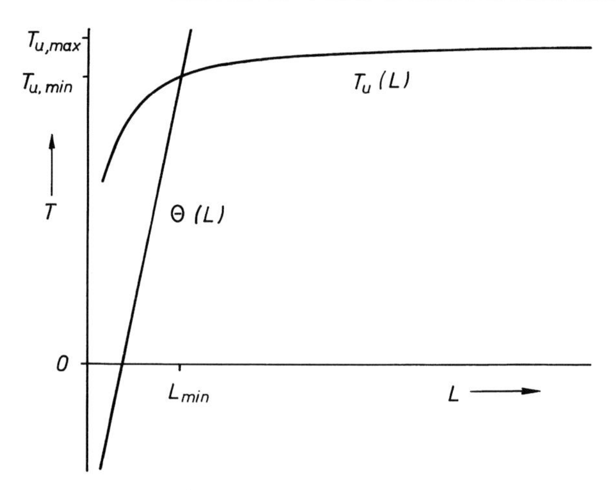

Fig. 8.18. Potential transition temperature T_u as a function of L according to (8.94) with (8.95). Temperature $\Theta(L)$ according to (8.98) with $\Delta v > 0$. A discontinuous transition can only appear in the regions $T_{u,\min} \leq T_u \leq T_{u,\max}$ and $L \geq L_{\min}$

to proceed from the G-representation of Gibbs fundamental equation. In addition, the G-representation requires that one has knowledge of the dependence of the volume on the concentrations of the conformational isomers, and possibly also of the dependence of the excess energy Δu and the interaction parameters w_{jk} on the pressure.

References

Adam G, Gibbs JH (1965) J Chem Phys 43:139
Alfrey T, Mark H (1942) J Phys Chem 46:112
Ambacher H, Kilian H-G (1992) The fundamental equations of Van Der Waals-networks. In: Mark E, Erman B (eds) Elastomeric Polymer Networks. Prentice Hall, Englewood Cliffs, NJ
Baughman RH (1973) J Chem Phys 58:2976
Baur H (1970) Kolloid-Z u Z Polymere 241:1057
Baur H (1971) Z Naturforschg 26 a:979
Baur H (1972) Kolloid-Z u Z Polymere 250:1000
Baur H (1972 a) Kolloid-Z u Z Polymere 250:289
Baur H (1975) Progr Colloid & Polymer Sci 58:1
Baur H (1977) Habilitationsschrift 1992 + a, Universität Hannover (Germany)
Baur H (1979) Progr Colloid & Polymer Sci 66:1
Baur H (1980) Pure & Appl Chem 52:457
Baur H (1984) Einführung in die Thermodynamik der irreversiblen Prozesse. Wissenschaftliche Buchgesellschaft, Darmstadt
Baur H (1986) Colloid & Polymer Sci 264:582
Baur H (1998) Z Naturforschg 53a:157
Blackman M (1955) The specific heat of solids. In: Encyclopedia of Physics, Flügge S (ed) Springer, Berlin Heidelberg New York
Blasenbrey S, Pechhold W (1967) Rheol Acta 6:174
Born M, Kármán Th v (1912) Physik Z 13:297
Born M, Huang K (1954) Dynamical theory of crystal lattices, Oxford University Press, London
Brandrup J, Immergut EH (eds) (1989) Polymer Handbook, Wiley, New York
Breuer H, Rehage G (1967) Kolloid-Z u Z Polymere 216/217:159
Brillouin L (1953) Wave propagation in periodic structures, Dover Publications, New York
Campell IA, Giovannella C (1990) Relaxation in complex systems and related topics, NATO ASI Series B: Physics, Vol 222, Plenum Press, New York
Casimir HBG (1945) Rev Mod Phys 17:343
Choy CL (1975) J Polymer Sci 13:1263
Christensen RM (1971) Theory of viscoelasticity, Academic Press, New York
Cloizeaux J des, Jannink G (1990) Polymer in solution (Their modelling and structure), Clarendon Press, Oxford
Courant R, Hilbert D (1965) Methods of mathematical physics, Vol 1, Interscience Publishers, New York
Curie P (1885) Bull Soc Minéralog France 8:145
Curie P (1887) Z Kristallogr 12:651
Davies RO (1952) Sur les soit-disant transitions de second ordre dans les milieux a relaxation, Comptes Rendus IUP, Société de Chimie Physique, Paris
Davies RO, Jones GO (1953) Adv Physics 2:370
Davies RO, Jones GO (1953) Proc Royal Soc A 217:26

Dean P (1960) Proc Royal Soc A 254:507
Dean P (1961) Proc Royal Soc A 260:264
Debye P (1912) Annalen der Physik 39:789
Dirac PMA (1958) The principles of quantum mechanics, Oxford University Press, London
Doi M, Edwards SF (1986) The theory of polymer dynamics, Clarendon Press, Oxford
Donth E-J (1992) Relaxation and thermodynamics in polymers, Akademie-Verlag, Berlin
Doolittle AK (1951) J Appl Phys 22:1471
Ehrenfest P (1933) Proc Kon Akad Amsterdam 36:153
Einstein A (1907) Annalen der Physik 22:180
Eirich FR (ed) (1978) Science and technology of rubber, Academic Press, New York
Elias HG (1984) Macromolecules, Vol 1, Plenum Publishing Corp, New York
Elyashevich GK, Baranov VG, Frenkel SYA (1977) Macromol Sci Phys B 13(2):255
Eyring H, Ree T (1961) Proc Nat Acad Sci (US) 47:526
Feller W (1957) An introduction to probability theory and its application, Vol 1, Wiley, New York
Ferry JD (1980) Viscoelastic properties of polymers, Wiley, New York
Fischer EW, Schmidt GF (1962) Angew Chemie 74:551
Fischer EW (1978) Pure & Appl Chem 50:1319
Flory PJ (1941) J Chem Phys 9:660
Flory PJ (1942) J Chem Phys 10:51
Flory PJ (1953) Principles of polymer chemistry, Cornell University Press, Ithaca, NY
Flory PJ (1962) J Am Chem Soc 84:2857
Flory PJ (1969) Statistical mechanics of chain molecules, Interscience Publ, New York
Fowler R, Guggenheim EA (1960) Statistical thermodynamics, Cambridge University Press, London
Fox TG, Flory PJ (1950) J Appl Phys 21:581
Frenkel JI (1946) Kinetic theory of liquids, Oxford University Press, London
Frith EM, Tuckett RF (1944) Trans Faraday Soc 40:251
Genensky SM, Newell GF (1957) J Chem Phys 26:486
Gennes P-G de (1979) Scaling concepts in polymer physics, Cornell University Press, Ithaca, NY
Gibbs JH, Di Marzio EA (1958) J Chem Phys 28:373
Gibbs JW (1928) Collected works, Longmans Green, New York, p 322
Glasstone S, Laidler KJ, Eyring H (1941) The theory of rate processes, McGraw-Hill, New York
Goldstein H (1962) Classical mechanics, Addison-Wesley, Reading, MA
Goldstein M (1963) J Chem Phys 39:3369
Green HS, Hurst CA (1964) Order-disorder phenomena, Interscience Publishers, New York
Gröbner W, Hofreiter N (1961) Integraltafeln, 2. Teil, Springer-Verlag, Wien
Groot, SR De (1963) Thermodynamics of irreversible processes, North-Holland, Amsterdam
Groot SR De, Mazur P (1962) Non-equilibrium thermodynamics, North-Holland, Amsterdam
Gross B (1953) Mathematical structure of the theories of viscoelasticity, Hermann, Paris (second impression 1968)
Guggenheim EA (1959) Thermodynamics, North-Holland, Amsterdam
Gupta PK, Moynihan CT (1976) J Chem Phys 65:4136
Gyarmati I (1970) Non-equilibrium thermodynamics, Springer, Berlin Heidelberg New York
Haken H (1973) Quantenfeldtheorie des Festkörpers, Teubner, Stuttgart
Hartwig G (1994) Polymer properties at room and cryogenic temperatures, Plenum Press, New York
Haward RN (ed) (1973) The physics of glassy polymers, Applied Science Publ, London
Hirai N, Eyring H (1958) J Appl Phys 29:810
Hirai N, Eyring H (1959) J Polymer Sci 37:51
Holliday L (ed) (1966) Composit materials, Elsevier, Amsterdam
Hopfinger AJ (1973) Conformational properties of macromolecules, Academic Press, New York
Hove L van (1953) Phys Rev 89:1189

Huang K (1963) Statistical Mechanics, Wiley, New York
Huggins ML (1941) J Chem Phys 9:440
Huggins ML (1943) Ann NY Acad Sci 44:431
Hunklinger S, Sussner H, Dransfeld K (1976) New dynamic aspects of amorphous dielectric solids. In: Treuch J (ed) Festkörperprobleme, Vieweg Verlagsgesellschaft, Braunschweig
Isihara A (1971) Statistical physics, Academic Press, New York
Jäckle J (1986) Rep Prog Phys 49:171
Jagodzinski H (1955) Kristallographie. In: Encyclopedia of Physics, Flügge S (ed), Vol VII/1, Springer, Berlin Heidelberg New York
Jou D, Casas-Vázquez J, Lebon G (1993) Extended irreversible thermodynamics, Springer, Berlin Heidelberg New York
Juilfs J, Kühne U (1971) Kolloid-Z u Z Polymere 244:304
Kanig G (1969) Kolloid-Z u Z Polymere 233:829
Kanig G (1991) Colloid & Polymer Sci 269:1118
Keller A (1957) Phil Mag 2:1171
Kirkwood JG (1939) J Chem Phys 7:506
Kittel C (1966) Quantum Theory of solids, Wiley, New York
Kohlrausch R (1847) Ann Phys (Leipzig) 12:393
Koningsveld R, Kleintjens LA (1977) J Polymer Sci, Polymer Symp 61:221
Kovacs AJ, Aklonis JJ, Hutchinson JM, Ramos AR (1979) J Polymer Sci Phys Ed 17:1097
Kubo R, Toda M, Hashitsume N (1985) Statistical physics II, Springer, Berlin Heidelberg, New York
Kuhn W (1934) Kolloid Z 68:2
Kuhn W, Kuhn H (1943) Helvet Chim Acta 26:1394
Kurata M (1982) Thermodynamics of polymer solutions, Harwood Academic Publishers, Chur, London, New York
Landau LD, Lifshitz EM (1986) Course of theoretical physics, vol 7: Theory of elasticity, Pergamon Press, Oxford
Landau LD, Lifshitz EM (1987) Course of theoretical physics, vol 6: Fluid mechanics, Pergamon Press, Oxford
Landsberg PT (1961) Thermodynamics, Interscience Publishers, New York
Laue M von (1943) Z Kristallogr 105:124
Lauritzen JI, Hoffman JD (1960) J Research NBS 64 A:73
Leibfried G (1955) Gittertheorie der mechanischen und thermischen Eigenschaften der Kristalle. In: Encyclopedia of Physics, Flügge S (ed), Springer, Berlin Heidelberg New York
Liebmann H (1914) Z Kristallogr 53:171
Lifshitz IM (1952) cited according to Perepechko (1980)
Lodge AS (1964) Elastic liquids, Academic Press, New York
Ludwig W (1967) Recent developments in Lattice theory. In: Springer tracts in modern physics, vol 42, Springer, Berlin
Mackenzie JD (ed) (1960, 1962, 1964) Modern aspects of the vitreous state, vol I–III, Butterworths, London
Mandelbrot BB (1983) The fractal geometry of nature, Freeman WH & Comp, New York
Mandelkren L (1964) Crystallization of polymers, McGraw–Hill, New York
Maradudin AA, Weiss GH (1958) J Chem Phys 29:631
Maradudin AA, Montroll EW, Weiss GH (1963) Theory of lattice dynamics in the harmonic approximation, Academic Press, New York
Margenau H, Murphy GM (1956) The mathematics of physics and chemistry, vol I, D Van Nostrand Comp, Princeton, NJ
Martin JL (1960) Proc Royal Soc London 254A:139
Martin JL (1961) Proc Royal Soc London 260A:139
Mattice WL, Suter UW (1994) Conformational theory of large molecules – The rotational isomeric state model, Wiley, New York
McCrum NG, Read BE, Williams G (1967) Anelastic and dielectric effects in polymeric solids, Wiley, New York

McMahon PE, McCullough RL, Schlegel AA (1967) J Appl Phys 38:4123
Meixner J (1949) Z Naturforschg 4a:594
Meixner J (1953) Kolloid-Z 134:3
Meixner J (1969) Z Physik 219:79
Meixner J, Reik HG (1959) Thermodynamik der irreversiblen Prozesse. In: Encyclopedia of physics, Flügge S (ed), vol III/2, Springer, Berlin Heidelberg New York
Moelwyn-Hughes EA (1961) Physical chemistry, Pergamon Press, London
Moynihan CT, Lesikar AV (1981) Ann NY Acad Sci 371:151
Münster A (1954) Z physik Chem NF 1:259
Nernst W, Lindemann FA (1911) Z Elektrochem 17:817
Newell GF (1953) J Chem Phys 21:1877
Nowick AS, Berry BS (1972) Anelastic relaxation in crystallin solids, Academic Press, New York
Nye JF (1975) Physical properties of crystals, Clarendon Press, Oxford
Onsager L (1931) Phys Rev 37:405 and 38:2265
O'Reilly JM (1962) J Polymer Sci 57:429
Pan R, Varma M, Wunderlich B (1989) J Thermal Anal 35:955
Pathria RK (1972) Statistical mechanics, Pergamon Press, Oxford
Pauling L, Wilson EB (1935) Introduction to quantum mechanics, McGraw–Hill, New York
Pechhold W (1968) Kolloid-Z u Z Polymere 228:1
Perepechko I (1980) Low-temperature properties of polymers, Pergamon Press, Oxford
Phillips WA (ed) (1981) Amorphous solids, low-temperature properties, Springer, Berlin
Pipkin AC (1972) Lectures on viscoelasticity theory, Springer, Berlin Heidelberg New York
Pitzer KS (1940) J Chem Phys 8:711
Prigogine I (1961) Introduction to thermodynamics of irreversible processes (revised ed), Interscience publishers, New York
Prigogine I, Defay R (1950) Traité de Thermodynamique, Tome I et II réunis, Éditions Desoer, Liège
Prigogine I, Defay R (1951) Traité de Thermodynamique, Tome III, Éditions Desoer, Liège
Ramsey NF (1956) Phys Rev 103:20
Rehage G (1980) J Macromol Sci-Phys B 18 (3):423
Roe RJ, Smith Jr KJ, Krigbaum WR (1961) J Chem Phys 35:1306
Sariban A, Binder K, Heermann DW (1987) Phys Rev B 35:6873
Schaefer C (1922) Einführung in die Theoretische Physik, Band I, Walter de Gruyter & Co, Berlin und Leipzig; see also: Love AEH (1892/93) A treatise on the mathematical theory of elasticity, Cambridge: At the University Press, and Landau, Lifshitz (1986)
Schelten J, Ballard DGH, Wignall GD, Longmann G, Schmalz W (1976) Polymer 17:751
Schelten J, Wignall GD, Ballard DGH, Longmann G (1977) Polymer 18:1111
Schrader E, Zachmann HG (1970) Kolloid-Z u Z Polymere 241:1015
Schuster HG (1984) Deterministic chaos, Physik-Verlag, Weinheim
Schwarzl FR (1990) Polymer-Mechanik, Springer, Berlin Heidelberg New York
Schwarzl FR, Staverman AJ (1952) J Appl Phys 23:383
Shannon CE (1962) The mathematical theory of communication, The University of Illinois Press, Urbana
Simha R, Boyer RF (1962) J Chem Phys 37:1003
Simon FE (1930) Ergeb Exakt Naturwiss 9:222
Stamm M, Fischer EW, Dettenmaier M, Convert P (1979) Faraday Discuss. Chem Soc 68:263
Stevels JM (1962) The structure and the physical properties of Glass. In: Encyclopedia of physics, Flügge S (ed), Vol XIII, Springer, Berlin Heidelberg New York
Stockmayer WH, Hecht CE (1953) J Chem Phys 21:1954
Stranski IN (1943) Z Kristallogr 105:91
Strickland-Constable RF (1968) Kinetics and mechanism of crystallization, Academic Press, London
Stuart HA (1967) Molekülstruktur, Springer, Berlin Heidelberg New York
Tarasov VV (1950, 1953) cited according to Perepechko (1980)

Tasumi M, Shimanouchi T, Miyazawa T (1962) J Molecular Spectroscopy 9:261
Tasumi M, Shimanouchi T (1965) J Chem Phys 43:1245
Tasumi M, Krimm S (1967) J Chem Phys 46:755
Taylor WJ (1947) J Chem Phys 15:412
Taylor WJ (1948) J Chem Phys 16:257
Telezhenko YaV, Sukharevskii BY (1982) J Low Temp Phys 8 (2):93
Tisza L (1951) On the general theory of phase transitions, Chap 1. In: Phase transformations in solids, Smoluchowski A, Mayer JE, Weyl WA (ed), Wiley, New York
Toda M, Kubo R, Saitô N (1983) Statistical physics I, Springer, Berlin
Tool AQ (1946) J Am Ceram Soc 29:240
Treloar LRG (1975) The physics of rubber elasticity, Clarendon Press, Oxford
Truell R, Elbaum C, Chick BB (1969) Ultrasonic methods in solid state physics, Academic Press, New York
Truesdell C (1969) Rational thermodynamics, McGraw-Hill, New York
Tsuan Wu Ting, Li JCM (1957) Phys Rev 106:1165
Tung LH, Buckser S (1958) J Phys Chem 62:1530
Ubbelohde AR (1952) Introduction to modern thermodynamical principles, Oxford University Press, London
Ubbelohde AR (1965) Melting and crystal structure, Claredon Press, Oxford
Volkenstein MV (1963) Configurational statistics of polymer Chains, Interscience Publishers, New York
Ward IM (1971) Mechanical properties of solid polymers, Wiley – Interscience Inc, New York
Williams G, Watts DC (1970) Trans Faraday Soc 66:80
Williams ML, Landel RF, Ferry JD (1955) J Amer Chem Soc 77:3701
Wrasidlow W (1974) Adv Polymer Sci 13:1
Wulff G (1901) Z Kristallogr 34:449
Wunderlich B, Baur H (1970) Adv Polymer Sci 7:151
Wunderlich B (1973) Macromolecular Physics, vol 1, Academic Press, New York
Wunderlich B (1976) Macromolecular Physics, vol 2, Academic Press, New York
Wunderlich B (1980) Macromolecular Physics, vol 3, Academic Press, New York
Wunderlich B, Grebowicz J (1984) Adv Polymer Sci 60/61:1
Wunderlich B, Möller M, Grebowicz J, Baur H (1988) Adv Polymer Sci 87:1
Yang CN, Lee TD (1952) Phys Rev 87:404 and 410
Zachmann HG (1964) Z Naturforschg 19a:1397
Zachmann HG, Spellucci P (1966) Kolloid-Z u Z Polymere 213:39
Zachmann HG (1967) Kolloid-Z u Z Polymere 216/217:180
Zachmann HG (1969) Kolloid-Z u Z Polymere 231:504
Zbinden R (1964) Infrared Spectroscopy of High Polymers, Academic Press, New York
Zener CM (1948) Elasticity and anelasticity of metals (Reprint 1960) University of Chicago Press, Chicago Ill
Ziman JM (1967) Electrons and phonons, Oxford University Press, London
Ziman JM (1969) Elements of advanced quantum theory, Cambridge University Press, London

Subject Index

Printing (computer to plate): Mercedes-Druck, Berlin
Binding: Buchbinderei Lüderitz & Bauer, Berlin